STRATEGY

Second Edition

STRATEGY

AN INTRODUCTION TO GAME THEORY

JOEL WATSON
University of California, San Diego

Second Edition

W. W. NORTON & COMPANY
NEW YORK • LONDON

W. W. Norton & Company has been independent since its founding in 1923, when William Warder Norton and Mary D. Herter Norton first published lectures delivered at the People's Institute, the adult education division of New York City's Cooper Union. The Nortons soon expanded their program beyond the Institute, publishing books by celebrated academics from America and abroad. By mid-century, the two major pillars of Norton's publishing program—trade books and college texts—were firmly established. In the 1950s, the Norton family transferred control of the company to its employees, and today—with a staff of four hundred and a comparable number of trade, college, and professional titles published each year—W. W. Norton & Company stands as the largest and oldest publishing house owned wholly by its employees.

Printed in the United States of America.

Composition by Integre Technical Publishing Company, Inc.
Manufacturing by Courier—Westford Division
Book design by Joan Greenfield
Project editor: Sandra Lifland
Production manager: Ben Reynolds

Library of Congress Cataloging-in-Publication Data
Watson, Joel.
Strategy : an introduction to game theory / Joel Watson. — 2nd ed.
p. cm.
Includes bibliographical references and index.
ISBN 978-0-393-92934-8 (hardcover)
1. Game theory. 2. Economics—Psychological aspects. 3. Strategic planning. I. Title.
HB144.W37 2008
330.01′5193—dc22

ISBN 13: 978-0-393-92934-8

W. W. Norton & Company, Inc., 500 Fifth Avenue, New York, NY 10110
www.wwnorton.com

W. W. Norton & Company Ltd., Castle House, 75/76 Wells Street, London W1T 3QT

3 4 5 6 7 8 9 0

This book is dedicated, with thanks, to Rosemary Watson.

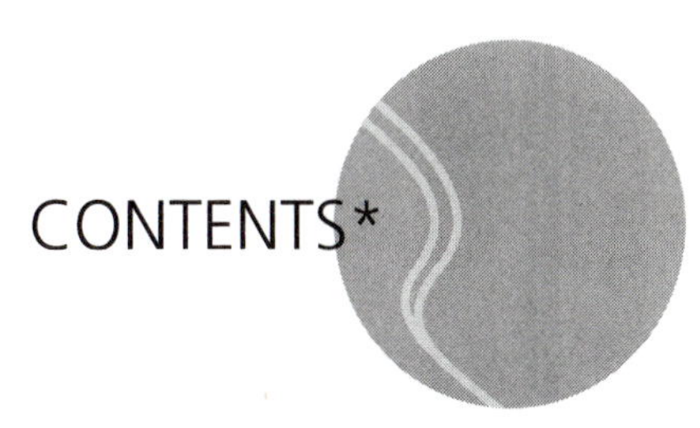

CONTENTS*

*Chapters in italics are applications chapters.

PREFACE

Game theory has become an enormously important field of study. It is now a vital methodology for researchers and teachers in many disciplines, including economics, political science, biology, and law. This book provides a thorough introduction to the subject and its applications, at the intermediate level of instruction. It is designed for the upper-division game theory courses that are offered in most undergraduate economics programs and also serves graduate students in economics, political science, law, and business who seek a solid and readable introduction to the theory. The book can be a primary or secondary source for graduate courses that have a significant game-theoretic component.

I have designed this book in pursuit of the following three ideals:

1. **Coverage of the Essential Topics, without Excess.** This book covers the basic concepts and insights of game-theoretic reasoning, along with classic examples and applications, without being overblown. The entire book can be digested in a semester course. Thus, instructors will not have to worry about having their students skipping or bouncing between chapters as much as is the case with most other textbooks. In terms of content, the book's major innovation is that it integrates an analysis of *contract* in the development of the theory. Research on contract has blossomed in the past few decades and, in my experience, students are keenly interested in this important game-theoretic topic. This book offers one of the first substantive treatments of contract at the intermediate level, without bending too far from standard game theory coverage.
2. **Proper Mathematics without Complication.** This book does not skimp on mathematical precision. However, each concept and application presented here is developed using the *simplest*, most straightforward model that I could find. In other words, the book emphasizes mathematically rigorous analysis of simple examples rather than complicated or incomplete analysis of general models. This facilitates the reader's understanding of key ideas without getting bogged down in unnecessary mathematical formalism.

3. **Relaxed Style.** This book is written in a style that is lively and not overly formal.

The book is also class-tested. It grew out of the undergraduate game theory course that I teach and has been developed and fine-tuned over the past decade. I used preliminary versions of the book for my game theory courses at the University of California, San Diego, and at Yale University. It was class-tested at Stanford University by Steven Tadelis, at the University of Michigan by Ennio Stacchetti, and at New York University by Charles Wilson. The first edition of the book was adopted for courses at numerous universities. I also have made it available as a supplemental textbook for my graduate game theory courses, including courses in the core microeconomics sequence. The journey to final, published book was carefully guided by comments from professors, reviewers, colleagues, and—to a significant extent—students. I am proud to report that the comments from students have been overwhelmingly positive.

The text is organized into four parts. Part I introduces the standard ways of representing games and carefully defines and illustrates the concepts of strategy and belief. Because the concept of strategy is so important, and because students who fail to grasp it inevitably sink deeper into confusion later on, I begin the formal part of the book with a nontechnical discussion of the extensive form. The material in Part I can be covered in just a couple of lectures, and it is an investment that pays great dividends later.

Part II presents the concepts used to analyze static settings, Part III surveys the refinements used to analyze dynamic settings, and Part IV looks at games with nature and incomplete information. In each part, I illustrate the abstract concepts and methodology with examples. I also discuss various "strategic tensions," and I remark on the institutions that may resolve them. Several chapters—those whose titles appear italicized in the table of contents—are devoted to applications; instructors will certainly want to supplement, emphasize, or de-emphasize some of these in favor of their preferred ones. Applications in the book may be skimmed or skipped without disruption. With a couple of exceptions, each chapter contains a guided exercise with a complete solution, followed by assorted exercises for the reader to complete.

Some instructors—especially those who teach on the quarter system—may wish to skip a bit of the material in the book. In my own experience, I can usually get through most of the book in a ten-week quarter, but I normally leave out a few applications. I think most instructors faced with a ten-week time constraint can easily do the same. As noted in the preceding paragraph, any application can be skipped without danger of causing the students to miss an important point that will later be needed. Instructors who do not want to cover contract can, for example, skip Chapters 20, 21, and 25. For more suggestions on how to plan a course, please see the Instructor's Manual posted on the publisher's Internet site at www.wwnorton.com/college/econ/watson.

I am grateful for the support of a great many people. I thank my theory colleagues at the University of California, San Diego—especially Vincent Crawford, Mark Machina, Garey Ramey, and Joel Sobel—whose instruction and advice have contributed significantly to this book and to my development as a scholar. My interest in game-theoretic models of contract, in particular, began with research I conducted with Garey Ramey.

With regard to the preparation of the original manuscript and second edition, I thank Chris Avery, Pierpaolo Battigalli, Jesse Bull, Hongbin Cai, Gavin Cameron, Vincent Crawford, Takako Fujiwara-Greve, Scott Gehlbach, Navin Kartik, David Miller, Ben Polak, Robert Powell, Ennio Stacchetti, Steven Tadelis, Rosemary Watson, Andrzej Wieczorek, Charles Wilson, Carina York, and anonymous reviewers for helpful comments and suggestions. Ben Polak also generously allowed me to use versions of exercises that he wrote for his courses at Yale University.

I am grateful to Ed Parsons and Jack Repcheck, the past and current economics editors at W. W. Norton & Company, who expertly guided me and the book through the publication process. Their professionalism and care are rarely seen in any business.

I also thank the following students and others who provided expert assistance and comments: Mike Beauregard, Jim Brennan, Jesse Bull, Clement Chan, Chris Clayton, Andrew Coors, Mei Dan, Travis Emmel, Cindy Feng, Lynn Fisher, Lauren Fogel, Shigeru Fujita, Alejandro Gaviria, Andreea Gorbatai, Michael Herron, Oliver Kaplan, Peter Lim, Emily Leppert, Magnus Lofstrom, Simone Manganelli, Andrew Mara, Barry Murphy, Herb Newhouse, Luis Pinto, Sivan Ritz, Heather Rose, Augusto Schianchi, Makoto Shimoji, Nick Turner, Bauke Visser, Ann Wachtler, Michael Whalen, and Wei Zhao. Furthermore, I thank the fall 2001 Economics 109 class at the University of California, San Diego; its students were the best and brightest that I have taught anywhere.

Finally, for their inspiring ways, I am grateful to the few people in this world who are both brilliant and modest.

Joel Watson
La Jolla, California
May 2007

INTRODUCTION 1

In all societies, people interact constantly. Sometimes the interaction is cooperative, such as when business partners successfully collaborate on a project. Other times the interaction is competitive, as exemplified by two or more firms fighting for market share or by several workers vying for a promotion that can be given to only one of them. In either case, the term *interdependence* applies—one person's behavior affects another person's well-being, either positively or negatively. Situations of interdependence are called *strategic settings* because, in order for a person to decide how best to behave, he must consider how others around him choose their actions. If partners want to successfully complete a project, they are well advised to coordinate their efforts. If a firm wishes to maximize its profit, it must estimate and analyze the stance of its rivals. If a worker wants to secure a promotion, she ought to consider the opposing efforts of her colleagues (so she can, for example, sabotage their work to her own benefit).

Even on a bad-cogitation day, we can easily discover the truth that strategy is fundamental to the workings of society. But this realization is just the beginning. For a greater challenge, we can try to develop an understanding of how people actually behave, and how they should be advised to behave, in strategic situations. A systematic study of this sort yields a *theory* of strategic interaction. The theory is useful in many ways. First, it identifies a language with which we can converse and exchange ideas about human endeavor. Second, it provides a framework that guides us in constructing models of strategic settings—a process that engenders insights by challenging us to be clear and rigorous in our thinking. Third, it helps us trace through the logical implications of assumptions about behavior.

Logical thinking about human behavior has proved useful for millennia. In ancient times, religious and civil code spoke directly to the methods and standards of negotiation, contract, and punishment, as they do today. The Babylonian Talmud, for example, established rules for the division of a man's estate that foreshadow modern theories of apportionment. Hundreds of years ago, mathematicians began studying parlor games in an attempt to formulate optimal strategies. In 1713, James Waldegrave communicated a solution to a particular card game to his colleagues Pierre-Remond de Montmort and Nicolas Bernoulli; Waldegrave's solution coincides with the conclusion of modern theory. In the 1800s, Augustin Cournot explored equilibrium in models of oligopoly, while Francis Ysidro Edgeworth tackled bargaining problems in the context of exchange in an economy. In 1913, the first formal theorem for games (a result about chess) was proved by Ernest Zermelo. Emile Borel followed with groundbreaking work on the concept of a strategy.[1]

Then, in the 1920s, 30s, and 40s, largely through the work of the brilliant scholar John von Neumann, a true, rigorous, general theory of strategic situations was launched. It is called the *theory of games*. Von Neumann and Oskar Morgenstern wrote the seminal game theory book, which proposed in great detail how to represent games in a precise mathematical way and also offered a general method of analyzing behavior.[2] But their method of analyzing behavior was quite limited in that it could only be applied to a small class of strategic settings. Game theory became truly activated with the mid-century work of John Nash, who made the key distinction between "noncooperative" and "cooperative" theoretical models and created concepts of rational behavior—so called "solution concepts"—for both branches.[3]

In the ensuing decades, mathematicians and economists enriched the foundation, gradually building one of the most powerful and influential toolboxes of modern social science. Currently, the theory is employed by practitioners from a variety of fields, including economics, political science, law, biology, international relations, philosophy, and mathematics. Research into the foundations and the proper application of game theory continues at a vigorous pace. Many components of the theory have yet to be developed; important settings await careful examination.

In this textbook, I introduce some of the building blocks of the theory and present several applications. Foremost, I wish to demonstrate the value of examining a strategic situation from the tight theoretical perspective. I believe structured, logical thinking can help anyone who finds herself in a strategic situation—and we all face strategic situations on a daily basis. More generally, game theory can help us understand how the forces of strategy influence the

[1] A history of game theory appears on the Web site of Paul Walker at the University of Canterbury (http://www.econ.canterbury.ac.nz/personal_pages/paul_walker/gt/hist.htm).

[2] J. von Neumann and O. Morgenstern, *Theory of Games and Economic Behavior* (Princeton, NJ: Princeton University Press, 1944).

[3] For more on von Neumann, Morgenstern, and Nash, see J. Watson, "John Forbes Nash, Jr.," in *The New Palgrave Dictionary of Economics* (2nd ed.), ed. L. Bloom and S. Durlauf (Hampshire, England: Palgrave Macmillan, 2007).

outcome of social—specifically economic—interaction. In other words, I hope you will find game theory useful in evaluating how the world works.

NONCOOPERATIVE GAME THEORY

Our tour will include a fairly complete presentation of "noncooperative game theory," which is a framework that treats strategic settings as *games* in the everyday sense of the word. Because the term *game* generally connotes a situation in which two or more adversaries match wits, games inherently entail interdependence in that one person's optimal behavior depends on what he or she believes the others will do. We also normally associate games with sets of rules that must be followed by the players. Baseball, for example, features myriad codified rules that govern play: how many players are on the field and what they can do; how runs are scored, outs recorded, and innings completed; what must be done if a fan interferes with play; and so forth. As another example, the game of Monopoly includes a formal specification of the rules of play, from the order of moves to the determination of a winner. Rules are synonymous with our ability to agree on exactly what game is actually being played, and thus the careful specification of rules is an important part of the formal theory.

One major feature distinguishes noncooperative game theory from other frameworks for studying strategy: the noncooperative framework treats all of the agents' actions as *individual actions*. An individual action is something that a person decides on his own, independently of the other people present in the strategic environment. Thus, it is accurate to say that noncooperative theories examine individual decision making in strategic settings. The framework does not rule out the possibility that one person can limit the options of another; nor is the theory incompatible with the prospect that players can make decisions in groups. In regard to group decision making, noncooperative models require the theorist to specify the procedure by which decisions get made. The procedure includes a specification of how agents negotiate options, which may include offers and counteroffers (taken to be individual actions). Indeed, there is a sense in which every decision that a person makes can be modeled as an individual action.

Thinking of actions as individual actions is perhaps the most realistic way of modeling the players' options in a strategic setting. It also implies how one must go about studying behavior. In a game, each player has to make his or her own decisions. A player's optimal decision depends on what he or she thinks the others will do in the game. Thus, to develop *solution concepts*—which are prescriptions or predictions about the outcomes of games—one must study how individual players make decisions in the presence of "strategic uncertainty" (not knowing for sure what other players will do).

One problem with noncooperative theory is that its tools are often very difficult to employ in applied settings. Creating a simple model necessarily entails leaving quite a bit of strategic interaction outside of view. Further, even the analysis of very simple games can be prohibitively complicated. There is an art to building game-theoretic models that are simple enough to analyze, yet rich enough to capture intuition and serve as sources of new insights. In fact, the process of devising interesting models is often the most rewarding and illuminating part of working with theory. Not only do models provide an efficient way of parceling and cataloging our understanding, but the struggle of fitting a simple mathematical model to a tremendously complex, real-world scenario often hastens the development of intuition. I try to gear the material in this book to highlight how useful modeling can be, in addition to developing most of the important theoretical concepts without becoming bogged down in the mathematics.

CONTRACT AND COOPERATIVE GAME THEORY

In some cases, it is helpful to simplify the analysis by abstracting from the idea that all decisions are treated as individual ones—that is, to depart from the strict noncooperative game theory mold. For example, instead of describing the complicated offers, counteroffers, and gestures available to people in a negotiation problem, one can sometimes find it useful to avoid modeling the negotiation procedure altogether and just think of the outcome of negotiation as a *joint action*. Analyzing behavior in models with joint actions requires a different set of concepts from those used for purely noncooperative environments; this alternative theory is called "cooperative game theory."

Cooperative game theory is often preferred for the study of *contractual relations,* in which parties negotiate and jointly agree on the terms of their relationship. Contractual relations form a large fraction of all strategic situations. In some cases, contracts are explicit and written, such as that between a worker and a manager, between a homeowner and a contractor, between countries that have agreed on a trade pact, or between husband and wife. In other cases, contracts are less formal, such as when fellow employees make a verbal agreement to coordinate on a project. This book considers contract as an integral part of strategic interaction. Contract selection and enforcement are therefore paid special attention.

For the sake of studying contract, this book presents some basic elements of cooperative game theory and provides a framework that generalizes noncooperative theory to incorporate joint actions.[4] Remember that incorporating joint actions into a strategic model is a shorthand device allowing one to represent agents negotiating over certain things, without explicitly modeling the negotiation process itself. The objects of negotiation are considered *spot-contractible* in the sense that an agreement on a joint action commits the agents to take this action.

THE MEANING OF "GAME"

Before we launch into the theory, a note on conflict and cooperation is in order. Most sports and leisure games are considered adversarial contests (in which someone wins and someone loses). However, as already noted, many settings of interdependence do not fit so neatly into this category. In fact, most situations contain elements of conflict as well as elements of potential cooperation or coordination or both. Consider a firm in which two managers interact to develop a new product. Their individual actions may affect each other's return from the project, so the setting involves interdependence. But must there be a winner and a loser? One can certainly imagine outcomes in which both managers "win" or "lose" to some degree. Perhaps if the managers cooperate with each other in developing the product, both will be likely to profit from the project. On the other hand, each of the managers may have an incentive to provide less effort than the other would prefer.

For another example having elements of both conflict and cooperation, consider a contracting problem between a worker and his employer. They may need to bargain over a wage contract prior to the production of an economic good. Although the interests of the parties may conflict regarding the worker's wage, their interests may be more aligned on another dimension. For instance, the parties may both prefer that the contract include a bonus for the worker to be granted in the event of exceptional performance on the job, because the bonus may give the worker the right incentive to generate profit that they can share. You may recognize this "concentrate on enlarging the pie" theme as the subject of some fluffy management-oriented books on bargaining. It is nonetheless a good example of how issues of conflict and cooperation arise simultaneously in many settings.

[4]To some extent, a rift has developed between the cooperative and the noncooperative approaches to studying games. As a result, "game theory" has developed somewhat independently of "contract theory," despite that the latter is, for the most part, an application of the former. (The study of contract also includes analysis of institutions, including law and politics.)

Keeping in mind that conflict and cooperation overlap, I take a broad view of what constitutes a game. In short, games are formal descriptions of strategic settings. Thus, game theory is a methodology of formally studying situations of interdependence. By "formally" I mean using a mathematically precise and logically consistent structure. With the right theoretical tools in place, we can study behavior in a variety of contexts and come to better understand economic and, more generally, social interaction.

This book is organized by conceptual topics, from the most basic (representations of games and simple models of behavior), to the more structured (equilibrium and institutions), to the most complicated (dynamic settings with incomplete information). Most chapters introduce a concept that builds from those developed in preceding chapters. In the first half of the book, I emphasize three major *tensions* of strategic interaction that are identified by the theory: (1) the conflict between individual and group interests, (2) strategic uncertainty, and (3) the specter of inefficient coordination. I highlight the institutions that help alleviate the tensions. I also attempt to cover a wide variety of applications of the theory. In each case, my presentation of applications is geared toward isolating their fundamental strategic and economic components. I always try to find the *simplest* model that brings out the essential intuition, rather than developing complicated and general applied models. Some of the chapters are solely dedicated to applied examples of the concepts developed previously; the titles of these chapters appear in italics in the table of contents. There also are two appendices; the first reviews some basic mathematical concepts that are used throughout the book, and the second elaborates on the material covered in Chapters 6, 7, and 9.

You will have already noticed that I prefer a relaxed style of writing. My aim is to engage you, rather than lecture to you. Thus, I use the pronouns "I," "we," and "you," with their standard meanings. I also toss in light-hearted comments here and there; I apologize if they rub you the wrong way. Regarding references, I cite relevant articles and books mainly to give credit to the originators of the basic concepts and to note historical events. If, after reading this book, you wish to probe the literature more deeply, I recommend consulting one of the fine graduate-level textbooks on game theory.[5] I view most graduate textbooks as good reference manuals; they are worth having nearby, but they are not much fun to read. Finally, to those who disapprove of my style, I can't

---

[5]Three popular books are D. Fudenberg and J. Tirole, *Game Theory* (Cambridge, MA: MIT Press, 1991); R. B. Myerson, *Game Theory* (Cambridge, MA: Harvard University Press, 1991); and M. J. Osborne and A. Rubinstein, *A Course in Game Theory* (Cambridge, MA: MIT Press, 1994).

resist taking liberties with a quip of Michael Feldman to say "write your own books."[6]

With all of these introductory comments now out of the way, let the fun begin.

[6]Michael Feldman hosts *Whad'Ya Know?*—a popular U.S. radio program produced by Wisconsin Public Radio.

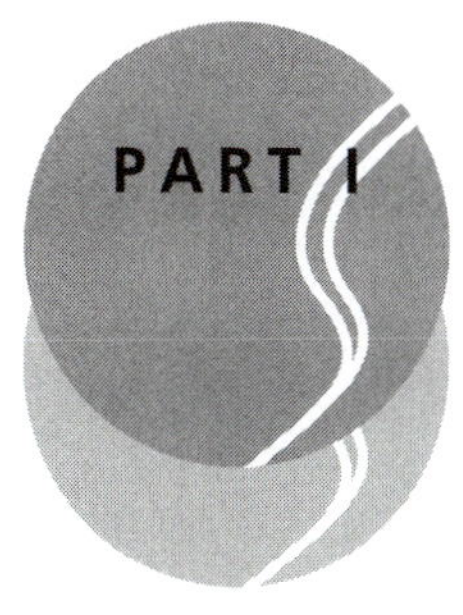

Representations and Basic Assumptions

There are several different ways of describing games mathematically. The representations presented here have the following formal elements in common:

1. a list of players,
2. a complete description of what the players can do (their possible actions),
3. a description of what the players know when they act,
4. a specification of how the players' actions lead to outcomes, and
5. a specification of the players' preferences over outcomes.

At this level of abstraction, the mathematical representation of a game is similar to the description of games of leisure. For example, the rules of the board game chess specify elements 1 through 4 precisely: (1) there are two players; (2) the players alternate in moving pieces on the game board, subject to rules about what moves can be made in any given configuration of the board; (3) players observe each other's moves, so each knows the entire history of play as the game progresses; (4) a player who captures the other player's king wins the game; in certain situations, a draw is declared. Although element 5 is not implied by the rules of chess, one generally can assume that players prefer winning over a draw and a draw over losing.

There are two common forms in which noncooperative games are represented mathematically: the *extensive form* and the *normal (strategic) form*. I begin with a nontechnical description of the extensive form.

2 THE EXTENSIVE FORM

In the fall of 1998, moviegoers had a generous selection of animated feature films from which to choose. Curiously, two of the films were about bugs: Disney Studio's *A Bug's Life* and DreamWorks SKG's *Antz*. Audiences marveled at the computer-generated, cuteness-enhanced critters, as many followed the rivalry that brought the twin films into head-to-head competition. Rumor has it that Disney executives pondered the idea of an animated bug movie in the late 1980s, during Jeffrey Katzenberg's term in charge of the company's studios.[1] However, *A Bug's Life* was not conceived until after Katzenberg left Disney in a huff (he was not given a promotion). Katzenberg resigned in August 1994; shortly thereafter, Pixar Animation pitched *A Bug's Life* to Disney, Michael Eisner accepted the proposal, and the film went into production.

At about the same time, Katzenberg joined with Steven Spielberg and David Geffen to form DreamWorks SKG, a new studio with great expectations. Shortly thereafter, SKG teamed with computer animation firm PDI to produce *Antz*. The two studios may have learned about each other's choices only after they had already made their own decisions. Disney chose to release *A Bug's Life* in the 1998 Thanksgiving season, when SKG's *Prince of Egypt* was originally scheduled to open in theaters. In response, SKG decided to delay the release of *Prince of Egypt* until the Christmas season and rushed to complete *Antz* so that it could open before *A Bug's Life* and claim the title of "first animated bug movie."

This story is intriguing because of the larger-than-life characters, the legal issues (did Katzenberg steal the bug idea from Disney?), and the complicated business strategy. In addition, there were bad feelings. Katzenberg sued Disney for unpaid bonuses. Eisner was embarrassed to admit in court that he may have said of Katzenberg, "I hate the little midget." Word on the street was that Pixar CEO Steve Jobs (also of Apple fame) believed Katzenberg stole the bug movie concept. (Someone ought to make a movie about this story.)

[1] Katzenberg was a major force in reviving the Disney animation department after being wooed from Paramount by Disney's boss, Michael Eisner. For a report on the events summarized here, see *Time*, September 28, 1998, p. 81.

FIGURE 2.1
Katzenberg's first move.

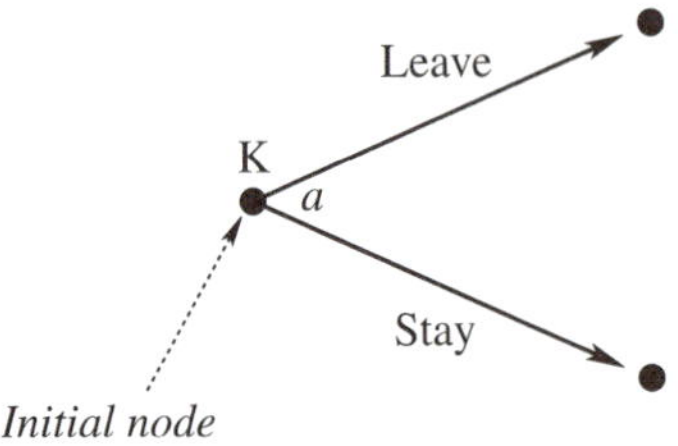

Let us use a mathematical model to tell the story of the bug films. To convert it into the abstract language of mathematics, we will have to abbreviate and stylize the story quite a bit. Our goal is to isolate just one or two of the strategic elements. This isolation will help us, as theorists, to understand the strategic setting and, as game players (putting ourselves in Eisner's or Katzenberg's shoes), to determine the best decisions. I expect that Eisner and Katzenberg have done this, as I expect have most successful business people.

I suggest that we focus on the competition between the two movies and, to keep things simple, concentrate on just some of the major decisions that influenced the market outcome. Think of it as a game between Katzenberg and Eisner, who are the players in our model. We can use a *tree* to graphically represent the strategic interaction between these two people. The tree is defined by *nodes* and *branches*. Nodes represent places where something happens in the game (such as a decision by one of the players), and branches indicate the various actions that players can choose. We represent nodes by solid circles and branches by arrows connecting the nodes. A properly constructed tree is called an *extensive-form representation*.[2]

To design the tree for the Katzenberg–Eisner game, it is useful to think about the chronological sequence of events as described in the story. (You will see, however, that not every extensive form represents the chronology of a strategic situation.) Let the game begin with Katzenberg's decision about whether to leave Disney. Node a in Figure 2.1 signifies where this decision is made. Because this decision is the start of the game, a is called the *initial node*. Every extensive-form game has exactly one initial node. Katzenberg's two options—stay and leave—correspond to the two branches, which are graphed as arrows from node a. Note that the branches are named and

[2]The extensive form was detailed in J. von Neumann and O. Morgenstern, *Theory of Games and Economic Behavior* (Princeton, NJ: Princeton University Press, 1944). Some of the material originally appeared in von Neumann, "Zur Theorie der Gesellschaftsspiele," *Mathematische Annalen* 100(1928):295–320, translated as "On the Theory of Games of Strategy," in *Contributions to the Theory of Games,* vol. IV (*Annals of Mathematics Studies*, 40), ed. A. W. Tucker and R. D. Luce (Princeton, NJ: Princeton University Press, 1959), pp. 13–42. More modern definitions have been provided by H. W. Kuhn, "Extensive Games and the Problem of Information," in *Contributions to the Theory of Games,* Vol. II (*Annals of Mathematics Studies*, 28), ed. H. W. Kuhn and A. W. Tucker (Princeton, NJ: Princeton University Press, 1953), pp. 193–216.

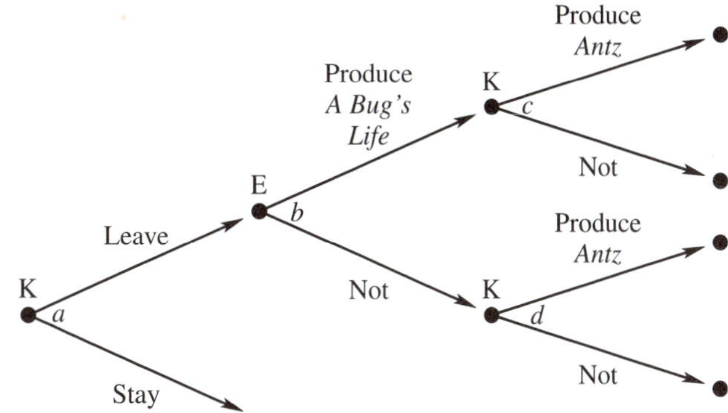

FIGURE 2.2
Adding the production decisions.

node *a* is labeled with Katzenberg's initial, signifying that it is his move in the game. These branches lead from node *a* to two other nodes.

If Katzenberg decides to stay at Disney (the lower branch from node *a*), assume that the game is over. On the other hand, if Katzenberg decides to leave, then other decisions have to be made. First, Eisner must decide whether to produce *A Bug's Life*. Figure 2.2 shows how the tree is expanded to include Eisner's choice. Note that, because Eisner has to make this decision only if Katzenberg has left Disney, Eisner's move occurs at node *b*. Eisner's two options—produce *A Bug's Life* or not—are represented by the two branches leading from node *b* to two other nodes, *c* and *d*.

After Eisner decides whether to produce *A Bug's Life,* Katzenberg must choose whether to produce *Antz*. Katzenberg's decision takes place at either node *c* or node *d*, depending on whether Eisner selected produce or not, as depicted in Figure 2.2. Note that there are two branches from node *c* and two from node *d*. Observe that Katzenberg's initial is placed next to *c* and *d*, because he is on the move at these nodes.

At this point, we have to address a critical matter: information. Specifically, does the tree specified so far properly capture the information that players have when they make decisions? With the extensive form, we can represent the players' information by describing whether they know where they are in the tree as the game progresses. For example, when Katzenberg decides whether to stay or leave, he knows that he is making the first move in the game. In other words, at node *a* Katzenberg *knows that he is at node a*. Further, because Eisner observes whether Katzenberg stays or leaves, when Eisner has to decide whether to produce *A Bug's Life* he knows that he is at node *b*.

However, as the story indicates, each player has to select between producing or not producing *without* knowing whether the other player has decided to produce. In particular, Katzenberg must choose whether to produce *Antz* before learning whether Eisner is producing *A Bug's Life*. The players learn

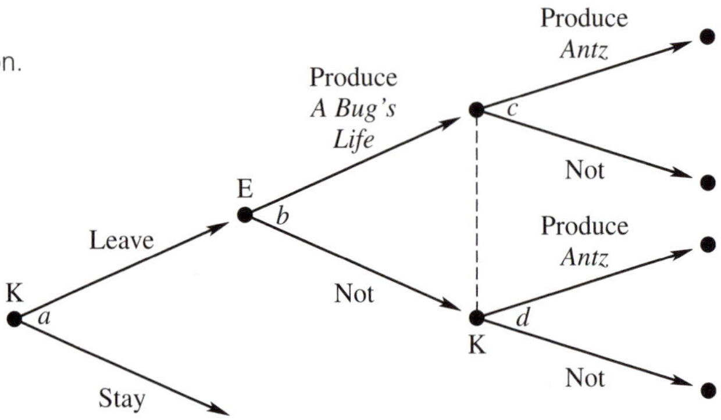

FIGURE 2.3
Capturing lack of information.

about each other's choices only after both are made. Referring again to the tree in Figure 2.2, we represent Katzenberg's lack of information by specifying that, during the game, he *cannot distinguish between nodes c and d*. In other words, when Katzenberg is on the move at either *c* or *d*, he knows that he is at one of these nodes but he does not know which one. Figure 2.3 captures this lack of information with a dashed line connecting nodes *c* and *d*. Because these two nodes are connected with the dashed line, we need to label only one of them with Katzenberg's initial.

Assume that, if either or both players chose not to produce his film proposal, then the game ends. If both players opted to produce, then one more decision has to be made by Katzenberg: whether to release *Antz* early (so it beats *A Bug's Life* to the theaters). Adding this decision to the tree yields Figure 2.4. Katzenberg makes this choice at node *e*, after learning that Eisner decided to produce *A Bug's Life*.

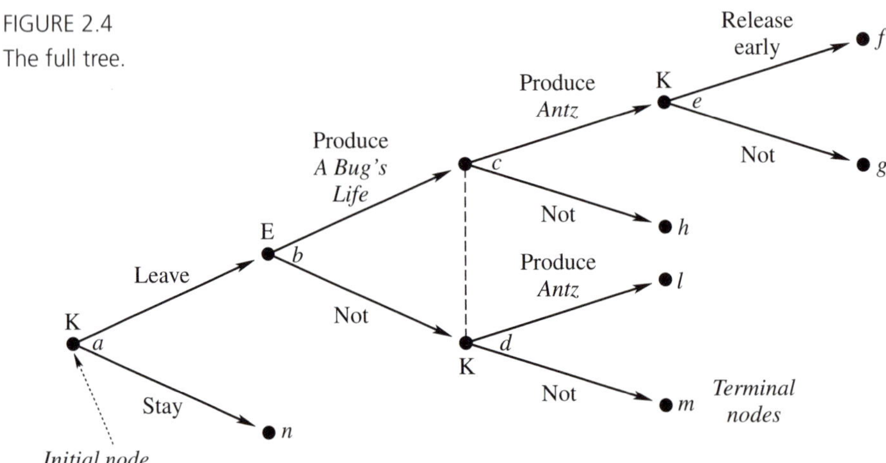

FIGURE 2.4
The full tree.

Figure 2.4 describes all of the players' actions and information in the game. Nodes a, b, c, d, and e are called *decision nodes,* because players make decisions at these places in the game. The other nodes (f, g, h, l, m, and n) are called *terminal nodes*; they represent *outcomes* of the game—places where the game ends. Each terminal node also corresponds to a unique *path* through the tree, which is a way of getting from the initial node through the tree by following branches in the direction of the arrows. In an extensive form, there is a one-to-one relation between paths and terminal nodes.[3]

It is common to use the term *information set* to specify the players' information at decision nodes in the game. An information set describes which decision nodes are connected to each other by dashed lines (meaning that a player cannot distinguish between them). Every decision node is contained in an information set; some information sets consist of only one node. For example, the information set for node a comprises just this node—Katzenberg can distinguish this node from his other decision nodes. Nodes b and e also are their own separate information sets. Nodes c and d, however, are in the same information set.

The information sets in a game precisely describe the different decisions that players have to make. For example, Katzenberg has three decisions to make in the game at hand: one at the information set given by node a, another at the information set containing nodes c and d, and a third at the information set given by node e. Eisner has one decision to make—at node b. Remember that only one decision is made at each information set. For example, because nodes c and d are in the same information set, Katzenberg makes the same choice at c as he does at d (produce or not). We always assume that all nodes in an information set are decision nodes for the same player.

You should check that the tree in Figure 2.4 delineates game elements 1 through 4 noted in the introduction to this part of the book (see page 9). We have just one more element to address: the players' preferences over the outcomes. To understand preferences, we must ask questions such as: Would Katzenberg prefer that the game end at terminal node f rather than at terminal node l? To answer such questions, we have to know what the players care about. We can then *rank* the terminal nodes in order of preference for each player. For example, Eisner may have the ranking h, g, f, n, l, m; in words, his favorite outcome is h, followed by g, and so on. It is usually most convenient to represent the players' preference rankings with numbers, which are called *payoffs* or *utilities*. Larger payoff numbers signify more preferred outcomes.

For many economic games, it is reasonable to suppose that the players care about their monetary rewards (profit). Katzenberg and Eisner probably

---

[3]I provide the technical details in Chapter 14.

FIGURE 2.5
The extensive form of the bug game.

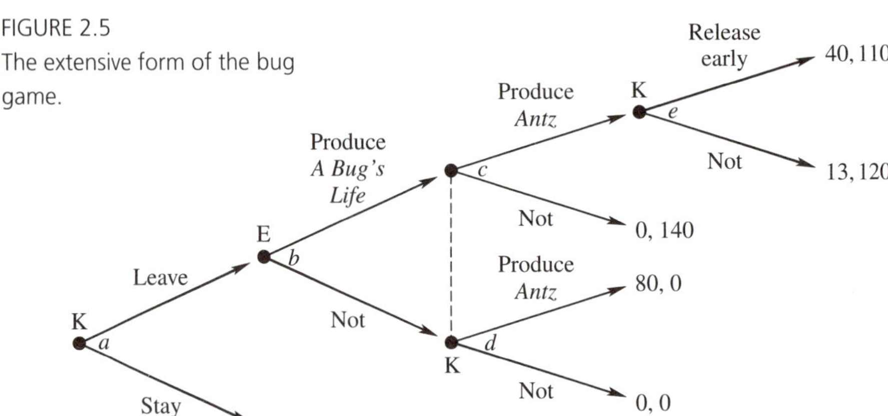

want to maximize their individual monetary gains. Therefore, let us define the payoff numbers as the profits that each obtains in the various outcomes. (The profits in this example are purely fictional—Who knows what a film's true profit is?) For example, in the event that Katzenberg stays at Disney, suppose he gets \$35 million and Eisner gets \$100 million. Then we can say that node n yields a payoff vector of (35, 100). This payoff vector, as well as payoffs for the other terminal nodes, is recorded in Figure 2.5. Note that, relative to Figure 2.4, the terminal nodes are replaced by the payoff vectors. In addition, we use the convention that one player's payoff is always listed first. Because Katzenberg is the first player to move in this game, I have listed his payoff first.[4]

Figure 2.5 depicts the full extensive form of the Katzenberg–Eisner game; it represents all of the strategic elements. A more compact representation (with all actions abbreviated to single letters) appears in Figure 2.6. This compact representation is usually the style in which I will present extensive forms to you. Observe that I have labeled one of Katzenberg's action choices N′ to differentiate it from the other "Not" action earlier in the tree. Then I can simply give you the name of a player and the name of an action, and you will be able to figure out where I am referring to on the tree. It will be important that you differentiate actions in this way whenever you draw an extensive-form game. Incidentally, you should always maintain conformity in labeling branches from nodes in the same information set. For example, Katzenberg's choices regarding whether to produce *Antz* are represented by actions P and N in Figure 2.6. The labels P and N are each used for two branches—from nodes

[4]We might also think that the players care about other things in addition to money, such as revenge. Although I have left them out of the current model, these considerations can be easily included in the specification of preferences.

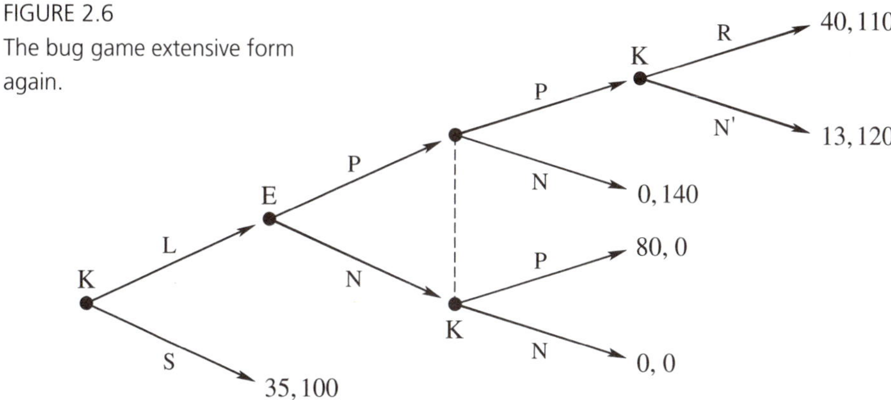

FIGURE 2.6
The bug game extensive form again.

c and *d,* which are in the same information set (refer to Figure 2.5). When Katzenberg finds himself at either node *c* or node *d* in the game, he knows only that he is at one of these two nodes and has to choose between P and N. In other words, this information set defines a single place where Katzenberg has to make a decision.

In the end, *Antz* produced a bit more than \$90 million in revenue for DreamWorks SKG, whereas *A Bug's Life* secured more than \$160 million. I guess these films each cost about \$50 million to produce and market, meaning that Katzenberg and Eisner faired pretty well. But did they make the "right" decisions? We will probably not be able to answer this question fully by using the simple game tree just developed, but analysis of our extensive form will yield some insights that are useful for instructing Katzenberg, Eisner, and other game players in how best to play. In addition, designing and analyzing game models will help you understand a wide range of strategic issues. Do not ignore the possibility that you may be advising the likes of Katzenberg and Eisner someday. One thing is obvious to me: I will not get \$160 million in revenue from this textbook.

OTHER EXAMPLES AND CONVENTIONS

So that you get a bit more exposure to the extensive form, and so that I can register some technical notes, consider a few abstract examples of games. Figure 2.7(a), on the next page, depicts a simple market game, where two firms compete by each selecting either a high (H) or low (L) price for a product that they both produce. Note that I have labeled the decision nodes with the *player numbers;* that is, firm 1 moves first, followed by firm 2. It is often useful to refer to players by number. Then, when we wish to speak of a generic

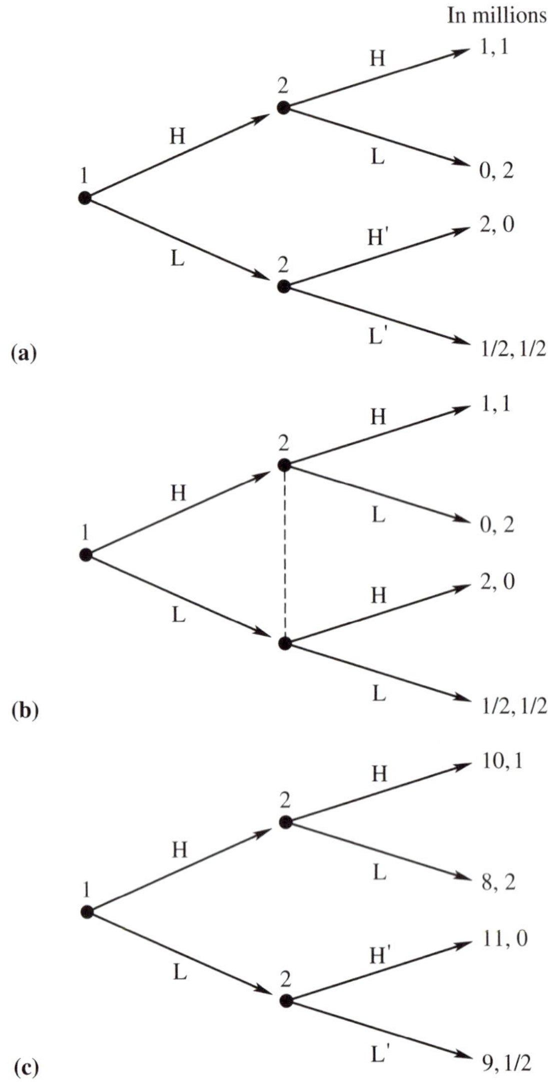

FIGURE 2.7
A price-competition game.

player, we can use the phrase "player i," where i stands for any one of the player numbers; that is, in a game with n players, $i = 1, 2, \ldots, n$. Remember to use the convention of listing the players' payoffs in the order of the players' identifying numbers. In general, player 1's payoff is written first, followed by player 2's payoff, and so on.

Observe that in Figure 2.7(a) I have used different action labels to describe firm 2's high and low price options in this firm's two information sets (the top and bottom nodes). This helps us avoid ambiguous statements such as, "Firm 2 chose a high price." In fact, firm 2 has two decisions to make in this game, one at its top node and another at its bottom node. By following the labeling convention, we are correctly forced to say, "Firm 2 chose a high price at its top information set" (action H) or "Firm 2 chose a high price at its bottom information set" (action H′).

Figure 2.7(b) depicts a game in which the firms select their prices simultaneously and independently, so neither firm observes the other's move before making its choice. In an extensive form, we must draw one player's decision before that of the other, but it is important to realize that this does not necessarily correspond to the *actual* timing of the strategic setting. In this example, the moves are simultaneous. The key to modeling simultaneous choice in the extensive form is to get the information sets right. Because firm 2 moves at the same time as does firm 1, firm 2 does not get to observe firm 1's selection before making its own choice. Thus, firm 2 cannot distinguish between its two decision nodes—they are in the same information set and therefore connected with a dashed line. To check your understanding, you might see how one can draw the extensive form in Figure 2.7(b) so that firm 2's action occurs at the initial node, followed by firm 1's decision. However the game is drawn, each firm has just one decision to make.

For games with monetary rewards, such as the Katzenberg–Eisner game and the one in Figure 2.7, it is convenient to use the monetary amounts as the payoffs themselves—as long as the players prefer more money to less. In nonrandom settings, in fact, any utility numbers will work as long as they preserve the players' preference *ranking*. For example, the extensive form in Figure 2.7(c) is the same as the extensive form in Figure 2.7(a), except that player 1's payoff numbers are different. Because the numbers follow the same ranking in these games, both correctly represent the preferences of player 1. Player 1 prefers the top terminal node to the bottom terminal node in both extensive forms, he prefers the top terminal node to the second terminal node, and so on. For now, either extensive form can be used. I say "for now" because, as noted earlier, these comments apply only to nonrandom outcomes. Later you will see that more than the order matters when players are uncertain of others' behavior or randomize in the selection of their own actions (this is briefly addressed in Chapter 4 and more thoroughly discussed in Chapter 25).

GUIDED EXERCISE

Problem: Represent the following game in extensive form. Firm A decides whether to enter firm B's industry. Firm B observes this decision. If firm A enters, then the two firms simultaneously decide whether to advertise. Otherwise, firm B alone decides whether to advertise. With two firms in the market, the firms earn profits of \$3 million each if they both advertise and \$5 million if they both do not advertise. If only one firm advertises, then it earns \$6 million and the other earns \$1 million. When firm B is solely in the industry, it earns \$4 million if it advertises and \$3.5 million if it does not advertise. Firm A earns nothing if it does not enter.

Solution: Let E and D denote firm A's initial alternatives of entering and not entering B's industry. Let a and n stand for "advertise" and "not advertise," respectively. Then the following extensive-form diagram represents the strategic setting.

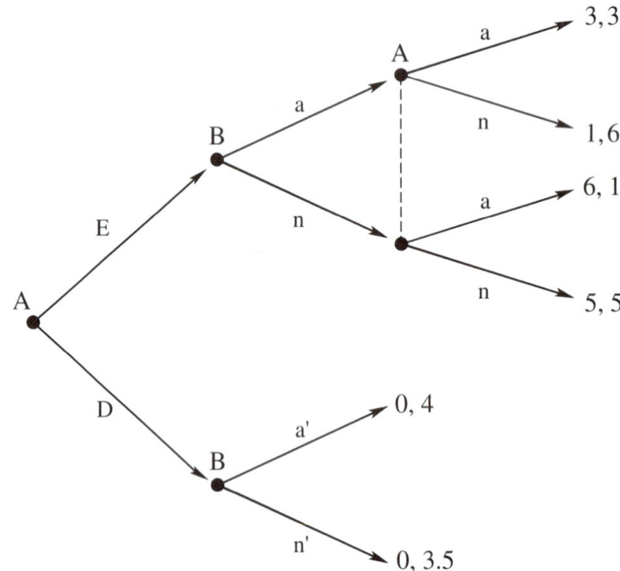

Note that simultaneous advertising decisions are captured by assuming that, at firm A's second information set, firm A does not know whether firm B chose a or n. Also note that primes are used in the action labels at firm B's lower information set to differentiate them from the actions taken at B's top information set.

EXERCISES

1. Represent the following strategic situation as an extensive-form game. Janet is a contestant on a popular game show and her task is to guess behind which door Liz, another contestant, is standing. With Janet out of the room, Liz chooses a

door behind which to stand—either door A or door B. The host, Monty, observes this choice. Janet, not having observed Liz's choice, then enters the room. Monty says to Janet either "Red" or "Green" (which sounds silly, of course, but it is a silly game show). After hearing Monty's statement, Janet picks a door (she says either "A" or "B"). If she picks the correct door, then she wins $100. If she picks the wrong door, then she wins nothing. Liz wins $100 if Janet picks the wrong door and nothing if she picks the correct door. (Thus, Liz would like to hide from Janet, and Janet would like to find Liz.) Monty likes the letter A. If Janet selects door A, then this selection makes Monty happy to the tune of 10 units of utility. If she selects door B, then Monty receives 0 utility units.

2. Consider the following strategic situation concerning the owner of a firm (O), the manager of the firm (M), and a potential worker (W). The owner first decides whether to hire the worker, to refuse to hire the worker, or to let the manager make the decision. If the owner lets the manager make the decision, then the manager must choose between hiring the worker or not hiring the worker. If the worker is hired, then he or she chooses between working diligently and shirking. Assume that the worker does not know whether he or she was hired by the manager or the owner when he or she makes this decision. If the worker is not hired, then all three players get a payoff of 0. If the worker is hired and shirks, then the owner and manager each get a payoff of -1, whereas the worker gets 1. If the worker is hired by the owner and works diligently, then the owner gets a payoff of 1, the manager gets 0, and the worker gets 0. If the worker is hired by the manager and works diligently, then the owner gets 0, the manager gets 1, and the worker gets 1. Represent this game in the extensive form (draw the game tree).

3. Draw the extensive form for the following game (invent your own payoff vectors, because I give you no payoff information). There is an industry in which two firms compete as follows: First, firm 1 decides whether to set a high price (H) or a low price (L). After seeing firm 1's price, firm 2 decides whether to set a high price (H) or a low price (L). If both firms selected the low price, then the game ends with no further interaction. If either or both firms selected the high price, then the attorney general decides whether to prosecute (P) or not (N) for anticompetitive behavior. In this case, the attorney general does not observe which firm selected the high price (or if both firms selected the high price).

4. The following game is routinely played by youngsters—and adults as well—throughout the world. Two players simultaneously throw their right arms up and down to the count of "one, two, three." (Nothing strategic happens as they do this.) On the count of three, each player quickly forms his or her hand into the shape of either a rock, a piece of paper, or a pair of scissors. Abbreviate these shapes as R, P, and S, respectively. The players make this choice at the same time. If the players pick the same shape, then the game ends in a tie. Otherwise, one of the players wins and the other loses. The winner is determined by the following rule: rock beats scissors, scissors beats paper, and paper beats rock. Each player obtains a payoff of 1 if he or she wins, -1 if he or she loses, and 0 if he or she ties. Represent this game in the extensive form. Also discuss the relevance of the order of play (which of the players has the move at the initial node) in the extensive form.

5. Consider the following strategic setting. There are three people: Amy, Bart, and Chris. Amy and Bart have hats. These three people are arranged in a room so that Bart can see everything that Amy does, Chris can see everything that Bart does, but the players can see nothing else. In particular, Chris cannot see what Amy does. First, Amy chooses either to put her hat on her head (abbreviated by H) or on the floor (F). After observing Amy's move, Bart chooses between putting his hat on his head or on the floor. If Bart puts his hat on his head, the game ends and everyone gets a payoff of 0. If Bart puts his hat on the floor, then Chris must guess whether Amy's hat is on her head by saying either "yes" or "no." This ends the game. If Chris guesses correctly, then he gets a payoff of 1 and Amy gets a payoff of -1. If he guesses incorrectly, then these payoffs are reversed. Bart's payoff is 0, regardless of what happens. Represent this game in the extensive form (draw the game tree).

6. Represent the following game in the extensive form. There are three players, numbered 1, 2, and 3. At the beginning of the game, players 1 and 2 simultaneously make decisions, each choosing between "X" and "Y." If they both choose "X," then the game ends and the payoff vector is $(1, 0, 0)$; that is, player 1 gets 1, player 2 gets 0, and player 3 gets 0. If they both choose "Y," then the game ends and the payoff vector is $(0, 1, 0)$; that is, player 2 gets 1 and the other players get 0. If one player chooses "X" while the other chooses "Y," then player 3 must guess which of the players selected "X"; that is, player 3 must choose between "1" and "2." Player 3 makes his selection knowing only that the game did not end following the choices of players 1 and 2. If player 3 guesses correctly, then both he and the player who selected "X" each obtains a payoff of 2, and the player who selected "Y" gets 0. If player 3 guesses incorrectly, then everyone gets a payoff of 0.

7. Draw the extensive-form diagram for the following strategic setting. There are three people: Amy, Bart, and Chris. Amy and Bart each have two cards, one of which has "K" (for King) written on it and the other has "Q" (for Queen) written on it; that is, Amy and Bart both have a King and a Queen. At the beginning of the game, Amy must place one of her cards (either K or Q) into an envelope and then give the envelope to Bart. Bart sees the card that Amy placed into the envelope, and then he places one of his cards (either K or Q) into the envelope as well. The envelope is then given to Chris, who has *not* observed the moves of Amy and Bart. Chris opens the envelope. Chris sees the two cards inside, but she does not know which card was placed there by Amy and which card was deposited by Bart. After observing the contents of the envelope, Chris selects "yes" (Y) or "no" (N). If Chris selects Y and Amy had put a King in the envelope, then Amy and Bart each get a payoff of 0 and Chris gets 1. If Chris selects N and Amy had put a Queen in the envelope then, again, Amy and Bart each get a payoff of 0 and Chris gets 1. In all other outcomes, Amy and Bart each get a payoff of 1 and Chris gets 0.

8. Consider the following strategic setting. There are three players, numbered 1, 2, and 3. Player 1 has two cards, labeled King and Ace. At the beginning of the game, player 1 deals one of the cards to player 2 and the other card to player 3; that is, player 1 either gives the Ace to player 3 and the King to player 2 (call this the action A) or the King to player 3 and the Ace to player 2 (action K). Player 2 observes the card dealt to him; player 3 does not get to see the card dealt to her. Player 2 then must decide between switching cards with player 3 (S) or not (N). Player 3 observes whether player 2 made the switch, but does not see her card. Finally, player 3 responds to the question "Is your card the Ace?" by saying either "yes" (Y) or "no" (N). If player 3 correctly states whether her card is the Ace, then she obtains a payoff of 1 and the other players get 0; otherwise, players 1 and 2 both get a payoff of 1 and player 3 obtains 0. Represent this game in the extensive form.

3 STRATEGIES AND THE NORMAL FORM

The most important concept in the theory of games is the notion of a *strategy*. The formal definition is simple:

> *A* **strategy** *is a complete contingent plan for a player in the game.*

The "complete contingent" part of this definition is what makes it powerful and, to some people, a bit confusing. By *complete contingent plan* I mean a full specification of a player's behavior, which describes the actions that the player would take at *each* of his possible decision points. Because information sets represent places in the game at which players make decisions, *a player's strategy describes what he will do at each of his information sets.*

Consider the Katzenberg–Eisner game depicted in Figure 2.5, and put yourself in Katzenberg's shoes. Your strategy must include what to do at the information set given by node *a*, what action you would pick at the *c–d* information set, and what your choice would be at the information set given by node *e*. Your strategy must specify all of these things *even* if you plan to select "Stay" at node *a*. (This point is elaborated later in this chapter.) As another example, in the game depicted in Figure 2.7(a), player 2's strategy tells us what this player will do at his top node (an information set) and at his bottom node (another information set). In words, in this game, a strategy for player 2 specifies what price to pick in response to H selected by player 1 as well as what price to pick in response to player 1's selection of L.

The easiest way of writing strategies is to put together the labels corresponding to the actions to be chosen at each information set. For example, for the game shown in Figure 2.7(a), one strategy for player 2 is to select the high price (H) in the top information set and the low price (L′) in the bottom information set; this strategy can be written HL′. Note that there are four strategies for player 2 in this game: HH′, HL′, LH′, and LL′. Also note that writing strategies in this way would be difficult if we did not use different labels for a player's various information sets. For example, if I wrote H and L instead of H′ and L′ in the bottom information set of player 2, "HL" would be ambiguous (is it H in the top information set and L in the bottom one, or vice versa?).

TERMINOLOGY AND NOTATION FOR STRATEGIES

It is now time to become more familiar with the theory. Formally, given a game, we let S_i denote the *strategy space* (also called the *strategy set*) of player i. That is, S_i is a set comprising each of the possible strategies of player i in the game. For the game discussed in the last paragraph, the strategy space of player 1 is $S_1 = \{\text{H}, \text{L}\}$ and the strategy space of player 2 is $S_2 = \{\text{HH}', \text{HL}', \text{LH}', \text{LL}'\}$. We use lowercase letters to denote single strategies (generic members of these sets). Thus $s_i \in S_i$ is a strategy for player i in the game. We could thus have $s_1 = \text{L}$ and $s_2 = \text{LH}'$, for instance. A *strategy profile* is a vector of strategies, one for each player. In other words, a strategy profile describes strategies for all of the players in the game. For example, suppose we are studying a game with n players. Then a typical strategy profile is a vector $s = (s_1, s_2, \ldots, s_n)$, where s_i is the strategy of player i, for $i = 1, 2, \ldots, n$. Let S denote the set of strategy profiles. Mathematically, $S = S_1 \times S_2 \times \cdots \times S_n$.[1]

Given a single player i, we often need to speak of the strategies chosen by all of the other players in the game. As a matter of notation, it will be convenient to use the term $-i$ to refer to all of the players except player i. Thus, s_{-i} is a strategy profile for everyone except player i:

$$s_{-i} = (s_1, s_2, \ldots, s_{i-1}, s_{i+1}, \ldots, s_n).$$

Separating a strategy profile s into the strategy of player i and the strategies of the other players, we write $s = (s_i, s_{-i})$.[2] Incidentally, sometimes I refer to the "$-i$" players as *player* i's *opponents,* but note that this expression is not appropriate for all strategic situations, because many games have cooperative elements. I will occasionally use "opponent" for adversarial settings or when it is the most convenient way of referring to "the other players."

Consider the examples pictured in Figures 3.1 and 3.2 on the next pages. The game in Figure 3.1(a) models a setting in which a firm may or may not exit a competitive industry. Firm 1 decides whether to be aggressive in the market (A), to be passive in the market (P), or to leave the market (O). If firm 1 leaves, then firm 2 enjoys a monopoly. Otherwise the firms compete and firm 2 selects whether or not to assume an aggressive stance. Furthermore, when firm 2 makes its decision, it knows only whether firm 1 is in or out of

[1]The symbol "×" denotes the Cartesian product. For example, if $S_1 = \{\text{A}, \text{B}\}$ and $S_2 = \{\text{X}, \text{Y}\}$ then $S = S_1 \times S_2 = \{(\text{A}, \text{X}), (\text{A}, \text{Y}), (\text{B}, \text{X}), (\text{B}, \text{Y})\}$.

[2]For example, in a three-player game with the strategy profile $s = (\text{B}, \text{X}, \text{Y})$, we have $s_{-2} = (\text{B}, \text{Y})$. Note that expressions such as $s = (s_2, s_{-2})$ can create some ambiguity because they reorder the players' individual components; the ambiguity can usually be avoided with clarifying remarks.

FIGURE 3.1 Some extensive-form games I.

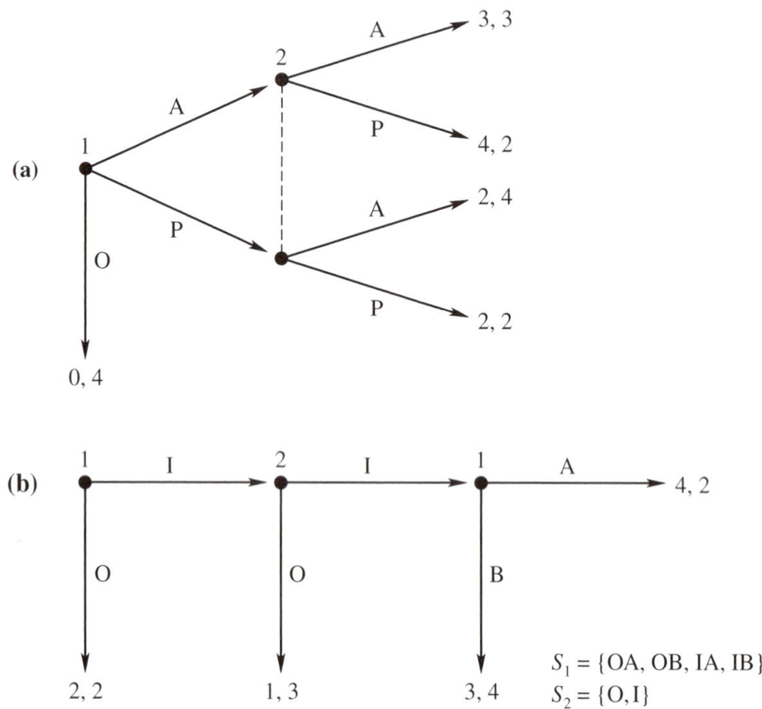

the market; firm 2 does not observe firm 1's competitive stance before taking its action. In this game, there is one information set for firm 1 (the initial node) and one for firm 2. The strategy sets are $S_1 = \{A, P, O\}$ and $S_2 = \{A, P\}$.

In the game in Figure 3.1(b), player 1 decides between "out" (O) and "in" (I). If he chooses O, then the game ends with a payoff vector of (2, 2). If he selects I, then player 2 is faced with the same two choices. If player 2 then chooses O, the game ends with the payoff vector (1, 3). If she picks I, then player 1 has another choice to make, between A and B (ending the game with payoffs (4, 2) and (3, 4), respectively). Player 1 has two information sets and player 2 has one information set. Note that, in this game, player 1's strategy specifies what he will do at the beginning of the game *and* what action he would take at his second information set. This specification may seem a bit nonsensical. Why, you may ask, should player 1 decide what to do at his second information set if he selects O at the beginning of the game? By selecting O, he guarantees that his second information set will not be reached. Doesn't player 1 have only three strategies, O, IA, and IB? Remember that our

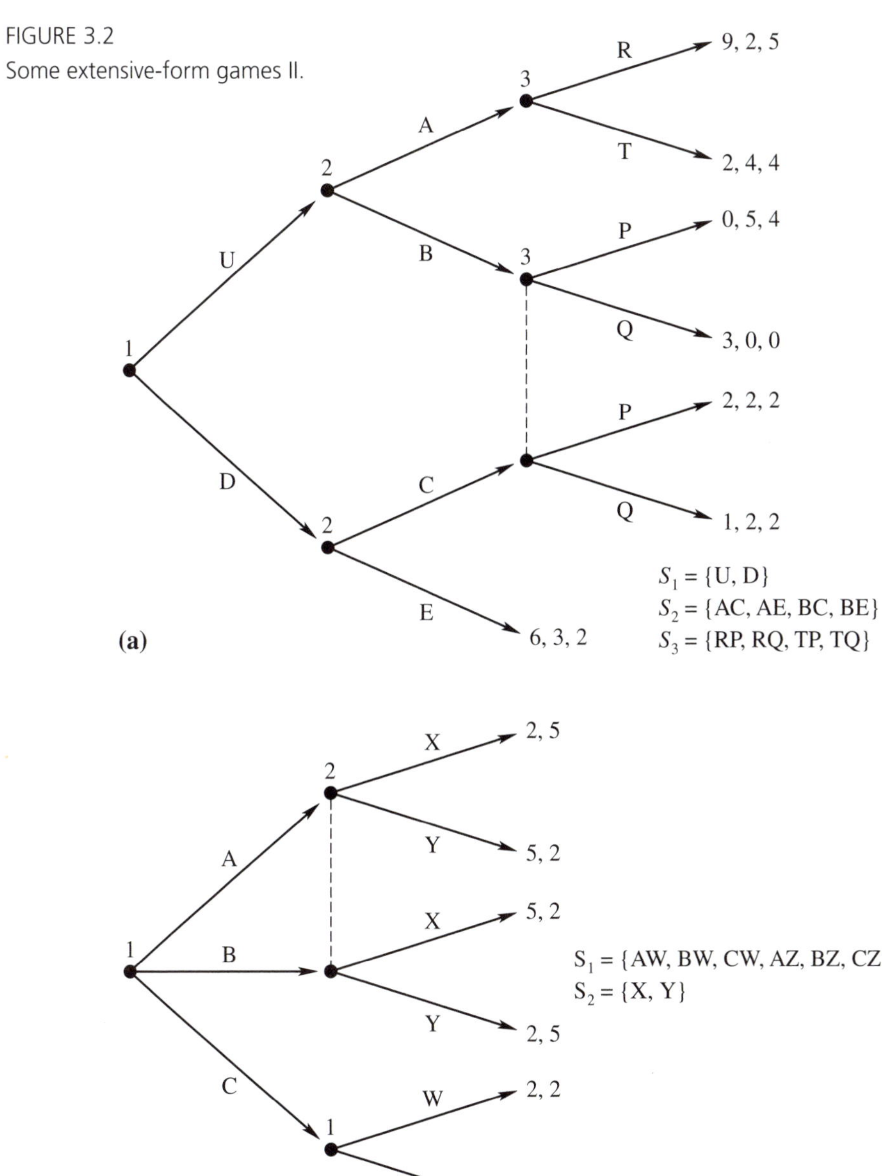

FIGURE 3.2
Some extensive-form games II.

definition of *strategy* is a complete specification of what a player does at each of his or her information sets. Because in the game shown in Figure 3.1(b) player 1 has two information sets (the first and third nodes), a strategy for player 1 prescribes actions to be taken at both of them, even if the action taken at the first information set precludes the second from being reached. Thus, player 1 has four strategies: OA, OB, IA, and IB.

At this point, you can see that "complete contingent plan" means more than just a "plan" for how to play the game. Indeed, why, you might wonder, should player 1 in the game shown in Figure 3.1(b) need a plan for his second information set if he selects O at the first? One reason is that our study of rationality will explicitly require the evaluation of players' optimal moves starting from arbitrary points in a game. This evaluation is connected to the beliefs that players have about each other. For example, in the game depicted in Figure 3.1(b), player 1's optimal choice at his first information set depends on what he thinks player 2 would do if put on the move. Furthermore, to select the best course of action, perspicacious player 2 must consider what player 1 would do at his second information set. Thus, player 2 must form a belief about player 1's action at the third node. A belief is a conjecture about what strategy the other player is using; therefore, player 1's strategy must include a prescription for his second information set, regardless of what that strategy prescribes for his first information set.

Another reason why it is important to specify a complete contingent plan is that players may need a contingency plan in case they make mistakes. For example, suppose that you are player 1 in the game shown in Figure 3.1(b). Further suppose that I am instructing you on how to play this game. I might write instructions on a piece of paper, hand the paper to you, and then leave town before the game begins. I could write "select O," but, even though that instruction seems to be enough information in advance, it may not give you enough guidance. You might make a mistake and select I instead of O at the beginning of the game. What if player 2 then picks I? There you are, red faced and indignant, yelling at the sky, "Now what am I supposed to do?"

Examine the other examples in Figure 3.2 to solidify your understanding of strategy. Note that, as required, I have used different labels to describe the actions that can be taken at different information sets.

THE NORMAL FORM

The extensive form is one straightforward way of representing a game. Another way of formally describing games is based on the idea of strategies. The alternative representation is more compact than the extensive form in some settings and may be preferred. As we develop concepts of rationality for games, you will notice the subtle differences between the two representations.

For a game in extensive form, we can describe the strategy spaces of the players. Furthermore, notice that each strategy profile fully describes how the game is played. That is, a strategy profile tells us what path through the tree is followed and, equivalently, which terminal node is reached to end the game. Associated with each terminal node (which we may call an *outcome*) is a payoff vector for the players. Notice, therefore, that each strategy profile implies a payoff vector.

For each player i, we can define a function $u_i : S \rightarrow \mathbf{R}$ (a function whose domain is the set of strategy profiles and whose range is the real numbers) so that, for each strategy profile $s \in S$ that the players choose, $u_i(s)$ is player i's payoff in the game. This function u_i is called player i's *payoff function.* As an example, take the game pictured in Figure 3.1(b). The set of strategy profiles in this game is

$$S = \{(\text{OA}, \text{O}), (\text{OA}, \text{I}), (\text{OB}, \text{O}), (\text{OB}, \text{I}), (\text{IA}, \text{O}), (\text{IA}, \text{I}), (\text{IB}, \text{O}), (\text{IB}, \text{I})\}.$$

Player i's payoff function u_i is defined over S so that $u_i(s)$ gives player i's payoff in the game when the strategy profile s is played. The easiest way to see how u_i is defined is to start at the initial node and trace through the tree according to the actions specified by the strategy profile. For instance, $u_1(\text{OA}, \text{O}) = 2$, $u_1(\text{IA}, \text{I}) = 4$, $u_2(\text{IA}, \text{O}) = 3$, and so forth.[3]

A convenient way of describing the strategy spaces of the players and their payoff functions for two-player games in which each player has a finite number of strategies is to draw a matrix. Each row of the matrix corresponds to a strategy of player 1 and each column corresponds to a strategy of player 2. Thus each cell of the matrix (which designates a single row and column) corresponds to a strategy profile. Inside a given cell, we write the payoff vector associated with the strategy profile. For example, the game shown in Figure 3.1(b) is described by the matrix in Figure 3.3 on the next page. In the matrix representation, we maintain the practice of putting player 1's payoff first.

The strategy sets and payoff functions of the players fully describe a strategic situation, without reference to an extensive form. In other words, strategies and payoffs can be taken as a fundamental representation of a game.[4] Here is

[3]Because we use notational conventions such as $s = (s_1, s_2)$, it is more proper to write $u_i(s) = u_i((s_1, s_2))$ than to write $u_i(s) = u_i(s_1, s_2)$. However, the latter is completely standard in mathematical discourse, so I refrain from using the double parentheses except where it helps prevent confusion. You will see their use a few times in the textbook.

[4]In fact, it is often easier to analyze a game without having to consider a tree structure.

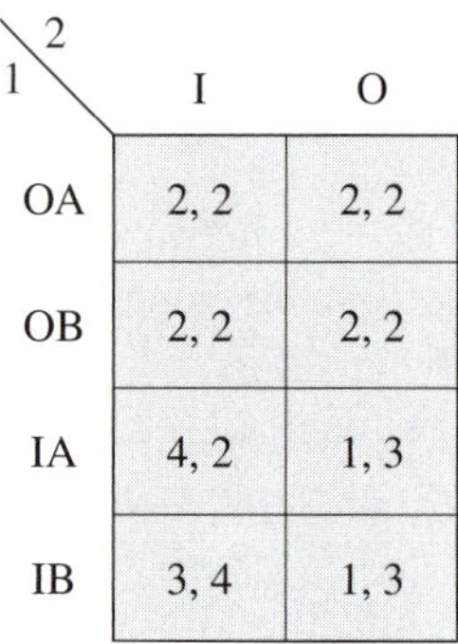

FIGURE 3.3
A game in normal form.

the formal definition of this representation. A game in *normal form* (also called *strategic form*) consists of a set of players, $\{1, 2, \ldots, n\}$, strategy spaces for the players, $S_1, S_2, \ldots, S_n$, and payoff functions for the players, $u_1, u_2, \ldots, u_n$.[5] As noted earlier, two-player normal-form games with finite strategy spaces can be described by matrices. Thus, such games are sometimes called "matrix games."[6]

CLASSIC NORMAL-FORM GAMES

Some classic normal-form games are depicted in Figure 3.4. In the game of matching pennies, two players simultaneously and independently select "heads" or "tails" by each uncovering a penny in his hand. If their selections match, then player 2 must give his penny to player 1; otherwise, player 1 gives his penny to player 2.

In the coordination game, both players obtain a positive payoff if they select the same strategy; otherwise they get nothing. The "Pareto coordination" game has the added feature that both players prefer to coordinate on strategy A rather than on strategy B.[7]

The prisoners' dilemma is a well-known example and is motivated by the following story. The authorities have captured two criminals who they know are guilty of a certain crime. However, the authorities have only enough evidence to convict them of a minor offense. If neither crook admits to the crime, then both will be charged with the minor offense and will pay a moderate

[5]The normal form was first defined by J. von Neumann and O. Morgenstern, *Theory of Games and Economic Behavior* (Princeton, NJ: Princeton University Press, 1944).

[6]Games with an infinite number of strategies are difficult to represent with a diagram, even in the extensive form. A recommendation appears in Chapter 14.

[7]Outcome (A, A) is said to "Pareto dominate" (B, B). This criterion is defined in Chapter 6.

FIGURE 3.4 Classic normal-form games.

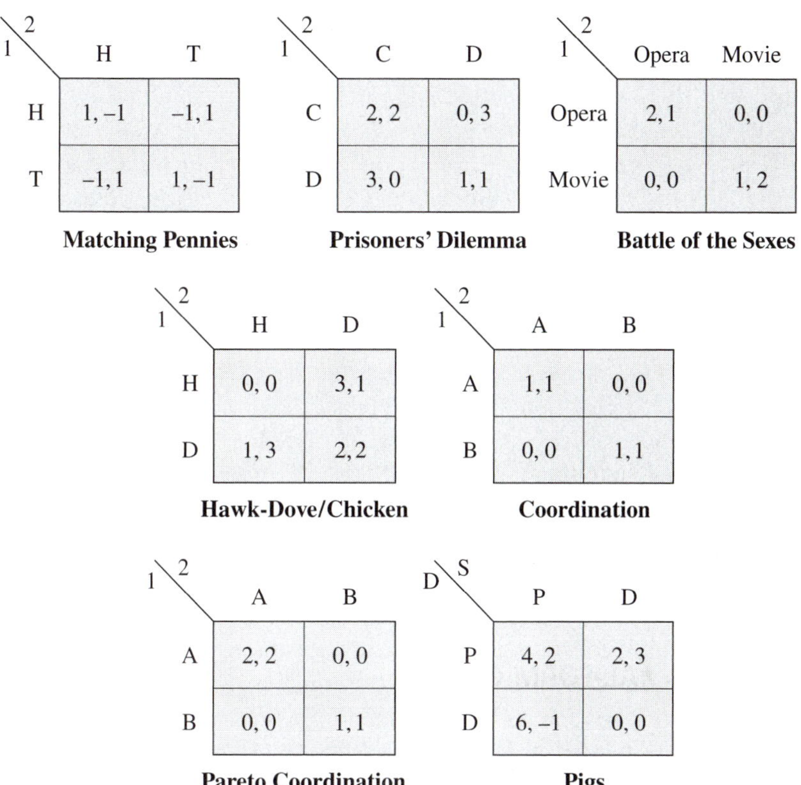

fine. The authorities have put the prisoners into separate rooms, where each prisoner is asked to squeal on the other. Squealing corresponds to strategy D (defect), and not squealing corresponds to strategy C (cooperate with the other prisoner). Each is told that if he squeals, he will be granted immunity and be released; his testimony, however, will be used to convict the other prisoner of the crime. If each squeals on the other, then they both get sent to jail, but their term is reduced because of their cooperation. The best outcome for a prisoner is to defect while the other cooperates (payoff 3); the next-best outcome occurs when neither defects (payoff 2); then comes the outcome in which both defect (payoff 1); the worst outcome for a prisoner is when he cooperates while the other defects.[8]

[8]Those who first described this game called it the "prisoner's dilemma." I prefer the plural form to highlight that it is a strategic situation. For an early account of this game and the battle of the sexes, see R. D. Luce and H. Raiffa, *Games and Decisions* (New York: Wiley, 1957).

The ill-titled "battle of the sexes" is a game in which two friends have to decide whether to see a movie or go to the opera. Unfortunately, they work in different parts of the city and, owing to a massive outage of the telephone system, find themselves incommunicado. They must simultaneously and independently select an event to attend. There is only one movie theater and only one opera venue, so the friends will meet each other if they manage to coordinate their decisions. Both prefer to be together, regardless of which event they attend. However, player 1 prefers the opera and player 2 prefers the movie.

The game of chicken you will recognize from several movies, in particular the 1955 James Dean film *Rebel Without a Cause*. Two players drive automobiles toward each other at top speed. Just before they reach each other, each chooses between maintaining course (H) and swerving (D). If both swerve, they both save face and are satisfied. If only one swerves, then he is proved to be a wimp, whereas the other is lauded as a tough guy with steely nerves. If both maintain course, they crash and are each horribly disfigured (and, needless to say, their girlfriends dump them).[9]

The pigs game refers to a situation in which a dominant and a submissive pig share a pen. On one side of the pen is a large button, which if pushed releases food into a dish at the other side of the pen. Each pig has the option of pushing the button (P) or not (D). If neither pushes, the pigs go hungry. If the submissive pig pushes the button and the dominant one does not, the released food is eaten entirely by the dominant pig because it gets to the food first. (Here the submissive pig is even worse off than if neither played P, because it expended the effort to push the button but got no food.) If the dominant pig pushes the button, then the submissive pig can enjoy some of the food before the dominant one reaches the dish.[10]

INTERPRETATION OF THE NORMAL FORM

One way of viewing the normal form is that it models a situation in which players simultaneously and independently select complete contingent plans for an extensive-form game. Theorists sometimes view selection of strategies in this fashion as equivalent to real-time play of the extensive form, but there are some subtleties to consider, which we will do in the third part of this book. On

[9]In *Rebel Without a Cause*, the players actually drive toward a cliff. My description of the chicken game more accurately describes the chicken scene in the 1984 Kevin Bacon film *Footloose*.

[10]The pigs game is inspired by B. A. Baldwin and G. B. Meese, "Social Behaviour in Pigs Studied by Means of Operant Conditioning," *Animal Behaviour* 27(1979):947–957.

FIGURE 3.5 Corresponding extensive and normal forms.

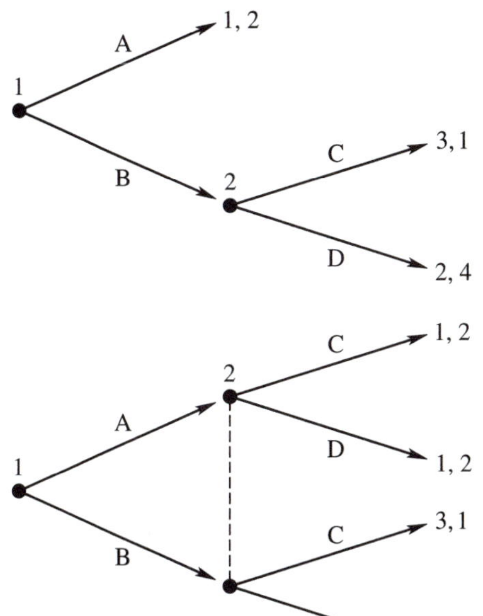

1 \ 2	C	D
A	1, 2	1, 2
B	3, 1	2, 4

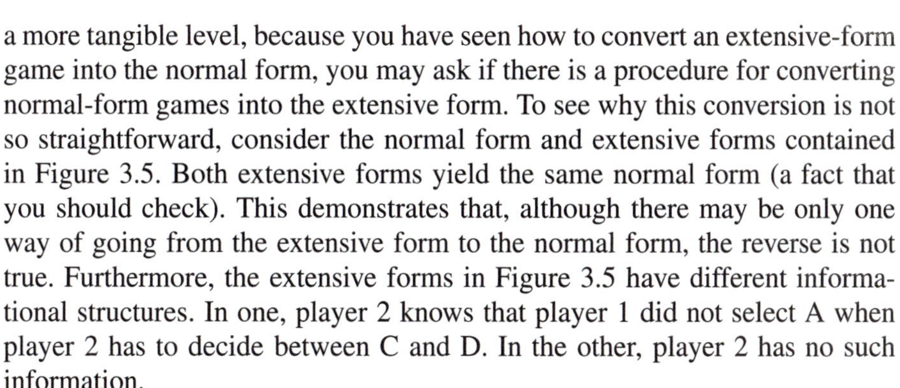

a more tangible level, because you have seen how to convert an extensive-form game into the normal form, you may ask if there is a procedure for converting normal-form games into the extensive form. To see why this conversion is not so straightforward, consider the normal form and extensive forms contained in Figure 3.5. Both extensive forms yield the same normal form (a fact that you should check). This demonstrates that, although there may be only one way of going from the extensive form to the normal form, the reverse is not true. Furthermore, the extensive forms in Figure 3.5 have different informational structures. In one, player 2 knows that player 1 did not select A when player 2 has to decide between C and D. In the other, player 2 has no such information.

Game theorists have been debating whether the normal form contains all of the relevant information about strategic settings. But you should realize that there is no discrepancy between the normal and extensive forms in settings in which the players make all of their decisions before observing what other players do (as is the case with simultaneous and independent moves). Such games are called "one-shot" or "static" games, and they are obviously well modeled in the normal form.

GUIDED EXERCISE

Problem: Describe the strategy spaces for the Katzenberg–Eisner game discussed in Chapter 2. Also draw the normal-form representation.

Solution: Recall that a strategy for player i must describe the action to be taken at each of player i's information sets. Examining Figure 2.6, which displays the extensive form of the Katzenberg–Eisner game, we see that Katzenberg (player K) has three information sets. At his first information set (the initial node), player K selects between actions L and S. At his second information set, he chooses between P and N. At his third information set, he chooses between R and N′. Thus, a strategy for player K is a combination of three actions, one for each of his information sets. An example of a strategy is LNR, which is a useful strategy to consider because it illustrates that even if player K is to select N at his second information set (which precludes his third information set from being reached), his strategy still must describe what he would do at his third information set. Because player K has two alternatives at each of three information sets, there are $2 \cdot 2 \cdot 2 = 8$ different combinations—thus eight different strategies. Player K's strategy space is

$$S_K = \{\text{LPR, LPN}', \text{LNR, LNN}', \text{SPR, SPN}', \text{SNR, SNN}'\}.$$

Observe that player E has just one information set. His strategy space is

$$S_E = \{\text{P, N}\}.$$

To draw the normal-form matrix, you should first note that it must have eight rows (for the eight different strategies of player K) and two columns (for the two strategies of player E). Looking at each individual strategy profile, you can trace through the extensive form to find the associated payoff vector. For example, consider strategy profile (LNN′, P). With this strategy profile, play proceeds from the initial node to node b, then node c (see Figure 2.5), and ends at the terminal node with payoff vector (0, 140). The complete normal-form matrix is:

K \ E	P	N
LPR	40, 110	80, 0
LPN′	13, 120	80, 0
LNR	0, 140	0, 0
LNN′	0, 140	0, 0
SPR	35, 100	35, 100
SPN′	35, 100	35, 100
SNR	35, 100	35, 100
SNN′	35, 100	35, 100

EXERCISES

1. Describe the strategy spaces of the players for the extensive-form games in Exercises 1 and 4 of Chapter 2. Also, draw the normal-form matrices.

2. Suppose a manager and a worker interact as follows. The manager decides whether to hire or not hire the worker. If the manager does not hire the worker, then the game ends. When hired, the worker chooses to exert either high effort or low effort. On observing the worker's effort, the manager chooses to retain or fire the worker. In this game, does "not hire" describe a strategy for the manager? Explain.

3. Draw the normal-form matrix of each of the following extensive-form games.

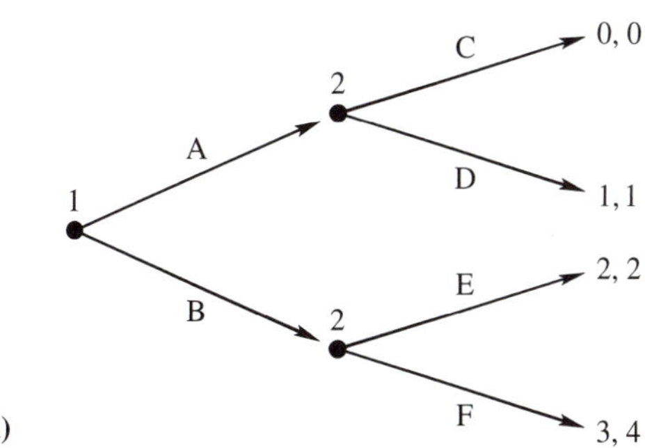

(a)

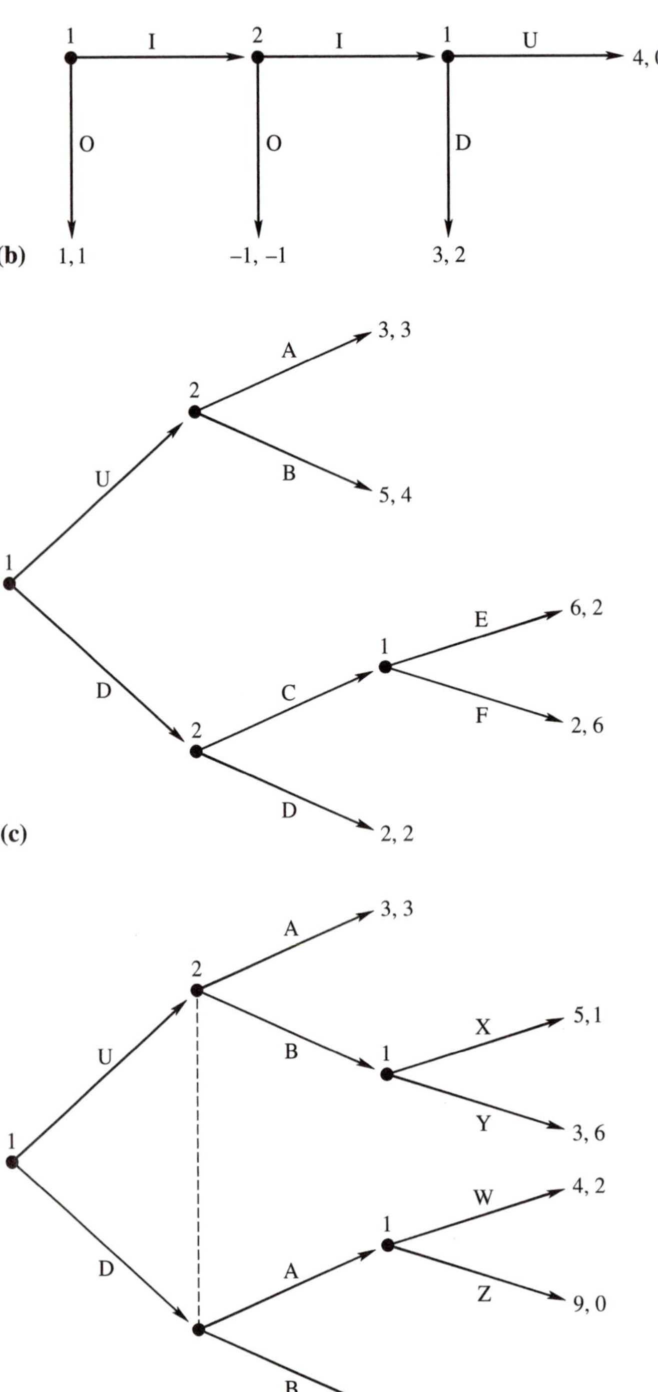
1
I
2
I
1
U
4, 0
O
O
D
(b)
1,1
−1, −1
3, 2
3, 3
A
2
U
B
5, 4
1
6, 2
E
1
D
C
F
2
2, 6
D
(c)
2, 2
3, 3
A
2
5,1
X
U
B
1
Y
1
3, 6
W
4, 2
1
D
A
Z
9, 0
B
(d)
2, 2

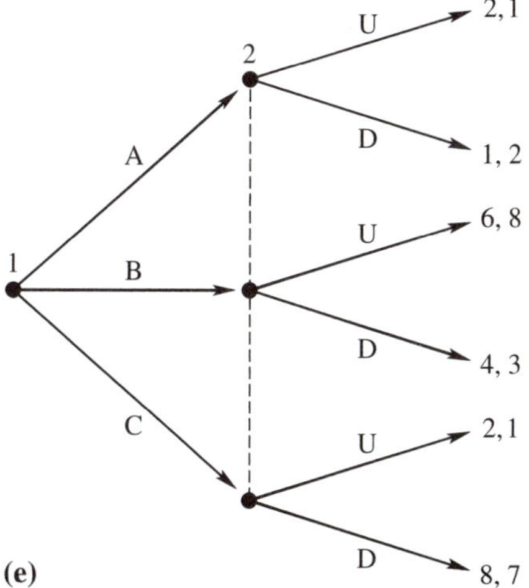

(e)

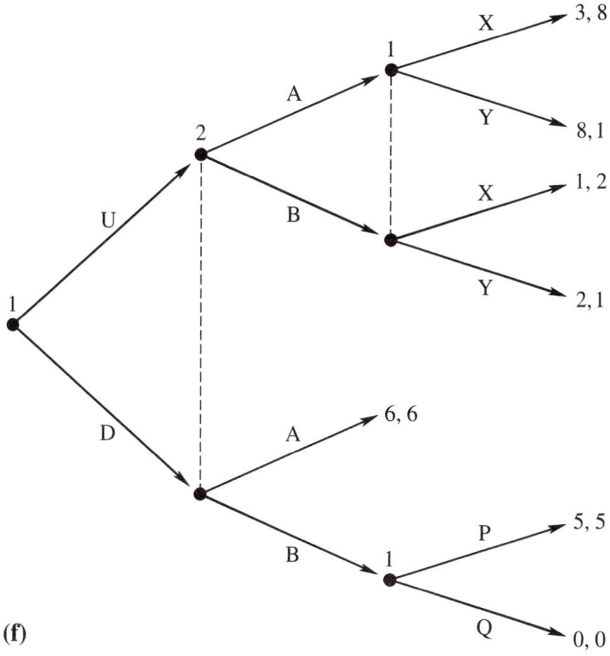

(f)

4. In the extensive-form game pictured at the top of the next page, how many strategies does player 2 have?

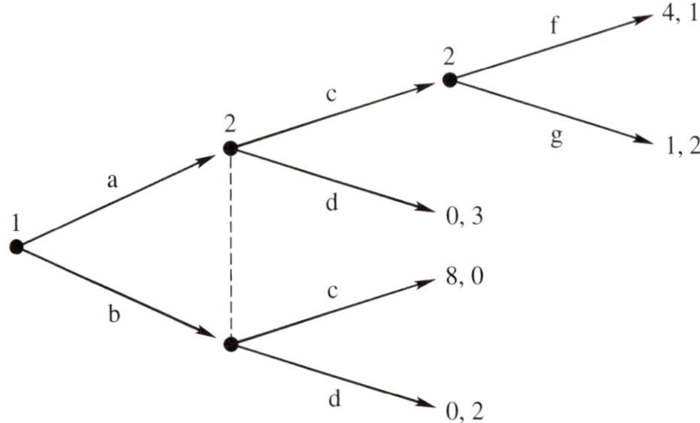

5. Consider a version of the *Cournot duopoly game*, which will be thoroughly analyzed in Chapter 10. Two firms (1 and 2) compete in a homogeneous goods market, where the firms produce exactly the same good. The firms simultaneously and independently select quantities to produce. The quantity selected by firm i is denoted q_i and must be greater than or equal to zero, for $i = 1, 2$. The market price is given by $p = 2 - q_1 - q_2$. For simplicity, assume that the cost to firm i of producing any quantity is zero. Further, assume that each firm's payoff is defined as its profit. That is, firm i's payoff is pq_i, where j denotes firm i's opponent in the game. Describe the normal form of this game by expressing the strategy spaces and writing the payoffs as functions of the strategies.

6. Consider a variation of the Cournot duopoly game in which the firms move sequentially rather than simultaneously. Suppose that firm 1 selects its quantity first. After observing firm 1's selection, firm 2 chooses its quantity. This is called the *von Stackelberg duopoly* model. For this game, describe what a strategy of firm 2 must specify.

7. Consider the normal-form games pictured here. Draw an extensive-form representation of each game. Can you think of other extensive forms that correspond to these normal-form games?

1 \ 2	C	D
AY	0, 3	0, 0
AN	2, 0	-1, 0
BY	0, 1	1, 1
BN	0, 1	1, 1

(a)

1 \ 2	HC	HD	LC	LD
H	3, 3	3, 3	0, 4	0, 4
L	4, 0	1, 1	4, 0	1, 1

(b)

4 BELIEFS, MIXED STRATEGIES, AND EXPECTED PAYOFFS

It is important for players to think about each other's strategic choices. We use the term *belief* for a player's assessment about the strategies of the others in the game. For example, consider the prisoners' dilemma game, in which each player chooses between strategies C and D. Player 1 may say to himself, "I think the other guy is likely to play strategy D" or "I think the other guy will surely play strategy C." These are two alternative beliefs that player 1 could have. Unfortunately, a statement such as "I think the other guy is likely to play strategy D" is somewhat ambiguous. Because we aim to model decision making mathematically, we need a precise way of representing a player's beliefs. The key is to use probabilities.

Here is an illustration. Continuing with the prisoners' dilemma example, suppose we let p represent the likelihood that player 1 thinks the other player will select strategy C in the game. Formally, p is a probability—a number between zero and one—where $p = 1$ means that player 1 is certain that player 2 will select strategy C and $p = 0$ means that player 1 is sure player 2 will *not* choose C. The probability that player 1 believes the other player will select D is $1 - p$. Thus, if $p = 1/2$, then player 1 believes that the other player is equally likely to select C and D. The numbers p and $1 - p$ constitute a *probability distribution* over the set {C, D}. For a more detailed review of the basic definitions of probability, please read Appendix A.

Note that if $p \in (0, 1)$, then player 1 thinks it is possible that player 2 will play strategy C and also thinks it is possible that player 2 will play D. Understand that player 1 might not believe that player 2 *actually* randomizes (by, say, flipping a coin). It's just that player 1 may be uncertain about what strategy player 2 will choose, and so he associates probabilities with player 2's strategies. Furthermore, player 1's belief may not be accurate; he could be certain that player 2 will select C when in fact player 2 selects D.

Let us formally define beliefs for general normal-form games. Mathematically, a belief of player i is a probability distribution over the strategies of the other players. We denote such a probability distribution θ_{-i} and write $\theta_{-i} \in \Delta S_{-i}$, where ΔS_{-i} is the set of probability distributions over the strategies of all the players except player i. (We often use Greek letters, such as theta, here, to represent probability distributions.) For example, take a two-player game (so that $-i = j$) and suppose each player has a finite number of strategies. The belief of player i about the behavior of player j is a function $\theta_j \in \Delta S_j$ such that, for each strategy $s_j \in S_j$ of player j, $\theta_j(s_j)$ is interpreted as the probability that player i thinks player j will play s_j. As a probability distribution, θ_j has the property that $\theta_j(s_j) \geq 0$ for each $s_j \in S_j$, and $\sum_{s_j \in S_j} \theta_j(s_j) = 1$.

Related to a belief is the notion of a *mixed strategy*. A mixed strategy for a player is the act of selecting a strategy according to a probability distribution. For example, if a player can choose between strategies U and D, we can imagine her selecting U with some probability and D with some probability. She might flip a coin and select U if the coin lands with the head up and D if the coin lands with the tail up. Formally, a mixed strategy is just like a belief in that they are both probability distributions. We will denote a generic mixed strategy of player i by $\sigma_i \in \Delta S_i$. To prevent confusion, we sometimes call a regular strategy a *pure strategy* to distinguish it from a mixed strategy. Notice that the set of mixed strategies includes the set of pure strategies (each of which is a mixed strategy that assigns all probability to one pure strategy).

If a player uses a mixed strategy and/or assigns positive probability to multiple strategies of the other player, then this player does not expect a particular payoff for sure. We can extend the definition of a payoff function to mixed strategies and beliefs by using the concept of *expected value*. When player i, for example, has a belief θ_{-i} about the strategies of the others and plans to select strategy s_i, then her *expected payoff* is the "average" payoff that she would get if she played strategy s_i and the others played according to θ_{-i}. Mathematically,

$$u_i(s_i, \theta_{-i}) = \sum_{s_{-i} \in S_{-i}} \theta_{-i}(s_{-i}) u_i(s_i, s_{-i}).$$

Take the game in Figure 4.1 and suppose that player 1 believes with probability 1/2 that player 2 will play strategy L, with probability 1/4 that she will play M,

FIGURE 4.1
Expected payoffs.

1 \ 2	L	M	R
U	8, 1	0, 2	4, 0
C	3, 3	1, 2	0, 0
D	5, 0	2, 3	8, 1

and with probability 1/4 that she will play R. That is, her belief θ_2 is such that $\theta_2(\text{L}) = 1/2$, $\theta_2(\text{M}) = 1/4$, and $\theta_2(\text{R}) = 1/4$. An often used, shorthand way of writing this belief is (1/2, 1/4, 1/4). If player 1 selects U, then she expects to get 8 with probability 1/2, 0 with probability 1/4, and 4 with probability 1/4. Thus, her expected payoff is

$$u_1(\text{U}, \theta_2) = (1/2)8 + (1/4)0 + (1/4)4 = 5.$$

When I began to formulate games in Chapter 2, payoff numbers were meant to describe the players' preferences over outcomes. I noted that any numbers can be used as long as they preserve the order describing the players' preferences over outcomes. At this point, we must observe that more is required. Payoff numbers also represent the players' preferences over probability distributions over outcomes. As an example, take the game in Figure 4.1. According to the payoff numbers in the matrix, player 1 is indifferent between the outcome in which he plays U and his opponent plays the mixed strategy (1/2, 1/4, 1/4) and the outcome of the strategy profile (D, L). Both yield an expected payoff of 5. If we raise or lower player 1's payoff from (D, L) to, say, 5.5 or 4.5, player 1's preferences over individual cells of the matrix do not change. But his preferences over uncertain outcomes do indeed change. Thus, there is more to the payoff numbers than simple order.[1]

For games with monetary outcomes, we will generally assume that the players seek to maximize their expected monetary gain. We can thus use the monetary amounts as the utility numbers themselves. Note that we can also multiply all of a player's payoffs by a positive number or add a number to all of them without affecting the player's preferences over certain or uncertain outcomes. We will take a closer look at various payoff functions when we get to the subject of risk in Chapter 25.

GUIDED EXERCISE

Problem: Consider the game in Figure 4.1. Suppose that player 2 believes that player 1 will select U with probability 1/2, C with probability 1/4, and D with probability 1/4. Also suppose that player 2 plans to randomize by picking M and R each with probability 1/2. What is player 2's expected payoff?

Solution: To calculate player 2's expected payoff, note that, according to player 2's belief, six strategy profiles occur with positive probability: (U, M),

[1]Two books that established this area of study are J. von Neumann and O. Morgenstern, *Theory of Games and Economic Behavior* (Princeton, NJ: Princeton University Press, 1944), and L. Savage, *The Foundations of Statistics* (New York: Wiley, 1954; revised and enlarged edition, New York: Dover Publications, 1972).

(U, R), (C, M), (C, R), (D, M), and (D, R). Profile (U, M) occurs with probability 1/4 (that is, 1/2 probability that player 1 selects U, times 1/2 probability that player 2 selects M). Profile (U, R) also occurs with probability 1/4. Each of the other four profiles occurs with probability 1/8. Note that the six probability numbers sum to one. Multiplying the individual probabilities by player 2's payoff in each case, we get a sum of

$$\frac{1}{4}\cdot 2+\frac{1}{4}\cdot 0+\frac{1}{8}\cdot 2+\frac{1}{8}\cdot 0+\frac{1}{8}\cdot 3+\frac{1}{8}\cdot 1=\frac{10}{8}=1.25.$$

Therefore, player 2's expected payoff is 1.25.

EXERCISES

1. Evaluate the following payoffs for the game given by the normal form pictured here. [Remember, a mixed strategy for player 1 is $\sigma_1 \in \Delta\{\text{U, M, D}\}$, where $\sigma_1(\text{U})$ is the probability that player 1 plays strategy U, and so forth. For simplicity, we write σ_1 as $(\sigma_1(\text{U}), \sigma_1(\text{M}), \sigma_1(\text{D}))$, and similarly for player 2.]

1 \ 2	L	C	R
U	10, 0	0, 10	3, 3
M	2, 10	10, 2	6, 4
D	3, 3	4, 6	6, 6

(a) $u_1(\text{U, C})$

(b) $u_2(\text{M, R})$

(c) $u_2(\text{D, C})$

(d) $u_1(\sigma_1, \text{C})$ for $\sigma_1 = (1/3, 2/3, 0)$

(e) $u_1(\sigma_1, \text{R})$ for $\sigma_1 = (1/4, 1/2, 1/4)$

(f) $u_1(\sigma_1, \text{L})$ for $\sigma_1 = (0, 1, 0)$

(g) $u_2(\sigma_1, \text{R})$ for $\sigma_1 = (1/3, 2/3, 0)$

(h) $u_2(\sigma_1, \sigma_2)$ for $\sigma_1 = (1/2, 1/2, 0)$ and $\sigma_2 = (1/4, 1/4, 1/2)$.

2. Suppose we have a game where $S_1 = \{\text{H, L}\}$ and $S_2 = \{\text{X, Y}\}$. If player 1 plays H, then her payoff is z regardless of player 2's choice of strategy; player 1's other payoff numbers are $u_1(\text{L, X}) = 0$ and $u_1(\text{L, Y}) = 10$. You may choose

any payoff numbers you like for player 2 because we will only be concerned with player 1's payoff.

(a) Draw the normal form of this game.

(b) If player 1's belief is $\theta_2 = (1/2, 1/2)$, what is player 1's expected payoff of playing H? What is his expected payoff of playing L? For what value of z is player 1 indifferent between playing H and L?

(c) Suppose $\theta_2 = (1/3, 2/3)$. Find player 1's expected payoff of playing L.

3. Evaluate the following payoffs for the game pictured here:

(a) $u_1(\sigma_1, \text{I})$ for $\sigma_1 = (1/4, 1/4, 1/4, 1/4)$

(b) $u_2(\sigma_1, \text{O})$ for $\sigma_1 = (1/8, 1/4, 1/4, 3/8)$

(c) $u_1(\sigma_1, \sigma_2)$ for $\sigma_1 = (1/4, 1/4, 1/4, 1/4)$, $\sigma_2 = (1/3, 2/3)$

(d) $u_1(\sigma_1, \sigma_2)$ for $\sigma_1 = (0, 1/3, 1/6, 1/2)$, $\sigma_2 = (2/3, 1/3)$

1 \ 2	I	O
OA	2, 2	2, 2
OB	2, 2	2, 2
IA	4, 2	1, 3
IB	3, 4	1, 3

4. For each of the classic normal-form games (see Figure 3.4), find $u_1(\sigma_1, \sigma_2)$ and $u_2(\sigma_1, \sigma_2)$ for $\sigma_1 = (1/2, 1/2)$ and $\sigma_2 = (1/2, 1/2)$.

5. Consider a version of the Cournot duopoly game, where firms 1 and 2 simultaneously and independently select quantities to produce in a market. The quantity selected by firm i is denoted q_i and must be greater than or equal to zero, for $i = 1, 2$. The market price is given by $p = 100 - 2q_1 - 2q_2$. Suppose that each firm produces at a cost of 20 per unit. Further, assume that each firm's payoff is defined as its profit. (If you completed Exercise 5 of Chapter 3, then you have already dealt with this type of game.) Suppose that player 1 has the belief that player 2 is equally likely to select each of the quantities 6, 11, and 13. What is player 1's expected payoff of choosing a quantity of 14?

5 GENERAL ASSUMPTIONS AND METHODOLOGY

The objective of game-theoretic modeling is to precisely sort through the logic of strategic situations—to explain why people behave in certain ways and to predict and prescribe future behavior. Game theory also gives us a language for recording and communicating ideas. You are now familiar with part of the language—that which is used to describe games.

In formulating an extensive form or normal form, we must balance the goal of realism with the need for manageable mathematics (as the Katzenberg–Eisner game illustrated; see Chapter 2). A simple model is more easily analyzed, whereas a complex model has room to incorporate more realistic features. The best theoretical models isolate a few key strategic elements and, by exploring their relation, develop new insights that can be broadly applied. Model development is often the most important part of game-theoretic analysis; sometimes it is a trial-and-error process.

Here is an example of the kind of trade-offs that we face in designing games. Suppose you are interested in modeling the interaction between my wife and me regarding how we will divide evening chores. You might think of this as a negotiation problem that you can describe with an extensive form. At the initial node, you might specify several branches to represent various verbal offers that I can make, such as, "Why don't you bathe the kids while I clean the kitchen?" or "I'll bathe the kids if you'll clean the kitchen and tidy the living room." In the extensive form, my offer would be followed by various alternative actions for my wife, such as saying "okay" or making a counteroffer. Because there are many such alternatives, you have a choice as to which to include in the extensive form and how precisely to describe them.

A coarse description of my wife's feasible responses would be simply "okay" or "no, you take care of everything." A more refined description might include whether my wife smiles or gives me the evil eye as she speaks. Whether you should adopt the more refined description in your model depends on your assessment of its strategic importance and how complicated it is to analyze. For instance, if you viewed my wife's countenance as something of no meaning that could be ignored, it would seem unnecessary to include it in your description of my wife's alternatives. On the other hand, you might rightly think that my wife communicates mainly by facial expression rather than with words. Further, you might know that my payoff depends on her affect and that I am terrified of the evil eye. Then perhaps you ought to use the refined description of her response in your model. I am starting to realize who has the upper hand in my marriage!

This book contains many examples of simple games that capture useful insights. Each example involves an abstraction from reality, but each also retains important realistic features that are illuminated with careful study. The examples will give you a taste of, and appreciation for, the exercise of modeling.

RATIONALITY

Once we have specified a game, our task becomes studying the players' behavior. Game-theoretic analysis generally rests on the assumption that each player behaves according to his preferences. More precisely, we can assume that, if a player's action will determine which of several outcomes will occur in a game, then this player will select the action that leads to the outcome he most prefers. This is the mathematical definition of *rationality*. The nice thing about payoff numbers is that, because they represent preferences (with larger numbers associated with more preferred outcomes), rationality simply means maximizing one's expected payoff. Thus, we should assume that each player acts to maximize his own expected payoff.

Rationality is a weak assumption in the sense that any coherent preferences can be accommodated by the theory. For instance, rationality does not necessarily imply that the players seek to maximize their own monetary gains. Consider, for example, an altruistic player who prefers to increase the amount of money that the *other* players get. The appropriate representation of this player's preferences calls for a payoff that is increasing in the others' money. Then rationality, in terms of payoff maximization, means acting to increase the other players' monetary gains. Incidentally, we shall generally assume that players care only about their own monetary gains because, for most settings, this is a reasonable approximation.[1]

[1]The phrase "coherent preferences" is not a formal term. The key property for preferences to have a meaningful mathematical representation is *transitivity*, which means that if a player prefers outcome A to outcome B and also prefers outcome B to outcome C, then this player must further prefer outcome A to outcome C.

COMMON KNOWLEDGE

As you have already seen from the development of the extensive-form representation, it is critical that we precisely model the players' information throughout the game being played. In particular, players may have different information at various points in the game. But conventional analysis of behavior requires that *players have a shared understanding of the entire game*. In other words, the players know the extensive-form (or normal-form) game that they are playing, which includes the recognition that, in the course of play, individual players will not be able to distinguish between nodes in the same information set.

To express the idea that the players have a shared understanding of the game being played, we can take advantage of the notion of *common knowledge*, which can be defined as follows:

> A particular fact F is said to be **common knowledge** between the players if each player knows F, each player knows that the others know F, each player knows that every other player knows that each player knows F, and so on.

A simpler way of stating this is that the players gather around a table where F is displayed, so that each player can verify that the others observe F and also can verify the same about everyone else.

Putting the notion of common knowledge to work, the assumption that underlies conventional analysis of games—an assumption that I maintain throughout this book—is that *the game is common knowledge between the players*. In other words, each player knows the game being played, each player knows that the others know this, each player knows that every other player knows that each player knows the game being played, and so on. It is as if, before the game is played, the players gather around a table on which the game has been depicted (in extensive or normal form, whichever is relevant).

Keep in mind that common knowledge of the game does not imply that, during the play of the game, players have common knowledge of where they are in the extensive form. There will still generally be asymmetric information between the players as represented by nontrivial information sets (that is, information sets comprising multiple nodes).

OVERVIEW OF SOLUTION CONCEPTS

The rest of this book is dedicated to the analysis of rational behavior in games. This endeavor separates roughly into two components that are interwoven throughout the book. The first component entails more precisely defining concepts of rational behavior, which leads to theories of how games are or should be played. The standard term for these theories is *solution concepts*. The second component involves applying the concepts to study specific examples and applications.

Part II focuses on a definition of rationality that operates on the normal form of a game. A player is thought to be rational if (1) through some cognitive process, he forms a belief about the strategies of the others, and (2) given this belief, he selects a strategy to maximize his expected payoff. The *rationalizability* solution concept combines this definition of rationality with the assumption that the game and the players' rationality are common knowledge. The *Nash equilibrium* solution concept adds the assumption that a social institution helps to coordinate the players so that their beliefs and actual behavior are consistent. Part III presents refinements of the rationality concept that are sensitive to individual information sets in the extensive-form representation of the game: *backward induction*, *subgame-perfect equilibrium*, and a variant for games with a negotiation component. Part IV expands the analysis to look at games with exogenous random events (moves of nature), where the solution concepts are called *Bayesian rationalizability*, *Bayesian Nash equilibrium*, and *perfect Bayesian equilibrium*.

THE ISSUE OF REALISM

Rational decision making may require complex calculations and a sophisticated understanding of the other players' motivations. Standard game theory assumes that the players are sophisticated and that they can handle whatever difficult calculations needed for payoff maximization. But in reality, behavior sometimes diverges from this rationality ideal and it does so in systematic ways. Sometimes real players do not have a shared understanding of the game they are playing. Sometimes a cognitive bias warps how they develop beliefs about each other. Sometimes real players are unable to perform all of the complicated calculations that would be required to determine an optimal strategy. Sometimes players do not fully understand each other's rationality or lack thereof.

For these reasons, we do not expect the outcome of our theoretical models to be perfect indicators of the real world. Thus, the relevant question is how closely the theory approximates reality. Where the approximation is poor, there might be an opportunity for the theory, in a normative sense, to play an instructive role in helping people to better understand the strategic settings they face (and to select better strategies). Alternatively, in a descriptive sense, there may be a need to enrich the theory to incorporate real bounds on rationality. Economists and game theorists have studied many theoretical enrichments in the category of cognitive biases, bounded rationality, and bounds on the degree to which players understand and internalize each other's rationality. Although this book concentrates on the conventional notion of full rationality, hints of how one might depart from this notion can be found here and there.

Analyzing Behavior in Static Settings

This part of the text introduces the basic tools for studying rational behavior. The tools treat strategies and payoffs as fundamental; that is, they address the *normal-form* specification of a game. Because extensive-form games can be easily translated into the normal form, the analysis developed here applies equally well to extensive-form representations. The theory presented in this part is most appropriately applied to games in which all of the players' actions are taken simultaneously and independently—so-called *one-shot,* or *static,* games. Most of the examples covered in this part are one-shot strategic settings. As noted earlier, normal and extensive forms model these games equally well.

DOMINANCE AND BEST RESPONSE 6

This chapter covers the two most basic concepts in the theory of strategic interaction: *dominance* and *best response*. These concepts form the foundation of most theories of rational behavior, so it is important to understand them well.[1]

DOMINANCE

Examine the game shown in Figure 6.1(a) on the next page, and suppose that you are player 1. Regardless of your decision-making process, player 2 selects a strategy independently of your choice. She will pick either L or R, and you cannot affect which she will choose. Of course, player 2 may base her choice on an assessment of what strategy you are likely to pick, but this is a matter of reason in the mind of player 2. Her decision-making process is independent of yours.[2] A good way of thinking about this is to imagine that player 2 has already chosen her strategy but you have not observed it. You must select your own strategy.

In the game shown in Figure 6.1(a) on the next page, strategy U has an interesting property. Regardless of player 2's choice, U gives you a strictly higher payoff than does D. If player 2 plays L, then you obtain 2 from playing U and 1 from playing D. Obviously, U is better in this case. Furthermore, if player 2 selects R, then you obtain 5 by playing U and 4 by playing D. Again, U is better. Technically, we say that strategy D is *dominated* by strategy U, and thus D should never be played by a rational player 1. Note that neither of player 2's strategies is dominated. Strategy L is better than R if player 1 selects U, but the reverse is true if player 1 selects D.

[1]These concepts were introduced by John Nash in "Non-Cooperative Games," *Annals of Mathematics* 54 (1951): 286–295; building from J. von Neumann and O. Morgenstern, *Theory of Games and Economic Behavior* (Princeton, NJ: Princeton University Press, 1944).

[2]There are some wacky theories of decision making that do not assume that players' decision-making processes are independent. To be precise, the wacky theories presume some metaphysical correlation between one's own decision and the decisions of others. For example, player 2 may think, "If I am the kind of person who would select R, then the world must be a kind place and, in such a world, there are only people who select D as player 1." We will not study such theories.

FIGURE 6.1 Examples of dominance.

(a)

1 \ 2	L	R
U	2, 3	5, 0
D	1, 0	4, 3

(b)

1 \ 2	L	C	R
U	8, 3	0, 4	4, 4
M	4, 2	1, 5	5, 3
D	3, 7	0, 1	2, 0

(c)

1 \ 2	L	R
U	4, 1	0, 2
M	0, 0	4, 0
D	1, 3	1, 2

Take another example, the game depicted in Figure 6.1(b). In this game, player 1's strategy D is dominated by strategy M. Regardless of what player 2 does, M yields a higher payoff for player 1 than does D. Strategy U is not dominated by M, however, because, if player 2 were to play L, then U would give player 1 a higher payoff than would M.

Finally, examine the game shown in Figure 6.1(c). This game has a more complicated dominance relation. Note that for player 1 no pure strategy is dominated by another pure strategy. Obviously, neither U nor M dominates the other, and D does not dominate these strategies. In addition, neither U nor M dominates D. For instance, although U is better than D when player 2 selects L, D performs better than U when player 2 selects R. However, a *mixed* strategy dominates D. Consider player 1's mixed strategy of selecting U with probability 1/2, M with probability 1/2, and D with probability zero. We represent this mixed strategy as (1/2, 1/2, 0). If player 2 selects L, then this mixed strategy yields an expected payoff of 2 to player 1; that is,

$$2 = 4(1/2) + 0(1/2) + 1(0).$$

Player 1 does worse by playing D. The same is true when player 2 selects R. Therefore, strategy D is dominated by the mixed strategy (1/2, 1/2, 0).

The formal definition of dominance is as follows:

> A pure strategy s_i of player i is **dominated** if there is a strategy (pure or mixed) $\sigma_i \in \Delta S_i$ such that $u_i(\sigma_i, s_{-i}) > u_i(s_i, s_{-i})$, for all strategy profiles $s_{-i} \in S_{-i}$ of the other players.

The best method for checking whether a strategy is dominated is to first decide whether it is dominated by another pure strategy. This procedure can be easily performed if the game is represented as a matrix. For example, to check if one of player 1's strategies dominates another, just scan across the two rows of the payoff matrix, column by column. You do not have to compare the payoffs of player 1 across columns. If the payoff in one row is greater than the payoff in another, and this is true for all columns, then the former strategy dominates the latter. If a strategy is not dominated by another pure strategy, you then must determine whether it is dominated by a mixed strategy.

Checking for dominance can be tricky where mixed strategies are concerned, but a few guidelines will help. First, note that there are many different mixed strategies to try. For example, in the game shown in Figure 6.1(c), $(1/2, 1/2, 0)$, $(3/5, 2/5, 0)$, $(2/5, 3/5, 0)$, and $(1/3, 1/3, 1/3)$ are four of the infinite number of mixed strategies for player 1. In fact, all four dominate strategy D. (Check this yourself for practice.) Second, when you are looking for a mixed strategy that dominates a pure strategy, look for alternating patterns of large and small numbers in the payoff matrix. This will help you find strategies that may be assigned positive probability by a dominating mixed strategy. Third, to demonstrate that a strategy is dominated, remember that you need to find only one strategy (pure or mixed) that dominates it. If, for example, you find that strategy X is dominated by pure strategy Y, your job is done; you do not need to check for mixed-strategy dominance. Finally, make sure that you check the correct payoff numbers. When evaluating the strategies of player i, you must look at player i's payoffs and no others.

An important component of the definition of dominance is the "strict" inequality. That is, in mathematical terms, we have the expression $u_i(\sigma_i, s_{-i}) > u_i(s_i, s_{-i})$ rather than $u_i(\sigma_i, s_{-i}) \geq u_i(s_i, s_{-i})$. Thus, for one strategy to dominate another, the former must deliver strictly *more* than does the latter. One strategy does not dominate another if they yield the same payoff against some strategy of the other players. As an example, take the game shown in Figure 6.1(b). We have already seen that D is dominated by M for player 1. Examine player 2's strategies and payoffs. Is R dominated by C? The answer is "no." For player 2, C yields a strictly higher payoff than does R against strategies M and D of player 1. However, against U, these strategies yield the same payoff. Because of this "tie," C does not dominate R. To emphasize the strict inequality in the definition, we sometimes use the expression "strict dominance."[3]

The most simple and compelling theory of behavior is that players do not play dominated strategies. As noted earlier, playing a dominated strategy is not rational in the sense that one's payoff can be increased by using another strategy, regardless of what the other players do in the game.

---

[3]By associating the term "dominance" with the strict inequality condition, I am utilizing John Nash's original terminology, which is common. But some people insist on saying "strict dominance."

THE FIRST STRATEGIC TENSION AND THE PRISONERS' DILEMMA

Before enriching the theory, let's use the concept of dominance to identify rational play in a simple application. Consider the prisoners' dilemma game pictured in Figure 3.4. For both players, strategy C is dominated by strategy D. We would therefore predict that neither player would select C. However, both players would be better off if they each selected C.

The prisoners' dilemma illustrates one of the major tensions in strategic settings: the clash between individual and group interests. The players realize that they are jointly better off if they each select C rather than D. However, each has the *individual* incentive to defect by choosing D. Because the players select their strategies simultaneously and independently, individual incentives win. One can even imagine the players discussing at length the virtues of the (C, C) strategy profile, and they might even reach an oral agreement to play in accord with that profile. But when the players go their separate ways and submit their strategies individually, neither has the incentive to follow through on the agreement. Strong individual incentives can lead to group loss.

While we're on the subject of conflicting interests, briefly consider two related issues. First, remember the meaning of payoff numbers. We take them to be utilities, as used generally in economics. As utilities, these numbers merely identify the players' preferences. They do not necessarily signify profit or money. For example, all we mean by the payoff numbers 2 and 5 is that the player in question prefers the outcome yielding the payoff 5 to the outcome yielding the payoff 2.

In the prisoners' dilemma, a player might be deterred from selecting D by the threat of his partner chastising him after play occurs. Certainly such considerations enter the minds of decision makers. As game theorists, we must insist that all such considerations be manifest in the payoffs. Suppose we have a setting like that portrayed by the prisoners' dilemma, except that the payoffs are in dollar terms. Further suppose that player 1 prefers not to play D for fear of retribution by his opponent after the game ends. If we were to draw the "actual" matrix describing this game, player 1's payoffs from selecting D should be less than those from selecting C (against each of player 2's strategies). The actual game, in this case, is not a prisoners' dilemma. Indeed, if retribution were possible after the players choose between C and D, then we ought to model the option for retribution formally as a part of the game.[4]

The second issue to recognize is that, in the real world, the players sometimes have the option of agreeing to legally binding contracts. For example, in

[4]Remember that modeling strategic situations by using game theory is an art. The best models capture the essence of the strategic environment without bowing so much to reality as to make analysis impossible.

a prisoners' dilemma-like situation, the players may have the option of writing a contract that binds them to select strategy C. As noted in the preceding paragraph, if such options are a critical part of the strategic setting, we should include them in our models. Theorists do in fact pursue this angle; we shall pick up the topic in Chapter 13. At this point, we can at least identify where binding contracts might be most valuable. Specifically, they are helpful when individual incentives interfere with group incentives.

The prisoners' dilemma is a widely discussed game and has proved to be a source of insight in the fields of economics, sociology, political science, international relations, and philosophy. In economics, it is ubiquitous. Settings in which workers interact to produce in a firm often have the same flavor, although the models are richer and more complicated. In the same vein are some models of international trade. So, too, are settings of industrial organization. You will recognize the Cournot model (analyzed later in Chapter 10) as having the basic prisoners' dilemma form. Firms have an individual incentive to "overproduce," but profits are low when all firms do so. When firms compete by selecting prices, the same kind of tension surfaces. Check your understanding by looking again at the game shown in Figure 2.7(b), which is just another prisoners' dilemma.

THE CONCEPT OF EFFICIENCY

The first strategic tension relates to the economic concept of *efficiency,* which is an important welfare criterion by which to judge behavior in a game. Suppose we wish to compare the outcomes induced by two strategy profiles, s and s'. We say that s is **more efficient** than s' if all of the players prefer the outcome of s to the outcome of s' and if the preference is strict for at least one player. In mathematical terms, s is more efficient than s' if $u_i(s) \geq u_i(s')$ for each player i and if the inequality is strict for at least one player.[5]

A strategy profile s is called **efficient** if there is no other strategy profile that is more efficient; that is, there is no other strategy profile s' such that $u_i(s') \geq u_i(s)$ for every player i and $u_j(s') > u_j(s)$ for some player j. The expression **Pareto efficient** is used to mean the same thing. Note that, in the prisoners' dilemma, (C, C) is more efficient than (D, D). Furthermore, (C, C), (C, D), and (D, C) are all efficient strategy profiles. In the game pictured in Figure 6.1(c), (D, R) is more efficient than both (M, L) and (U, R). In this game, (U, L) and (D, L) are the only efficient strategy profiles.

[5]In this case, we can also say that strategy s' is *Pareto dominated,* invoking the name of Wilfredo Pareto (who first introduced this concept of efficiency). I shall avoid using the expression "Pareto dominate," because I do not want it to be confused with the individual dominance criterion defined in this chapter.

BEST RESPONSE

It seems reasonable that rational people refrain from using dominated strategies. Dominance is a good descriptive and prescriptive concept. But it is just the beginning of our development of a theory of behavior. Indeed, in most games, players have more than one undominated strategy. Take, as an example, some of the simple games in Figure 3.4. Matching pennies, the battle-of-the-sexes, and the coordination games have no dominated strategies, so one cannot predict how people should or will play these games on the basis of the dominance criterion. We must move on to explore the process by which players actually select their strategies, at least among those that are not dominated.

Rational folks think about the actions that the other players might take; that is, people form beliefs about one another's behavior. In games, it is wise to form an opinion about the other players' behavior before deciding your own strategy. For example, if you were to play the coordination game pictured in Figure 3.4 and you thought that the other player would definitely play strategy B, it would be prudent for you to play B as well. If you thought he would select A, then you should follow suit. To maximize the payoff that you expect to obtain—which we assume is the mark of rational behavior—you should select the strategy that yields the greatest expected payoff against your belief. Such a strategy is called a *best response* (or *best reply*). Formally,

> Suppose player i has a belief $\theta_{-i} \in \Delta S_{-i}$ about the strategies played by the other players. Player i's strategy $s_i \in S_i$ is a **best response** if $u_i(s_i, \theta_{-i}) \geq u_i(s_i', \theta_{-i})$ for every $s_i' \in S_i$.

As the example described next shows, there may be more than one best response to a given belief. It is not difficult to show that, in a finite game, every belief has at least one best response. For each belief θ_{-i} of player i, we denote the set of best responses by $BR_i(\theta_{-i})$.

As an example, take the game in Figure 6.2. Suppose player 1, on deliberation, believes with probability 1/3 that player 2 will play L, with probability 1/2 that player 2 will play C, and with probability 1/6 that player 2 will play R. Recall that we can write this belief as (1/3, 1/2, 1/6). Then if player 1 selects strategy U, he expects a payoff of

$$(1/3)2 + (1/2)0 + (1/6)4 = 8/6.$$

If he plays M, he expects

$$(1/3)3 + (1/2)0 + (1/6)1 = 7/6.$$

FIGURE 6.2
An example of best response.

1 \ 2	L	C	R
U	2, 6	0, 4	4, 4
M	3, 3	0, 0	1, 5
D	1, 1	3, 5	2, 3

If he plays D, he obtains

$$(1/3)1 + (1/2)3 + (1/6)2 = 13/6.$$

Thus, his best response is strategy D, the strategy that yields the greatest expected payoff given his belief. Strategy D is his only best response. We thus have

$$BR_1(1/3, 1/2, 1/6) = \{D\}.$$

Remember that, because BR_i is a *set* of strategies, we enclose its elements in brackets, even if it has only one element.[6]

To continue with the game of Figure 6.2, suppose that player 2 has the belief (1/2, 1/4, 1/4) regarding the strategy that player 1 employs. That is, player 2 believes with probability 1/2 that player 1 will select U and that player 1's other two strategies are equally likely. Player 2 then expects

$$(1/2)6 + (1/4)3 + (1/4)1 = 4$$

from playing L. He expects

$$(1/2)4 + (1/4)0 + (1/4)5 = 13/4$$

from playing C. He expects

$$(1/2)4 + (1/4)5 + (1/4)3 = 4$$

from playing R. With this belief, player 2 has *two* best responses, strategies L *and* R. Thus,

$$BR_2(1/2, 1/4, 1/4) = \{L, R\}.$$

[6] We will depart from this rule for a few examples in which it is useful to think of BR_i as a function from strategies to strategies.

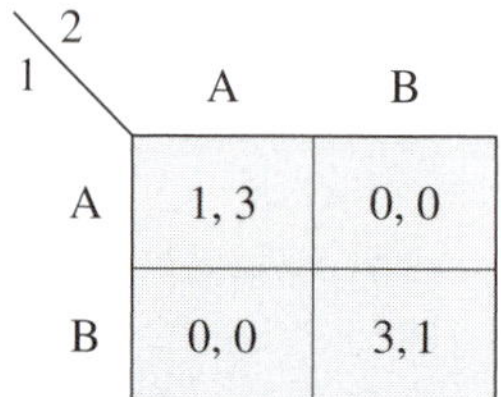

FIGURE 6.3
A battle of the sexes game.

Understand that playing a best response to one's belief about the moves of others is not in itself a strategic act. One can think of it as merely a computational exercise associated with rationality. Forming one's belief is the more important component of strategy. For example, suppose you are to play the game pictured in Figure 6.3 as player 2 and you are discussing your intentions with a friend. You might explain to your friend that, on the basis of your knowledge of player 1, you think that player 1 will certainly choose strategy B in the game. Therefore you plan to select B as well. Upset with your statement, he criticizes you for "giving in." You should demand the best outcome for yourself by playing A, he asserts. Should you follow the advice of your friend?

In fact, you should not. Tell your friend he is just plain crazy, moonstruck perhaps. Your actual choice in the game has *no* direct effect on the other player's thought process or action. Remember, players select their strategies simultaneously and independently. You cannot directly influence how the other player will act in a game such as this one. The best you can do is to form a well-reasoned opinion about what strategy he will take and then play a best response to this belief. The rules of the game do not allow you to signal your intentions to the other player by your choice of strategy.[7] So perform whatever mental gymnastics are necessary for you to settle on a belief about player 1's behavior and then play a best response. Of course, if your friend has information about player 1's tendencies, then you might incorporate your friend's information into your belief; but, when your belief has been set, then a best response is in order. Incidentally, don't be too hard on your friend; he is just trying to lend some support.

I am trying to convince you that the most substantive component of behavior is the formation of beliefs. Indeed, herein lies the real art of game theory. Success in games often hinges on whether you understand your opponent better than he understands you. We often speak of someone "outfoxing" others, after all. In fact, beliefs are subject to scientific study as well. The bulk of our work in the remainder of this text deals with determining what beliefs are

[7]If such a signal were physically possible, it would have to be formally included in the specification of the game (yielding a game that is different from that pictured in Figure 6.3).

rational in games. With the important concepts of dominance and best response under our belts, we can continue our study of strategic behavior by placing the beliefs of players at the center of attention. Before doing so, however, it is important to understand the relation between dominance and best response.

DOMINANCE AND BEST RESPONSE COMPARED

There is a precise relation between dominance and best response, the latter of which underpins the theories of behavior to come. For a given game, let UD_i be the set of strategies for player i that are not strictly dominated. Let B_i be the set of strategies for player i that are best responses, over *all* of the possible beliefs of player i. Mathematically,

$$B_i = \{s_i \mid \text{there is a belief } \theta_{-i} \in \Delta S_{-i} \text{ such that } s_i \in BR_i(\theta_{-i})\}.$$

That is, if a strategy s_i is a best response to *some* possible belief of player i, then s_i is contained in B_i. As heretofore noted, the notion of best response will be of primary importance. Unfortunately, determining the set B_i is sometimes a greater chore than determining UD_i. In fact, the two sets are closely related.

To build your intuition, examine the game in Figure 6.4. Note first that R is dominated for player 2.[8] Thus, $UD_2 = \{\text{L}\}$ in this game. Also note that strategy R can never be a best response for player 2, because L yields a strictly higher payoff regardless of what player 1 does. In other words, for any belief of player 2 about player 1's behavior, player 2's only best response is to select L.

FIGURE 6.4
Dominance and best response.

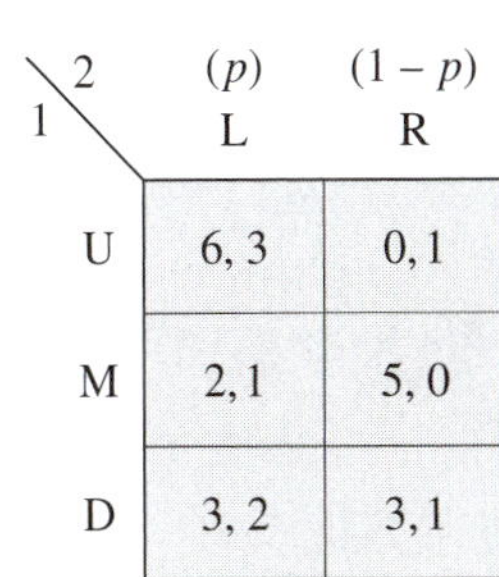

1 \ 2	(p) L	(1 − p) R
U	6, 3	0, 1
M	2, 1	5, 0
D	3, 2	3, 1

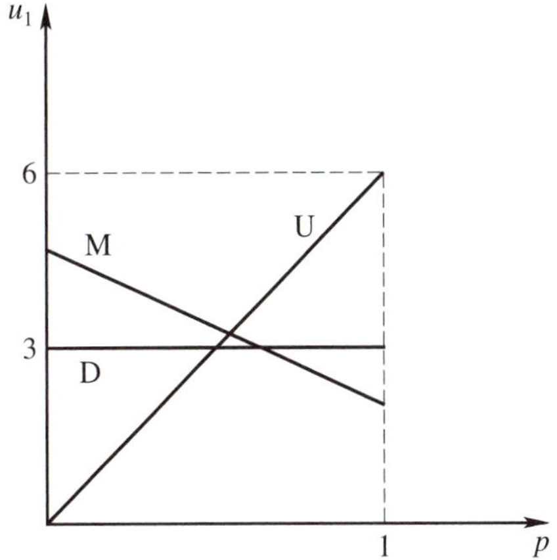

[8]Remember that when evaluating player 2's strategies you should look at player 2's payoffs.

Therefore $B_2 = \{L\}$. Obviously, $B_2 = UD_2$ and, for this game, dominance and best response yield the same conclusion about rational behavior for player 2.

To continue with the example, consider player 1's strategies. Obviously neither U nor M is dominated. Furthermore, strategy D is dominated by neither U nor M. But is D dominated by a mixed strategy? The answer is "yes"; D is dominated by the mixed strategy that puts probability 1/3 on U, 2/3 on M, and 0 on D.[9] Thus the set of undominated strategies for player 1 is $UD_1 = \{U, M\}$. To find the set of best responses, examine the diagram at the right of the matrix in Figure 6.4, where the expected payoffs of player 1's three pure strategies are graphed as functions of the probability that player 1 believes player 2 will select L. That is, with player 1's belief $(p, 1 - p)$, the upward sloping line is $6p + 0(1 - p)$, the expected payoff of playing strategy U. The downward sloping line, $2p + 5(1 - p)$, is the expected payoff from selecting M, and the flat line is the payoff from choosing D. There are values of p making U and M best responses, but for no value of p is D a best response. Thus $B_1 = \{U, M\}$ and we have $B_1 = UD_1$ in this game.

You may have concluded by now that strategies are best responses if and only if they are not strictly dominated. For two-player games, this is true.[10]

Result: In a finite two-player game, $B_1 = UD_1$ and $B_2 = UD_2$.

For games with more than two players, the relation between dominance and best response is a bit more subtle and requires a better understanding of the kinds of beliefs that players may hold. This is addressed in Appendix B. The key to the analysis has to do with whether a given player's belief is thought to exhibit correlation between the strategies of this player's multiple opponents. One can define B_i so that correlated conjectures are not allowed. Then one can let B_i^c be the version in which correlation is allowed. The general result is:

Result: For a finite game, $B_i \subset UD_i$ and $B_i^c = UD_i$ for each $i = 1, 2, \ldots, n$.

That is, if one imagines that players hold only uncorrelated beliefs, then strictly dominated strategies are never best responses. If players are allowed to hold correlated beliefs, then a strategy is undominated if and only if it is a best response to some belief. Thus, to find the set of strategies that a player may adopt as a rational response to his belief, you can simply find the set of undominated strategies. Because correlation in a player's belief is not an issue

[9] The mixed strategy (1/2, 1/2, 0) does not dominate D, demonstrating that you sometimes have to search a bit harder to find a dominance relation. Of course, to show that D is dominated, you need to find only *one* pure or mixed strategy that does the job.

[10] The results reported here are from the work of D. Pearce, "Rationalizable Strategic Behavior and the Problem of Perfection," *Econometrica* 52(1984):1029–1050.

FIGURE 6.5
The procedure for calculating B_1 and UD_1.

1 \ 2		(3/5) L	(2/5) R
	W	2, 4	2, 5
(p)	X	2, 0	7, 1
$(1-p)$	Y	6, 5	1, 2
	Z	5, 6	3, 0

when the player has just one opponent, the preceding result is a special case of this result. I provide a partial proof of the result in Appendix B, which technically oriented readers having more than one competitor or more than one friend should definitely read.

The relation between dominance and best response comes in handy when calculating the sets B_i and UD_i. For two-player, matrix games, where $B_i = UD_i$, you can most easily compute this set by using the following procedure.

Procedure for calculating $B_i = UD_i$:

1. Look for strategies that are best responses to the simplest beliefs—those beliefs that put all probability on just one of the other player's strategies. These best responses are obviously in the set B_i, so they are also in UD_i.
2. Look for strategies that are dominated by other *pure* strategies; these dominated strategies are *not* in UD_i and thus they are also *not* in B_i.[11]
3. Test each remaining strategy to see if it is dominated by a mixed strategy. This final step is the most difficult, but you will rarely have to perform it.

For an illustration of the procedure, consider the game in Figure 6.5. Let us calculate the set $B_1 = UD_1$. The first step is to find the strategies that are best responses to the simplest beliefs. Note that, against L, player 1's best response is Y. Against R, X is the best response. Thus, we know that $X \in B_1$ and $Y \in B_1$. The second step is to check whether any strategy is dominated by another pure strategy. You can quickly see that W is dominated by Z. Therefore, we know that $W \notin B_1$.

One question remains: Is strategy Z an element of B_1? This question is answered by the third step of the procedure. Note that, if a mixed strategy

[11] Note that at this point in the procedure you will have found a set of strategies that are definitely in B_i and UD_i and you will have found a set of strategies that are definitely not in B_i and UD_i.

were to dominate Z, then this strategy would be a mixture of X and Y. Let us assume that X is played with probability p and Y is played with probability $1 - p$. In order for this mixed strategy to dominate Z, it must be that it delivers a higher expected payoff for player 1 than does Z. Against player 2's strategy L, this comparison requires

$$2p + 6(1 - p) > 5.$$

Against R, it must be that

$$7p + 1(1 - p) > 3.$$

The first inequality simplifies to $p < 1/4$, whereas the second inequality simplifies to $p > 1/3$. Because there is no value of p that simultaneously satisfies $p < 1/4$ and $p > 1/3$, we know that Z is not dominated. Therefore, $Z \in B_1$. In fact, you can verify that Z is a best response to the belief $\theta_2 = (3/5, 2/5)$.

GUIDED EXERCISE

Problem: Suppose that two people decide to form a partnership firm. The revenue of the firm depends on the amount of effort expended on the job by each person and is given by:

$$r(e_1, e_2) = a_1 e_1 + a_2 e_2,$$

where e_1 is the effort level of person 1 and e_2 is the effort level of person 2. The numbers a_1 and a_2 are positive constants. The contract that was signed by the partners stipulates that person 1 receives a fraction t (between 0 and 1) of the firm's revenue and person 2 receives a $1 - t$ fraction. That is, person 1 receives the amount $tr(e_1, e_2)$, and person 2 receives $(1 - t)r(e_1, e_2)$. Each person dislikes effort, which is measured by a personal cost of e_1^2 for person 1 and e_2^2 for person 2. Person i's utility in this endeavor is the amount of revenue that this person receives, minus the effort cost e_i^2. The effort levels (assumed nonnegative) are chosen by the people simultaneously and independently.

(a) Define the normal form of this game (by describing the strategy spaces and payoff functions).

(b) Using dominance, compute the strategies that the players rationally select (as a function of t, a_1, and a_2).

(c) Suppose that you could set t before the players interact. How would you set t to maximize the revenue of the firm?

Solution:

(a) The game has two players. Each player selects an effort level, which is greater than or equal to zero. Thus, $S_i = [0, \infty)$ for $i = 1, 2$. As described, each player's payoff is the amount of revenue he receives, minus his effort cost. Thus, the payoff functions are

$$u_1(e_1, e_2) = t[a_1 e_1 + a_2 e_2] - e_1^2$$

and

$$u_2(e_1, e_2) = (1 - t)[a_1 e_1 + a_2 e_2] - e_2^2.$$

(b) In this game, each player has a strategy that dominates all others. To see this, observe how player 1's payoff changes as e_1 is varied. Taking the derivative of u_1 with respect to e_1, we get $ta_1 - 2e_1$. Setting this equal to zero and solving for e_1 reveals that player 1 maximizes his payoff by selecting $e_1^* = ta_1/2$. Similar analysis for player 2 yields $e_2^* = (1-t)a_2/2$. Note that, although each player's payoff depends on the strategy of the other player, a player's optimal strategy does not depend on the other's strategy. The set of undominated strategies is therefore

$$UD_1 = \{ta_1/2\} \quad \text{and} \quad UD_2 = \{(1 - t)a_2/2\}.$$

(c) Because they depend on a_1, a_2, and t, let us write the optimal strategies e_1^* and e_2^* as functions of these parameters. The revenue of the firm is then given by

$$a_1 e_1^*(a_1, a_2, t) + a_2 e_2^*(a_1, a_2, t).$$

Plugging in the values e_1^* and e_2^* from part (b), the revenue is

$$a_1 \cdot \frac{ta_1}{2} + a_2 \cdot \frac{(1 - t)a_2}{2} = t\frac{a_1^2}{2} + (1 - t)\frac{a_2^2}{2}.$$

Note that the objective function is linear in t. Thus, maximization occurs at a "corner," where either $t = 0$ or $t = 1$. If $a_1 > a_2$, then it is best to set $t = 1$; otherwise, it is best to set $t = 0$.

Incidentally, one can also consider the problem of maximizing the firm's revenue minus the partners' effort costs. Then the problem is to maximize

$$a_1 \cdot \frac{ta_1}{2} + a_2 \cdot \frac{(1-t)a_2}{2} - \left(\frac{ta_1}{2}\right)^2 - \left(\frac{(1-t)a_2}{2}\right)^2$$

and, using calculus, the solution is to set $t = a_1^2/(a_1^2 + a_2^2)$.

EXERCISES

1. Determine which strategies are dominated in the following normal-form games.

1 \ 2	L	R
A	3, 3	2, 0
B	4, 1	8, –1

(a)

1 \ 2	L	C	R
U	5, 9	0, 1	4, 3
M	3, 2	0, 9	1, 1
D	2, 8	0, 1	8, 4

(b)

1 \ 2	W	X	Y	Z
U	3, 6	4, 10	5, 0	0, 8
M	2, 6	3, 3	4, 10	1, 1
D	1, 5	2, 9	3, 0	4, 6

(c)

1 \ 2	L	R
U	1, 1	0, 0
D	0, 0	5, 5

(d)

2. For the game in Exercise 1 of Chapter 4, determine the following sets of best responses.

(a) $BR_1(\theta_2)$ for $\theta_2 = (1/3, 1/3, 1/3)$

(b) $BR_2(\theta_1)$ for $\theta_1 = (0, 1/3, 2/3)$

(c) $BR_1(\theta_2)$ for $\theta_2 = (5/9, 4/9, 0)$

(d) $BR_2(\theta_1)$ for $\theta_1 = (1/3, 1/6, 1/2)$

3. Consider a version of the Cournot duopoly game (described in earlier exercises), where firms 1 and 2 simultaneously and independently select quantities to produce in a market. The quantity selected by firm i is denoted q_i and must be greater than or equal to zero, for $i = 1, 2$. The market price is given by $p = 100 - 2q_1 - 2q_2$. Suppose that each firm produces at a cost of 20 per unit. Further, assume that each firm's payoff is defined as its profit. Is it ever a best response for player 1 to choose $q_1 = 25$? Suppose that player 1 has the belief that player 2 is equally likely to select each of the quantities 6, 11, and 13. What is player 1's best response?

4. For the game of Figure 6.2, determine the following best-response sets.

(a) $BR_1(\theta_2)$ for $\theta_2 = (1/6, 1/3, 1/2)$

(b) $BR_2(\theta_1)$ for $\theta_1 = (1/6, 1/3, 1/2)$

(c) $BR_1(\theta_2)$ for $\theta_2 = (1/4, 1/8, 5/8)$

(d) $BR_1(\theta_2)$ for $\theta_2 = (1/3, 1/3, 1/3)$

(e) $BR_2(\theta_1)$ for $\theta_1 = (1/2, 1/2, 0)$

5. Represent in the normal form the rock–paper–scissors game (see Exercise 4 of Chapter 2 to refresh your memory) and determine the following best-response sets.

(a) $BR_1(\theta_2)$ for $\theta_2 = (1, 0, 0)$

(b) $BR_1(\theta_2)$ for $\theta_2 = (1/6, 1/3, 1/2)$

(c) $BR_1(\theta_2)$ for $\theta_2 = (1/2, 1/4, 1/4)$

(d) $BR_1(\theta_2)$ for $\theta_2 = (1/3, 1/3, 1/3)$

6. In the game pictured here, is it ever rational for player 1 to select strategy C? Why?

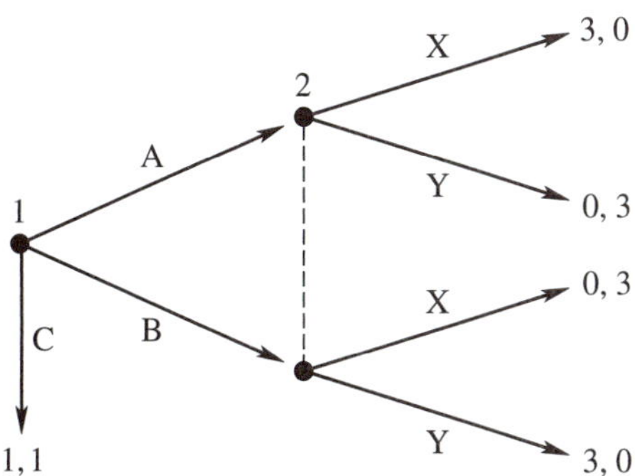

7. In the normal-form game pictured below, is player 1's strategy M dominated? If so, describe a strategy that dominates it. If not, describe a belief to which M is a best response.

1 \ 2	X	Y
K	9, 2	1, 0
L	1, 0	6, 1
M	3, 2	4, 2

8. For the Cournot game described in Exercise 3, calculate UD_1.

7 RATIONALIZABILITY AND ITERATED DOMINANCE

The concepts of dominance and best response are the basis of theories of rational behavior. We must recognize, however, that rational behavior may consist of more than avoiding dominated strategies. Indeed, sophisticated players understand the fundamental notion of a game—that the actions of one player may affect the payoff of another—and can shrewdly estimate the other players' actions by putting themselves in the others' shoes. In other words, mind games can be the art of game playing. At this point, we must take our analysis to another level. Please continue to breathe normally.

Suppose you are player 1 in the game pictured in Figure 7.1. You will notice that neither of your strategies is dominated. Therefore, you can rationally play either A or B, depending on your belief about the action of your opponent. Let us perform the mathematics behind this statement. Let p denote the probability that you think the opponent will play X, and let q be the probability that you think she will play Y. Obviously, $1 - p - q$ is the probability that she will play strategy Z, and these numbers make sense only if $p \geq 0$, $q \geq 0$, and $p + q \leq 1$. Given your belief, your expected payoff from playing A is

$$3p + 0q + 0(1 - p - q) = 3p.$$

Your expected payoff from playing B is

$$0p + 3q + 1(1 - p - q) = 1 - p + 2q.$$

Thus, your best response is to play A if $3p > 1 - p + 2q$ (which simplifies to $4p > 1 + 2q$) and to play B if $4p < 1 + 2q$. Both A and B are best responses if $4p = 1 + 2q$.

FIGURE 7.1
Playing mind games.

1 \ 2	X	Y	Z
A	3, 3	0, 5	0, 4
B	0, 0	3, 1	1, 2

FIGURE 7.2 Iterative removal of strictly dominated strategies.

1 \ 2	X	Y	Z
A	3, 3	0, 5	0, 4
B	0, 0	3, 1	1, 2

(a)

1 \ 2	X	Y	Z
A	3, 3	0, 5	0, 4
B	0, 0	3, 1	1, 2

(b)

1 \ 2	X	Y	Z
A	3, 3	0, 5	0, 4
B	0, 0	3, 1	1, 2

(c)

Are we finished with the analysis? Not by a long shot. Suppose it is *common knowledge* between you and your opponent that both of you are rational and understand exactly the game that is being played. That is, you and she both know the matrix that describes the game; she knows that you know this; you know that she knows this; she knows that you know that she knows this; and so forth.[1] Can you rationalize playing strategy A against a rational opponent? In fact, you cannot. Put yourself in the position of player 2. Her strategy X is strictly dominated by Y. Therefore, she will not play X; there is no belief that she could have about your strategy that would cause her to play X. Knowing this, you should assign *zero* probability to her strategy X (your belief should have $p = 0$).

You might as well strike the column of the matrix corresponding to strategy X, as demonstrated in Figure 7.2(a), because your opponent would never play this strategy. Striking X leads to a new "game" in which you choose between A and B and your opponent chooses between Y and Z. In this reduced game, your strategy A is strictly dominated by B. That is, knowing that your opponent will not play X, your only rational strategy is to play B. To see this in terms of beliefs, note that, if $p = 0$, it is not possible that $4p \geq 1 + 2q$ and so, regardless of your belief about whether your opponent will play Y or Z, the only best response is to play B. Thus, as in Figure 7.2(b), we can strike strategy A from consideration.

[1] Recall the definition of common knowledge from Chapter 5.

Interestingly enough, we can take this logic one step further in this game. Your opponent knows that you know that she will not play X. Thus, recognizing that you are rational, she deduces that you will not play A. Putting probability 1 on the event that you will play B, her only best response is to play Z. That is, in the reduced game where you play only B, strategy Y is dominated, as represented in Figure 7.2(c), where Y is stricken.

Simple logic led to a unique prediction in the example. With rational players, the only outcome that can arise is the strategy profile (B, Z), yielding the payoff vector (1, 2). Note that, as in the prisoners' dilemma, our prediction here is a decidedly inefficient outcome. Both players could fare better if the strategy profile (A, X) were played, but neither player finds it in his or her interest to do so.

The procedure just illustrated is called *iterative removal of strictly dominated strategies* (or **iterated dominance**, for short). We can apply the procedure to any normal-form game as follows. First, delete all of the dominated strategies for each player, because no rational player would adopt one of them. When these strategies have been deleted, a new "smaller" game is formed. Then, delete strategies that are dominated in this smaller game, forming an even smaller game on which the deletion process is repeated. Continue this process until no strategies can be deleted.

Remember that, at least in two-player games, dominance and best response imply the same restrictions. Therefore, iterated dominance is identical with the procedure in which strategies that are never best responses are removed at each round. Thus, the procedure identifies strategies for each player that can be rationalized in the following way. Player 1 selects a strategy that is a best response to his belief about the behavior of player 2. Furthermore, his belief is consistent with the rationality of player 2 in that every strategy to which he assigns positive probability is a best response for player 2 to some belief of hers. For instance, suppose player 1's belief assigns positive probability to some strategy s_2 of player 2. That is, $\theta_2(s_2) > 0$. Then there must be a belief θ_1 of player 2 over the strategies of player 1 to which s_2 is a best response. Furthermore, each strategy assigned positive probability by θ_1 must be a best response to some belief about the strategy of player 2. The logic continues endlessly.

Your first impression of these mind games may be less than favorable, but a very simple and precise theory of behavior is represented. Each player should play a best response to her belief. In this way, the belief *rationalizes* playing the strategy. Furthermore, each player should assign positive probability only to strategies of the other players that can be similarly rationalized. The set of strategies that survive iterated dominance is therefore called the

FIGURE 7.3
Another example of rationalizability/iterated dominance.

1 \ 2	L	C	R
U	5, 1	0, 4	1, 0
M	3, 1	0, 0	3, 5
D	3, 3	4, 4	2, 5

rationalizable strategies.[2] In precise terms, rationalizability refers to strategies that remain after one iteratively removes those that are never best responses. But, because this process is equivalent to iterated dominance, we can just perform the latter.[3]

Remember that the logic of rationalizability depends on common knowledge of rationality and the game. If the players have this information, then we shall predict that the strategy profile is rationalizable. Understand, though, that it may be a stretch to assume that players have this kind of information in some settings. For instance, you can certainly imagine situations in which one agent does not know whether the other players want to maximize anything in particular or what the payoffs of the others are or both. In such a case, this person may not be able to put himself in the other players' shoes with confidence, and rationalizability may be too strong a behavioral concept.

To further illustrate the concept of rationalizability, examine the game in Figure 7.3. Realize first that player 2's strategy L is strictly dominated, so we can strike it from consideration. If you do not see this right away, note that strategy L is dominated by the mixed strategy (0, 1/2, 1/2), the strategy that puts probability 1/2 on both C and R. After striking L, player 1's strategy U is strictly dominated by D, so we can strike U. Then player 2's strategy C is dominated by R, leaving her with only one rational strategy, R. Finally, player 1's strategy D is dominated by M. The set of rationalizable strategy profiles thus consists of only strategy M for player 1 and only strategy R for player 2; it is the set {(M, R)}.

---

[2] Rationalizability was invented, and its relationship to iterated dominance characterized, by D. G. Pearce, "Rationalizable Strategic Behavior and the Problem of Perfection," *Econometrica* 52(1984):1029–1050, and D. Bernheim, "Rationalizable Strategic Behavior," *Econometrica* 52(1984):1007–1028.

[3] Recall that in games with more than two players, best response and dominance are not exact complements. Thus, in these games, the set of rationalizable strategies may be different from the set of strategies that survive iterated dominance. The difference has to do with whether correlated conjectures are allowed, as briefly discussed in Appendix B, but is not important for our study. Thus, the terms *iterated dominance* and *rationalizability* are used interchangeably.

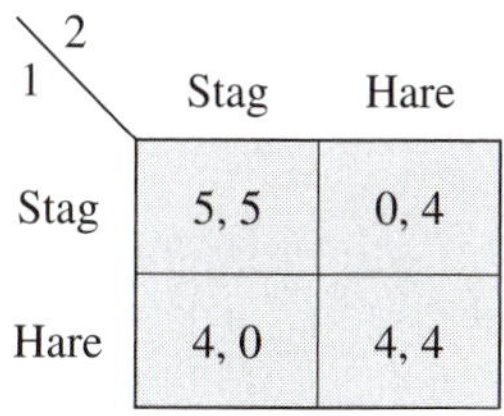

1 \ 2	Stag	Hare
Stag	5, 5	0, 4
Hare	4, 0	4, 4

FIGURE 7.4
Stag hunt.

THE SECOND STRATEGIC TENSION

Observe that rationalizability does not always lead to a unique strategy profile. In fact, it often has little bite in games. For example, neither player has a dominated strategy in the battle of the sexes game (in Figure 6.3). Player 1 might rationally believe that player 2 will select A and thus play A in response. But player 2 may believe that player 1 will select B and therefore will play B herself. Rationalizability merely requires that the players' beliefs and behavior be consistent with common knowledge of rationality. It does not require that their beliefs be correct. We have thus identified another tension in games, which is called *strategic uncertainty*. One manifestation of strategic uncertainty is the *coordination problem,* which players often face in games such as the battle of the sexes.

Indeed, lack of strategic coordination seems entirely reasonable and an accurate description of some actual events. You might have had the experience of losing a friend or family member at Disneyland and being without a cell phone to contact this person. Seeking to find each other, you and your friend each decided to stand near a prominent site at the theme park. You thought it obvious to meet at Cinderella's Castle. Your friend thought it equally obvious to meet at the foot of Main Street. You each behaved in a rational way in response to rational conjectures about the other. But your beliefs in the end were not consistent. Perhaps you finally caught up to her at the car at the end of the day, at which point you were both fuming.

In some games, there is also a direct tension between strategic uncertainty and efficiency. Consider a version of the *stag hunt* game discussed by French philosopher Jean-Jacques Rousseau and pictured in Figure 7.4.[4] As the story goes, two players go on a hunt. Simultaneously and independently, they each decide whether to hunt stag or hare. If a player hunts for hare, he will definitely catch one (worth 4 units of utility), regardless of whether the other player joins him in the hunt for this kind of animal. On the other hand, two people are required to catch a stag (which yields 5 to both players). Thus, if only one player hunts for stag, then he will get nothing. Note that both players

[4] J.-J. Rousseau, *A Discourse on Inequality* (New York: Penguin, 1984. Originally published in 1755).

would like to coordinate on the efficient (stag, stag) strategy profile. But suppose player i has some doubt about whether the other player will hunt stag. If player i is sufficiently uncertain about the other player's strategy—in particular if player i figures that the other player will hunt stag with a probability less than 4/5—then player i should hunt for hare. In other words, strategic uncertainty sometimes hinders attainment of an efficient outcome. To vegetarians, I apologize.

Strategic uncertainty is a part of life, but there are devices in the world that help us coordinate our behavior and avoid inefficiency. Institutions, rules, norms of behavior, and, in general, culture often facilitate coordination in society. Examples are ubiquitous. Disneyland, and most other places of dense tourism, have lost-and-found centers. These centers act as a "focal point" for people in search of lost items or lost people; that is, they are prominent in the minds of those in need.[5] Security on our motorways is buoyed by traffic laws and conventions, which keep drivers on one side of the road. (If you thought the drivers of oncoming cars would "drive on the left," then you should as well.) Furthermore, we empower our governments to police the roads, because drivers often have the incentive to bend the rules.

Communication provides a simple way of coordinating behavior as well. For example, suppose Louise and Wayne are planning a trip to a theme park. Before entering the grounds, they might agree on a suitable meeting place in the event that they become separated and cannot contact each other by phone. This agreement will save them from the undesired outcome in which they meet in the parking lot at the end of the day. However, this miscoordinated outcome might be the *most* preferred by Louise's boyfriend Chris. Perhaps it is Louise and Chris who ought to communicate better. As therapists say, communication is *essential* to maintaining a healthy relationship.

Communication and many social institutions serve to coordinate beliefs and behavior. They create durable systems of beliefs in which economic agents can have confidence. They align our expectations and give us the security of knowing that what we expect will actually take place. I pick up the analysis of coordinating institutions in Chapter 9.

GUIDED EXERCISE

Problem: The normal-form game pictured below represents a situation in tennis, whereby the server (player 1) decides whether to serve to the opponent's forehand (F), center (C), or backhand (B) side. Simultaneously, the re-

[5] Nobel laureate Thomas Schelling discussed the idea of a focal point in his enormously influential book titled *The Strategy of Conflict* (London: Oxford University Press, 1960).

ceiver (player 2) decides whether to favor the forehand, center, or backhand side. Calculate the set of rationalizable strategies for this game.

1 \ 2	F	C	B
F	0, 5	2, 3	2, 3
C	2, 3	0, 5	3, 2
B	5, 0	3, 2	2, 3

Solution: First note that player 1's strategy F is dominated by a mixture of C and B. You can get a sense of this by noticing that

(a) if player 2 selects F, both C and B yield a strictly higher payoff than does F for player 1.

(b) if player 2 selects B, then a mixture of C and B yields a higher payoff for player 1 than does F, as long as player 1 puts strictly positive probability on C.

(c) if player 2 selects C, then a mixture of C and B yields a higher payoff for player 1 than does F, as long as player 1 puts high enough probability on B.

To quantify (c), consider the mixed strategy in which player 1 puts probability q on B and $1 - q$ on C. Then, in the event that player 2 selects C, player 1's mixed strategy yields an expected payoff of $3q$, whereas strategy F yields the payoff 2. The mixed strategy is better if $q > 2/3$. Note that (b) requires $q < 1$. Thus, for any $q \in (2/3, 1)$, the mixed strategy dominates F.

After removing player 1's strategy F from consideration, we see that player 2's strategy F is dominated by C, and so F is removed from player 2's strategy set in the second round of the iterated-dominance procedure. This leaves C and B for both players. You can quickly confirm that no further deletion of strategies can occur, which means the set of rationalizable strategy profiles is

$$S = \{\text{C}, \text{B}\} \times \{\text{C}, \text{B}\}.$$

EXERCISES

1. Determine the set of rationalizable strategies for each of the following games.

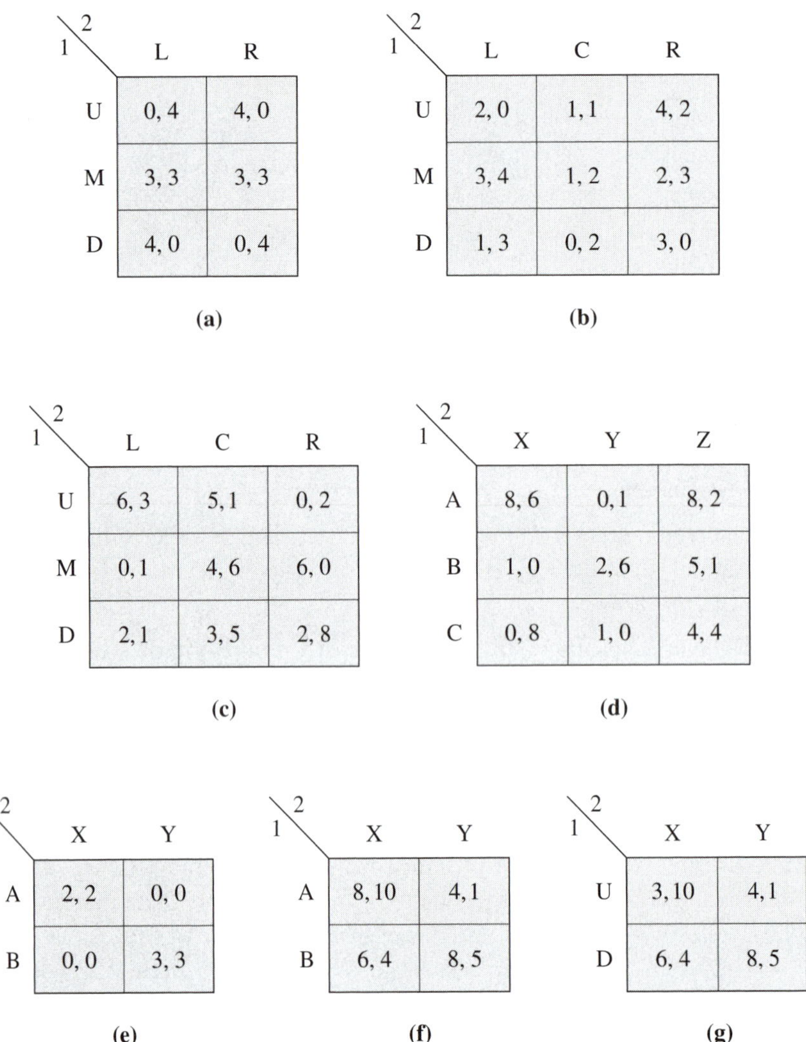

1 \ 2	L	R
U	0, 4	4, 0
M	3, 3	3, 3
D	4, 0	0, 4

(a)

1 \ 2	L	C	R
U	2, 0	1, 1	4, 2
M	3, 4	1, 2	2, 3
D	1, 3	0, 2	3, 0

(b)

1 \ 2	L	C	R
U	6, 3	5, 1	0, 2
M	0, 1	4, 6	6, 0
D	2, 1	3, 5	2, 8

(c)

1 \ 2	X	Y	Z
A	8, 6	0, 1	8, 2
B	1, 0	2, 6	5, 1
C	0, 8	1, 0	4, 4

(d)

1 \ 2	X	Y
A	2, 2	0, 0
B	0, 0	3, 3

(e)

1 \ 2	X	Y
A	8, 10	4, 1
B	6, 4	8, 5

(f)

1 \ 2	X	Y
U	3, 10	4, 1
D	6, 4	8, 5

(g)

2. Suppose that you manage a firm and are engaged in a dispute with one of your employees. The process of dispute resolution is modeled by the following game, where your employee chooses either to "settle" or to "be tough in negotiation," and you choose either to "hire an attorney" or to "give in."

Employee \ You	Give in	Hire attorney
Settle	1, 2	0, 1
Be tough	3, 0	x, 1

In the cells of the matrix, your payoff is listed second; x is a number that both you and the employee know. Under what conditions can you rationalize selecting "give in"? Explain what you must believe for this to be the case.

3. Find the set of rationalizable strategies for the following game.

1 \ 2	a	b	c	d
w	5, 4	4, 4	4, 5	12, 2
x	3, 7	8, 7	5, 8	10, 6
y	2, 10	7, 6	4, 6	9, 5
z	4, 4	5, 9	4, 10	10, 9

Note that each player has more than one dominated strategy. Discuss why, in the iterative process of deleting dominated strategies, the order in which dominated strategies are deleted does not matter.

4. Imagine that there are three major network-affiliate television stations in Turlock, California: RBC, CBC, and MBC. All three stations have the option of airing the evening network news program live at 6:00 P.M. or in a delayed broadcast at 7:00 P.M. Each station's objective is to maximize its viewing audience in order to maximize its advertising revenue. The following normal-form representation describes the share of Turlock's total population that is "captured" by each station as a function of the times at which the news programs are aired. The stations make their choices simultaneously. The payoffs are listed according to the order RBC, CBC, MBC. Find the set of rationalizable strategies in this game.

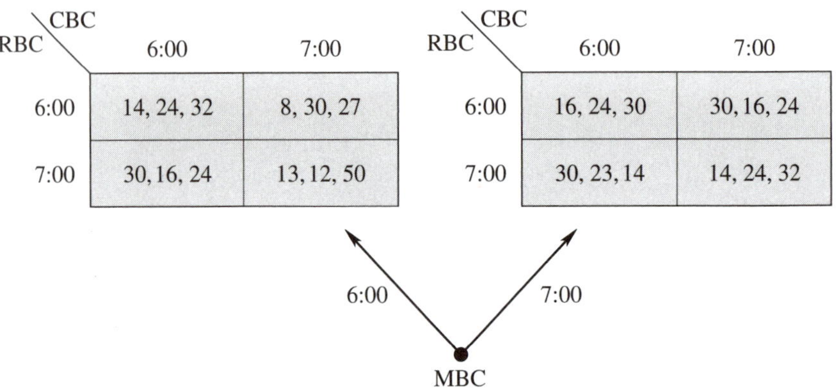

5. Suppose that in some two-player game, s_1 is a rationalizable strategy for player 1. If s_2 is a best response to s_1, is s_2 a rationalizable strategy for player 2? Explain.

6. Suppose that in some two-player game, s_1 is a rationalizable strategy for player 1. If, in addition, you know that s_1 is a best response to s_2, can you conclude that s_2 is a rationalizable strategy for player 2? Explain.

7. Consider a guessing game with ten players, numbered 1 through 10. Simultaneously and independently, the players select integers between 0 and 10. Thus player i's strategy space is $S_i = \{0, 1, 2, 3, 4, 5, 6, 7, 8, 9, 10\}$, for $i = 1, 2, \ldots, 10$. The payoffs are determined as follows: First, the average of the players' selections is calculated and denoted a. That is,

$$a = \frac{s_1 + s_2 + \cdots + s_{10}}{10},$$

where s_i denotes player i selection, for $i = 1, 2, \ldots, 10$. Then, player i's payoff is given by $u_i = (a - i - 1)s_i$. What is the set of rationalizable strategies for each player in this game?

LOCATION AND PARTNERSHIP 8

You have seen in some abstract examples how the concepts of iterated dominance and rationalizability constrain behavior. Our next step is to analyze some models that more closely represent real economic situations. As always, we have to keep our models simple; otherwise they can become unwieldy. Good models capture some of the important features of real-world settings while maintaining tractability; they represent and enrich our intuition in digestible doses of analytics.

A LOCATION GAME

Where firms locate in a city is a matter of strategy. Often a firm's proximity to its customers and its competitors determines how profitably it can operate. Consider a very simple model of location choice by competitors.[1] Suppose two people—call them Pat and Chris (P and C)—work for the same large soft drink company. Their job is to sell cans of the company's soda at a popular beach. They will be working at the same beach and, by company policy, must charge the same price. The company has committed to give each salesperson a commission of 25 cents per can sold. Because of the quantity of drinks that will be sold, as well as the coolers and refrigeration needed, each salesperson must work from a booth that is stationary on the beach. The only decision that each person has to make is where on the beach to set his booth at the beginning of the day. Pat and Chris select the locations for their booths independently and simultaneously. (Assume that salespeople have to call the city in the morning to obtain a permit, at which point they must commit to a location.)

The beach is divided into nine regions of equal size, as pictured in Figure 8.1 on the next page. Booths may be located in any one of these regions (and it is possible for the salespeople to locate in the same region). On any given day, fifty people in each region will each wish to purchase a soda. Thus, if a salesperson serves all of the customers in a single region, he will earn

[1]The model analyzed here is of the type first studied by H. Hotelling, "Stability in Competition," *The Economic Journal* 39(1929):41–57.

FIGURE 8.1 A location game.

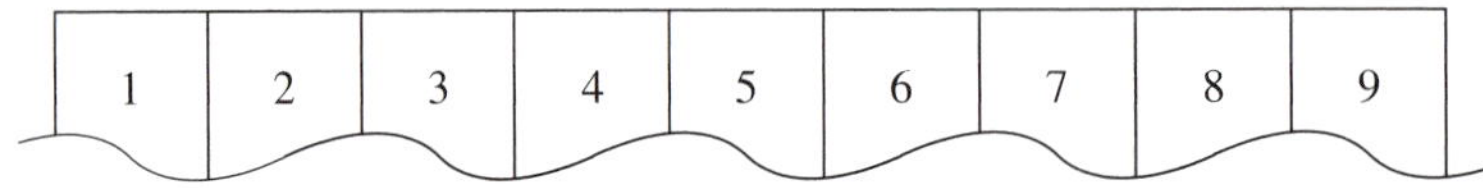

\$12.50. Assume that fifty customers in every region each purchase a drink, regardless of where Pat and Chris locate their booths. However, customers walk to the *nearest* booth to make their purchases. For example, if Pat locates in region 3 and Chris locates in region 8, then Pat sells drinks to all of the customers in regions 1 through 5 and Chris sells to all of those in regions 6 through 9. If the customers of a given region are indifferent between the two booths, assume that half go to one booth and half go to the other. Pat and Chris seek to maximize their incomes.

What should Pat and Chris rationally do? To analyze this game formally, first represent the game in normal form. Each player's strategy is a description of where this player locates on the beach. Thus, there are nine different strategies and player i's strategy space is $S_i = \{1, 2, 3, 4, 5, 6, 7, 8, 9\}$, for $i = \text{C, P}$. We could draw a matrix to neatly describe the payoff functions for this game; I leave it to you to do so and instead just note a few of the payoff numbers. When describing strategy profiles, put Chris's strategy first and Pat's second; that is, (x, y) is the strategy profile in which Chris selects x and Pat selects y. If Chris chose 2 and Pat selected 5, Chris's payoff would be $u_\text{C}(2, 5) = 3 \times 12.50 = 37.50$ and Pat would earn $u_\text{P}(2, 5) = 6 \times 12.50 = 75.00$. If Chris chose 1 and Pat selected 9, then the payoffs would be $u_\text{C}(1, 9) = 56.25$ and $u_\text{P}(1, 9) = 56.25$. In this case, Chris sells to regions 1 through 4, Pat to regions 6 through 9, and they split region 5.

To compute the set of rationalizable strategies, we need to perform iterated dominance. Begin by evaluating strategies 1 and 2 against all of the opponent's strategies. If Chris plays 1 and Pat plays 1, then Chris gets 56.25; they split all of the regions. If Chris plays 2 when Pat chooses 1, then Chris obtains 100; she captures regions 2 through 9 in this case. Thus, when Pat plays 1, Chris's strategy 2 yields a greater payoff than does strategy 1. Comparing these two strategies for Chris against Pat's strategy 2 yields the same conclusion: $u_\text{C}(2, 2) = 56.25 > u_\text{C}(1, 2) = 12.50$. Continuing, we find that strategy 2 yields a greater payoff than does strategy 1 against *all* of Pat's strategies. That is, $u_\text{C}(2, x) > u_\text{C}(1, x)$ for $x = 1, 2, \ldots, 9$. You should verify this.

We have demonstrated that strategy 2 strictly dominates strategy 1, and it is obviously true for both players. By the same reasoning, strategy 9 is dominated for both players. In fact, strategies 1 and 9 are the only dominated strategies in the game. To see this, note that 2 is a best response to the opponent selecting 1, 3 is a best response to the opponent selecting 2, and so on. By deleting strategies 1 and 9 for the players, we have reduced the strategy space $S_i = \{1, 2, 3, 4, 5, 6, 7, 8, 9\}$ to the set of undominated strategies $\{2, 3, 4, 5, 6, 7, 8\}$. And we can go farther by iterating the dominance criterion.

Given that the opponent's strategy is contained in $\{2, 3, 4, 5, 6, 7, 8\}$, strategies 2 and 8 are now strictly dominated. That is, $u_i(3, s_j) > u_i(2, s_j)$ and $u_i(7, s_j) > u_i(8, s_j)$ for $s_j = 2, 3, 4, 5, 6, 7, 8$, where s_j is the strategy of player i's opponent. If it is not obvious, you should check this. In the second round of deleting strictly dominated strategies, we thus remove 2 and 8, forming the undominated strategy set $\{3, 4, 5, 6, 7\}$ for each player. The same analysis requires us to remove 3 and 7 in the third round and 4 and 6 in the fourth. We are left with only one rational strategy for each player: 5.

In this game, rationalizability implies that the players, in their quest for a large share of the market, both locate in the same region at the center of the beach. The model solidifies some intuition that can be applied more generally. For example, in a community with two supermarkets, we should expect them to be located close together and in the geographical center of the population. Furthermore, "location" can be more broadly interpreted, allowing the model to address, among other things, product variety and politics. In the first case, think of consumers differing in their taste for sweet cereal. A producer can "locate" its cereal anywhere in the sweetness spectrum, by choosing how much sugar to include. Our model suggests that, in a small market, firms produce cereals of similar flavor that appeal to the average taste.

Regarding politics, note that the voters in a country are generally a diverse bunch. Some desire an open trade policy, whereas others prefer isolationism. Some people want government regulation in certain industries, whereas other people oppose any government intervention. One can categorize citizens according to their preferences. In the United States and in other countries, this categorization often takes the form of a one-dimensional *political spectrum,* with liberal-minded people on the left and conservatives on the right. Candidates for public office may try to appease certain groups of people to get elected. We suspect that an individual citizen votes for the candidate who is "closest" to him or her in regard to policy.

The location model suggests that candidates will try to locate themselves toward the center of the spectrum—that is, they will try to argue that they are in the mainstream, that they are moderate, and so forth. Lo and behold, candidates actually behave this way. The model also explains why candidates' positions shift over time (such as between a primary election, when a candidate wants to appeal to the party faithful, and a general election). I further discuss voting in Chapter 10.

The simple location model is clearly limited. Among others, there are four obvious criticisms. First, in the context of market competition, it does not include the firms' specification of prices. Firms might be able to charge higher prices when they are separated, and it is not clear how the predictions of our model would change with this extension. Second, agents may not have to move simultaneously in the real world. Firms can delay the construction of stores to observe where their competitors locate. Firms may even be able to relocate in response to their rivals' positions. Third, our model does not apply to larger markets where there are more than two firms or more than two products. Fourth, our model is one-dimensional, whereas interesting applications have more than one dimension.

The cereal industry illustrates a more elaborate economic setting. There are many cereal producers (although the market is dominated by General Mills, Kellogg's, and Post) and, more important, firms produce more than one kind of cereal. In fact, the dominant firms produce many different varieties, which, in reference to our model, is like locating at several different points on the spectrum. Firms simultaneously offer various cereals to suit different tastes. The technology of cereal production offers one explanation for the plethora of varieties on the market. Different varieties of cereal can be produced in the same plant, and companies can change from one variety to another at relatively low cost. If there are large segments of the taste spectrum at which no variety is located, a firm can easily locate a cereal in this region and earn a profit. Thus, we expect all such opportunities to be exhausted, leading to a market with a large number of varieties positioned at regular intervals along the spectrum of tastes.

As with the topic of firm competition, the analysis of politics can benefit from an enriched location model. In fact, intuition garnered from the basic model can guide us in extending the model and developing an understanding of complicated political processes. You might think about what happens if the citizens are not uniformly arranged on the political spectrum but, instead, are distributed in a different way. What if there are more than two candidates? What if the candidates cannot so easily remake themselves (locate freely)? What if there is more than one dimension to policy decisions? Why are there only two strong parties in the United States? These and many other questions can be addressed by using extensions of the basic model.

A PARTNERSHIP GAME: STRATEGIC COMPLEMENTARITIES

Most jobs entail interaction between different people. Whether in a small firm with two partners or in a major conglomerate, the success of the enterprise requires cooperation and shared responsibility. Because one person's effort affects another person's prosperity—perhaps through the profit that they generate together—people may not always have the incentive to work in the most efficient way. In other words, the nature of the workplace can create "distortions" whereby self-interested behavior spoils the cause of joint productivity. Games are played between friends and colleagues just as they are played between adversaries.[2]

To construct an extremely simple model of such distortions, think of a partnership between two people. The firm's profit, which the partners share, depends on the effort that each person expends on the job. Suppose that the profit is $p = 4(x + y + cxy)$, where x is the amount of effort expended by partner 1 and y is the amount expended by partner 2. The value c is a positive constant, which is assumed to be between 0 and $1/4$; it measures how complementary the tasks of the partners are. Partner 1 incurs a personal cost x^2 of expending effort, which is measured in monetary terms. Partner 2's cost of effort is y^2. Assume that x and y have to be set between 0 and 4.

The business environment is such that the partners cannot write a contract that dictates how much effort each must expend. The reason could be that the courts have no way of verifying effort on the job and so cannot enforce such contracts. Therefore partner 1 selects x and partner 2 selects y independently. Assume that they do so simultaneously as well. The partners seek to maximize their individual share of the firm's profit net of the cost of effort. In mathematical terms, partner 1 cares about $p/2 - x^2$ and partner 2 cares about $p/2 - y^2$.

To be complete, let us translate this model into the terminology of game theory using the normal form. The strategy space for player 1 is the set of numbers between zero and four, because he selects an effort level in this range. Player 2 has the same strategy space. Mathematically, $S_1 = [0, 4]$ and $S_2 = [0, 4]$. The payoff function for player 1 is $2(x + y + cxy) - x^2$ and the payoff function for player 2 is $2(x + y + cxy) - y^2$. Note that, because the players have an infinite number of strategies, we cannot represent this game with a matrix.

In the partnership game, the distortion in the interaction between the two partners arises in that each partner does not fully "internalize" the value of his effort. A partner knows that, if he works to increase the firm's profit by one dollar, he obtains only one-half of this amount. He is therefore less willing to provide effort. The firm suffers here because the partner does not incorporate the benefit of his labor to the rest of the firm.

Let us analyze the game and see whether this intuition is captured by the theory of rationalizability. Because the game has an infinite strategy space, looking for dominated strategies is a bit of a challenge. Rather than trying for a complete mathematical analysis, we can construct the set of rationalizable strategies by examining a graph of the players' best-response functions. This type of graph provides a great deal of intuition and will be useful later as well.

---

[2]Remember that game theory is the study of interdependence, not necessarily pure conflict.

Player 1 constructs a belief about the strategy of player 2, which is a probability distribution over y, player 2's effort level. Player 1 then selects a best response to her conjecture, which maximizes her expected payoff in the game. Let us compute this best response. Given her belief, if player 1 selects effort level x, then her share of the firm's profit is the expected value of $2(x + y + cxy)$. This expected value is $2(x + \overline{y} + cx\overline{y})$, where $\overline{y}$ is the mean (or "average") of her belief about y.[3] Player 1's expected payoff is thus $2(x+\overline{y}+cx\overline{y}) - x^2$. To compute player 1's best response, take the derivative of this expression with respect to x and set the derivative equal to zero (to find where the slope of this function of x is zero). This yields

$$2 + 2c\overline{y} - 2x = 0.$$

Solving for x, we have $x = 1 + c\overline{y}$. We thus have computed player 1's best response as a function of $\overline{y}$. In terms of the notation introduced earlier, $BR_1(\overline{y}) = 1 + c\overline{y}$. Player 2's best-response function is computed in the same way and is $BR_2(\overline{x}) = 1 + c\overline{x}$.

To be consistent with earlier definitions, we should actually write $BR_1(\overline{y}) = \{1 + c\overline{y}\}$ and $BR_2(\overline{x}) = \{1 + c\overline{x}\}$, because BR_i is a *set* of strategies (which, in this case, consists of a single strategy). However, I will be a bit loose with the nature of BR_i for games in which BR_i contains a single strategy for each belief. For such games, we can treat BR_i as a function. Thus, I will write $x = BR_1(y)$ instead of $x \in BR_1(y)$.

Figure 8.2 depicts the best-response functions of the two players. The strategy of player 1 is on the x-axis, and the strategy of player 2 is on the y-axis. The best-response function of player 2 is a function of x. Observe that, at $x = 0$, $BR_2(0) = 1$; at $x = 4$, $BR_2(4) = 1 + 4c$. The best-response function for player 1 has the same properties, although it is a function of player 2's strategy y.

To compute the set of rationalizable strategies, first observe that, regardless of player 2's belief about player 1, player 2's best response is a number between 1 and $1+4c$. To see this, remember that player 1 selects an effort level between 0 and 4, which means that $\overline{x}$ must be between these numbers as well. Thus, the greatest that $1 + c\overline{x}$ can be is $1 + 4c$, and the least that it can be is 1. Strategies of player 2 below 1 or above $1 + 4c$ can never be best responses; they are dominated. The same argument establishes that these same strategies are dominated for player 1 as well. Thus, if we start with the strategy space $[0, 4]$ for each player, the undominated strategies are in the set $[1, 1 + 4c]$. Note that, because $c < 1/4$, we know that $1 + 4c < 2$.

[3] For example, maybe she believes that, with probability 1/4, player 2 will select $y = 1/2$ and, with probability 3/4, player 2 will select $y = 2$. Then the mean of her conjecture is $\overline{y} = (1/4)(1/2) + (3/4)2 = 13/8$.

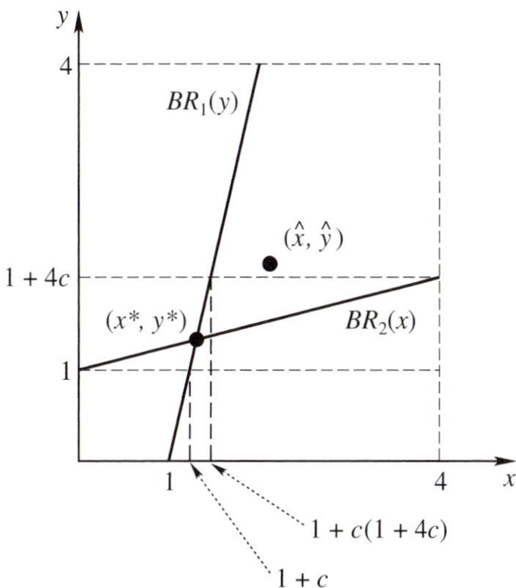

FIGURE 8.2
Partnership game best-response functions.

On to the next level of dominance. Understand that player 1 knows that player 2's strategy will be contained in the set $[1, 1 + 4c]$, and so player 1's average expectation about player 2's strategy has the same bounds. That is, $\overline{y} \in [1, 1 + 4c]$, which implies that player 1's best response to his belief must be a strategy in the set $[1 + c, 1 + c(1 + 4c)]$. We get this set by plugging the extreme values 1 and $1 + 4c$ into player 1's best-response function, which corresponds to the two vertical dashed lines in the center of Figure 8.2. Observe that $1+c > 1$ and $1+c(1+4c) < 1+4c$; so this new set of undominated strategies is "smaller" than the one described in the preceding paragraph. Again, the same argument can be used to narrow the set of rational strategies for player 2.

We obtain the set of rationalizable strategies by continuing to delete dominated strategies in stages, disregarding strategies that we have deleted in prior stages. You can tell from the preceding analysis that at each stage there will be an upper and a lower bound on the set of strategies that have survived so far. To get the upper bound for the next stage, plug the upper bound of the current stage into a player's best-response function. To get the lower bound for the next stage, plug in the current lower bound. From looking at the graph, it is not difficult to see where this process leads: as we remove dominated strategies, both the lower and upper bounds converge to the point at which the players' best-response functions cross. They cross at the point where

$$x = \frac{1}{1 - c} \quad \text{and} \quad y = \frac{1}{1 - c},$$

which jointly solve the equations $x = 1 + cy$ and $y = 1 + cx$. Therefore, $x^* = 1/(1 - c)$ is the only rationalizable strategy for player 1, and $y^* = 1/(1 - c)$ is the only rationalizable strategy for player 2.[4]

Given that the players select their only rationalizable strategies, the firm's profit is

$$4\left[\frac{1}{1-c} + \frac{1}{1-c} + \frac{c}{(1-c)^2}\right],$$

which simplifies to $(8 - 4c)/(1 - c)^2$. Splitting the profit and including the effort cost, each partner gets a payoff of $(3 - 2c)/(1 - c)^2$.

Rationalizability yields a unique prediction in this model of a partnership and, as we have seen before, the outcome is not the one that the players like best. To find the efficient outcome, maximize the total net profit of the firm,

$$4(x + y + cxy) - x^2 - y^2,$$

by choosing x and y. Taking the derivatives of this expression with respect to x and y and setting these derivatives equal to zero, we get $4 + 4cy = 2x$ and $4 + 4cx = 2y$. Solving these equalities simultaneously yields the *jointly optimal* effort levels:

$$\hat{x} = \frac{2}{1-2c} \quad \text{and} \quad \hat{y} = \frac{2}{1-2c}.$$

Note that

$$\frac{1}{1-c} < \frac{2}{1-2c}.$$

By playing their rationalizable strategies, the partners provide *less* effort than they jointly prefer. The jointly optimal strategy profile is denoted $(\hat{x}, \hat{y})$ in Figure 8.2.[5]

We have seen such a conclusion before, the prisoners' dilemma being the standard example of how individual incentives can interfere with group

---

[4]For the benefit of those who are mathematically inclined, the convergence property can be proved in the following way: Take any number z and define $w = z - 1/(1 - c)$. The number w measures the distance between z and $1/(1 - c)$. Plugging z into the best-response function yields $1 + cz = 1 + c[w + 1/(1 - c)] = cw + 1/(1 - c)$. In words, if $|w|$ is the distance between z and $1/(1 - c)$, then $c|w|$ is the distance between $BR_i(z)$ and $1/(1 - c)$. Because $0 < c < 1$, the distance shrinks every time the number is plugged into the best-response function.

[5]The term *joint* refers to the "total" welfare of all agents in the model—in this case, the two partners. Sometimes the expression *social optimal* is used, although it can be ambiguous if we consider people outside the game.

gain. But here the process of iteratively deleting strictly dominated strategies is about as complicated as the title of the procedure. There is also a bit more economics at work in the partnership example. As already noted, the reason our predicted outcome differs from the joint-optimum strategy is that the partners' individual costs and benefits differ from joint costs and benefits. The joint optimum balances joint benefits (the firm's gross profit) with joint costs (the sum of the partners' effort costs). Because each partner cares only about his own cost and benefit and because his individual benefit is only a fraction of the joint benefit, his calculations understate the value of his effort. Thus, he expends less effort than is best from the social point of view. Because both players do so, they both end up worse off.

The term cxy was included in the partners' profit function for two reasons. First, it makes iterated dominance more interesting than if the term were absent, which allows us to build intuition that will be useful later. Incidentally, you will notice that one of the problems at the end of this chapter is a simplified version of the game heretofore presented; it is much easier to solve.

The second reason for including the cxy term is that it represents an important feature of many economic settings—a feature called *complementarity*. In the partnership game, increases in effort by one of the partners is beneficial to him, up to a point. For example, raising x from 0 to $1/2$ will increase partner 1's payoff. Furthermore, increasing a partner's effort is more valuable *the greater is the other partner's effort level*. To see this, note that the first derivative of player 1's payoff function with respect to his own effort, x, increases as y increases. The same is true for player 2's payoff. The relation between the partners' strategies is complementary here; in formal terms, the game is one of *strategic complementarities*. In general, strategic complementarity is easy to detect in games with continuous strategy spaces. It exists if $\partial^2 u_i(s)/\partial s_i \partial s_j \geq 0$ for all $s \in S$, each player i, and each other player j. That is, the cross partial derivative of player i's payoff function with respect to player i's strategy and any other player's strategy is not negative.[6]

Because of the strategic complementarity, each player's best-response function is increasing in the mean belief about the partner's strategy. That is, as $\overline{y}$ increases, so does the optimal x in response. In terms of the graph in Figure 8.2, the players' best-response functions are positively sloped. In the partnership game, as c increases, so does the slope of the best-response functions. (Because player 1's best response is a function of $\overline{y}$, this means the graph in Figure 8.2 flattens.) As a result, the rationalizable effort levels

[6]Strategic complementarity was discussed by J. Bulow, J. Geanakoplos, and P. Klemperer, "Multimarket Oligopoly: Strategic Substitutes and Complements," *The Journal of Political Economy* 93(1985):488–511. The concept was analyzed more generally by P. Milgrom and J. Roberts, "Rationalizability, Learning, and Equilibrium in Games with Strategic Complementarities," *Econometrica* 58(1990):1255–1277.

increase, too. Intuitively, they do so because each player's optimal strategy depends "more" on the effort of the other.

Rationalizability does not always lead to a unique prediction in games such as this one, even when there are strategic complementarities. The force of iterated dominance depends on the relative slopes of the best-response functions. To develop your understanding, it might be helpful to draw a few pictures like Figure 8.2 with various different best-response functions. A best-response function for player 1 must assign a value for every y, so it must stretch from the bottom of the picture to the top. A best-response function for player 2 must stretch from the left to the right. One result of which you may convince yourself is that rationalizability leads to a unique prediction in two-player games with three properties: (1) the strategy spaces are intervals with lower and upper bounds, (2) there are strategic complementarities, and (3) the slope of the best-response functions is less than one. These conditions are not required for rationalizability to yield a unique strategy profile, but they are sufficient. You should verify that they hold in the partnership game.

GUIDED EXERCISE

Problem: Consider the location game analyzed in this chapter, but with different preferences for the players. Instead of each player seeking to sell as *many* cans of soda as possible, suppose that each wants to sell as *few* cans as possible. To motivate such preferences, think of the players being paid a fixed wage for their work, and imagine that sales entail effort that they would rather not exert. In this new game, are there any dominated strategies for the players? Compute the rationalizable strategies for the players, and explain the difference between this game and the original game.

Solution: Intuitively, players in this game want to do the opposite of what was shown to be rational in the standard location game. Whereas in the standard game a player benefits from being closer to the middle location than is her opponent, in this game the players gain (that is, sell fewer sodas) by moving away from the middle. Indeed, you can easily check that the strategy of locating in region 1 dominates each of the strategies 2 through 5, because locating in region 1 yields strictly lower sales regardless of where the opponent locates. Furthermore, the strategy of locating in region 9 dominates each of the strategies 5 through 8. Thus, strategies 2 through 8 for both players are removed in the first round of the iterated-dominance procedure. In the second round, we find that neither strategy 1 nor strategy 9 dominates the other. Therefore, the set of rationalizable strategies for each player i is $R_i = \{1, 9\}$, $i = 1, 2$.

EXERCISES

1. Consider a location game like the one discussed in this chapter. Suppose that, instead of the nine regions located in a straight line, the nine regions form a box with three rows and three columns. Two vendors simultaneously and independently select on which of the nine regions to locate. Suppose that there are two consumers in each region; each consumer will walk to the nearest vendor and purchase a soda, generating a \$1.00 profit for the vendor. Consumers *cannot* walk diagonally. For example, to get from the top-left region to the middle region a consumer has to walk through either the top-center region or the middle-left region. (This means that the middle region is the same distance from top-left as top-right is from top-left.) Assume that, if some consumers are the same distance from the two vendors, then these consumers are split equally between the vendors. Determine the set of rationalizable strategies for the vendors.

2. Consider a location game with nine regions like the one discussed in this chapter. But instead of having the customers distributed uniformly across the nine regions, suppose that region 1 has a different number of customers than the other regions. Specifically, suppose that regions 2 though 9 each has ten customers, whereas region 1 has x customers. For what values of x does the strategy of locating in region 2 dominate locating in region 1?

3. Consider our nine-region location game to be a model of competition between two political candidates. Fifty voters are located in each of the nine regions of the political spectrum; each voter will vote for the closest candidate. Voters in regions equidistant from the two candidates will split evenly between them.

   (a) Assume that each candidate wants to maximize the number of votes he or she receives. Are the candidates' preferences the same as those modeled in the basic location game? Discuss how this setting differs from the setting in which candidates want to maximize the *probability* of winning (with the rule that the candidate with the most votes wins the election and, in case of a tie, each candidate is awarded the office with probability $1/2$).

   (b) Suppose, contrary to the basic location model, that voters are not uniformly distributed across the nine regions. Instead, whereas regions 1, 2, 3, 4, 5, 8, and 9 each contains fifty voters, regions 6 and 7 each contains x voters. Analyze the candidate location game for the case in which $x = 75$. In a rationalizable outcome, can one of the candidates win with probability 1?

   (c) What happens if $x > 75$ or $x < 75$?

4. Consider the partnership game analyzed in this chapter, but assume that $c < 0$. Graph the players' best-response functions for this case, and explain how they differ from those in the original model. If you can, find the rationalizable

strategies for the players. Use the graph to find them. Repeat the analysis under the assumption that $c > 1/4$.

5. Consider a duopoly game in which two firms simultaneously and independently select prices. Assume that the prices are required to be greater than or equal to zero. Denote firm 1's price as p_1 and firm 2's price as p_2. The firms' products are differentiated. After the prices are set, consumers demand $10 - p_1 + p_2$ units of the good that firm 1 produces. Consumers demand $10 - p_2 + p_1$ units of the good that firm 2 produces. Assume that each firm must supply the number of units demanded. Also assume that each firm produces at zero cost. The payoff for each firm is equal to the firm's profit.

(a) Write the payoff functions of the firms (as a function of their strategies p_1 and p_2).

(b) Compute firm 2's best-response function (as a function of p_1).

(c) Can you determine the rationalizable strategies in this game by inspecting the graph of the best-response functions? What are the rationalizable strategies?

6. Consider a location game with five regions on the beach in which a vendor can locate. The regions are arranged on a straight line, as in the original game discussed in the text. Instead of there being two vendors, as with the original game, suppose there are *three* vendors who simultaneously and independently select on which of the five regions to locate. There are thirty consumers in each of the five regions; each consumer will walk to the nearest vendor and purchase a soda, generating a \$1.00 profit for the vendor. Assume that, if some consumers are the same distance from the two or three nearest vendors, then these consumers are split equally between these vendors.

(a) Can you rationalize the strategy of locating in region 1?

(b) If your answer to part (a) is "yes," describe a belief that makes locating at region 1 a best response. If your answer is "no," find a strategy that strictly dominates playing strategy 1.

7. Consider a game in which, simultaneously, player 1 selects a number $x \in [2, 8]$ and player 2 selects a number $y \in [2, 8]$. The payoffs are given by:

$$u_1(x, y) = 2xy - x^2$$

$$u_2(x, y) = 4xy - y^2.$$

Calculate the rationalizable strategy profiles for this game.

9 NASH EQUILIBRIUM

The concept of rationalizability embodies just a few simple assumptions about what players know about each other and how players respond to their own beliefs. The assumptions are that (1) people form beliefs about others' behavior, (2) people best respond to their beliefs, and (3) these facts are common knowledge among the players. The concept is quite *weak* in that no additional assumptions about behavior are made. In particular, we do not assume that each player's beliefs are consistent with the strategies actually used by the other players. We saw at the end of Chapter 7 how strategic uncertainty can lead to very uncoordinated outcomes; in the battle of the sexes, for example, both players may receive their lowest possible payoffs if the expectation of one player is not synchronized with the behavior of the other.

There are many settings in which rationalizability is the appropriate behavioral concept. In particular, it is justified in studying situations in which people have little history in common on which to coordinate their beliefs. Locating a friend on your first trip to Disneyland may fit into this category, as long as you have not communicated extensively with your friend (presumably because you lost your cell phone or the battery ran down) and you have not agreed in advance on places to meet under certain contingencies. In a more traditional economic vein, bargaining between the buyer and the seller of a house or car often takes place with strategic uncertainty. The parties may not have met each other previously and do not know what strategies each will adopt in the negotiation. They may understand that they are both rational and sophisticated thinkers, which can restrict how they act, but there is strategic uncertainty nonetheless.

In other settings, strategic uncertainty is resolved through a variety of social institutions, such as norms, rules, or communication between the players. For instance, if people can communicate before playing a game, they may discuss what strategies each will take, and they may even agree on how the game will be played. Communication therefore can reduce strategic uncertainty by bringing beliefs into harmony with actual behavior. We can imagine how this is accomplished in coordination games, where the players can all benefit from agreeing to a course of play. In more adversarial games, communication may not be quite as effective. In general, communication that aligns the players' expectations should be considered a form of agreement, or contract. This interpretation is elaborated in later chapters.

Historical factors also may coordinate beliefs and behavior. For example, consider the game of deciding how to pass someone who is walking toward you on the sidewalk. If neither of you changes direction, you will collide. At least one of you must alter your path to prevent a collision. But if you both move toward the street or you both move away from the street, then you will also collide and be embarrassed to boot. You are better off if one person moves a bit *toward* the street and the other moves a bit *away* from the street. You need to coordinate your actions. One way of doing so is to yell to each other commands such as "move to your right please," but yelling seems as embarrassing as colliding.

In fact, the "avoiding people on the street" game is played every day in society, and historical precedent has served to align our beliefs and behavior. At some point in the United States, people began to favor the practice of organizing the sidewalk in much the same way as streets are organized for automobile travel. People generally move to their right to avoid others. How this transpired may not be known; perhaps it was the result of random behavior. Regardless of its origins, people have come to *expect* that others will prevent collisions by moving to their right. It is a social convention, one reinforced every day as people conform to it.

History, rules, and communication are just as effective in coordinating beliefs and behavior in economic settings as they are in walking games. Firms that compete over time often settle into a routine in which the manager of each firm has learned to accurately predict the strategies employed by his rivals each week. Business partners who work together on similar projects over and over again learn what to expect from each other. They can also communicate to coordinate their actions. Bargaining over the price of a house can be moderated by social norms, even when the parties have little personal experience with the market. The parties' beliefs and behavior are often guided by how others have played the game in the past.

The point of this chapter is to begin to explore rational behavior when actions and beliefs are coordinated by social institutions. The underlying idea is that through some social force (such as those just discussed) behavior in a game is coordinated, or *congruous*. Congruity can refer to consistent and regular behavior in a game that is played over and over in society or by the same parties who interact repeatedly. Congruity can also refer to behavior in a one-shot game in which communication or social history has aligned the beliefs of each player with the strategies of the others. Here are three versions of the congruity concept, geared toward different settings:

1. A game is repeatedly played in society or by a group of agents. The behavior of the players "settles down" in that the same strategies are used each time the game is played.
2. The players meet before playing a game and reach an agreement on the strategy that each will use. Subsequently, the players individually honor the agreement.
3. An outside mediator recommends to the players that they adopt a specific strategy profile in a game. Each player, expecting that the others will follow the mediator's suggestion, has the incentive to follow the mediator as well.

Instead of studying these interpretations separately, we shall examine their common element by defining congruity through using the notion of best response.

CONGRUOUS SETS

Inherent in "congruity" is that players have no reason to deviate from their prescribed course of play. In other words, each player selects a best response. We can then think of sets of strategies as congruous if each strategy prescribed for a given player can be rationalized as a best response to some belief over the strategies prescribed for the other players. To be precise:

> Consider a set of strategy profiles $X = X_1 \times X_2 \times \cdots \times X_n$, where $X_i \subset S_i$ for each player i. The set X is called **weakly congruous** if, for each player i and each strategy $s_i \in X_i$, there is a belief $\theta_{-i} \in \Delta X_{-i}$ (putting probability only on strategies in X_i) such that $s_i \in BR_i(\theta_{-i})$. The set X is called **best response complete** if, for each player i and each belief $\theta_{-i} \in \Delta X_{-i}$, it is the case that $BR_i(\theta_{-i}) \subset X_i$. Finally, X is called **congruous** if it is both weakly congruous and best response complete.

In words, X is weakly congruous if each of the strategies in X_i can be rationalized with respect to X_{-i}. It is best response complete if X_i contains *all* of the strategies that can be rationalized with respect to X_{-i}. It is congruous if X_i contains *exactly* those strategies that can be so rationalized.[1]

Consider the game in Figure 9.1 on the next page. Observe that $X = \{U, M\} \times \{L, C\}$ is a congruous set of strategies. To understand why, note that,

[1] Note that in the formal definitions, "ΔX_{-i}" denotes the set of probability distributions on S_{-i} that put positive probability only on strategies in X_{-i}.

FIGURE 9.1
Congruity.

1 \ 2	L	C	R
U	10, 0	0, 10	3, 3
M	2, 10	10, 2	6, 4
D	3, 3	4, 6	6, 6

if player 1's belief puts probability 0 on player 2's strategy R, then his strategy D is never a best response. You should verify this by computing player 1's best response to various beliefs. In addition, U and M are both rationalized as best responses to beliefs putting probability only on L and C. The same is true for player 2's strategies L and C, with respect to beliefs over U and M. Therefore, X is both weakly congruous and best response complete.

In this example, {(D, R)} is also weakly congruous, because D is a best response to R and vice versa. Note, though, that {(D, R)} is not best response complete, because M is also a best response to R. An example of a set that is best response complete but not weakly congruous is the entire strategy space of the prisoners' dilemma. In fact, the entire strategy space of *any* game is best response complete, because players cannot select anything outside of this set. For finite games, the set of rationalizable strategies is congruous. In fact, one can prove that the rationalizable set is the largest congruous set, in that every other congruous set is contained in it.

NASH EQUILIBRIUM DEFINED

As Figure 9.1 demonstrates, congruity does not imply that strategic uncertainty is erased. In the example, $\{U, M\} \times \{L, C\}$ is congruous, but it leaves open the possibility that the players' strategies are uncoordinated. For instance, player 1 might select U under the firmly held belief that player 2 will choose L. Player 2 may select C believing that player 1 plays U and M each with probability 1/2. The players' beliefs and actual choices are inconsistent in this case.

One simple way of capturing the idea of strategic certainty is to assume that the players have coordinated on a *single strategy profile*. From this perspective, the components of congruity heretofore sketched verbally have a straightforward and intuitive interpretation. For example, an agreement between people about how to play a game often takes the simple form of an agreement to play a specific strategy profile. Combining coordination on a strategy profile with the notion of weak congruity leads to a simple, but extremely powerful, theory of behavior. It is one of the many contributions of Nobel laureate John Nash to the field of game theory. Nash defined an *equilibrium* concept for games—we now call it a Nash equilibrium—which is precisely a weakly congruous strategy profile.[2] In a meaningful way, a Nash equilibrium has no strategic uncertainty: each player's belief about another's strategy is concentrated on the actual strategy that the other player uses.

A strategy profile is a Nash equilibrium if and only if each player's prescribed strategy is a best response to the strategies of the others. To be formal:

> A strategy profile $s \in S$ is a **Nash equilibrium** if and only if $s_i \in BR_i(s_{-i})$ for each player i. That is, $u_i(s_i, s_{-i}) \geq u_i(s_i', s_{-i})$ for each $s_i' \in S_i$ and each player i.

For example, consider a game that is repeated in society, and suppose that over time behavior settles to the point that it conforms to a single strategy profile, s. If player i believes that the behavior of the others today will correspond to historical behavior, then player i should play a best response to s_{-i}. If s_i is not a best response to s_{-i}, then s is not a good description of what will happen today. That is, s can be stable over time only if s_i is a best response to s_{-i}.

As another example, consider the setting in which the players meet before playing a game and agree on a strategy profile s that should be played. The agreement is weakly congruous—that is, the players each have an individual incentive to abide by it—only if each player's prescribed strategy is a best response to the prescription for the others. If the agreement is to play strategy profile s and $s_i \notin BR_i(s_{-i})$ for some player i, then this player has no incentive to abide by the agreement and will choose a strategy that is different from s_i.

To illustrate the concept of a Nash equilibrium, consider the classic normal-form games pictured in Figure 3.4. The games are redrawn in Figure 9.2 at the top of the next page, and the Nash equilibria are designated by circles around the relevant cells of the matrices. The set of rationalizable strategies also is indicated for each game (in that strategies removed by the iterated dominance procedure are stricken). Note that all strategies are rationalizable for every game except the pigs game and the prisoners' dilemma.

In the prisoners' dilemma, regardless of the other player's strategy, the only best response is to play strategy D. We have $\text{D} \in BR_i(\text{D}) = \{\text{D}\}$ for both players, and so the strategy profile (D, D) is the Nash equilibrium of the prisoners' dilemma. The battle of the sexes has two equilibria, one in which the couple coordinates on the opera and one in which both players go to the

[2] Nash reported his equilibrium concept in "Non-Cooperative Games," *Annals of Mathematics* 51(1951):286–295.

FIGURE 9.2 Equilibrium and rationalizability in the classic normal forms.

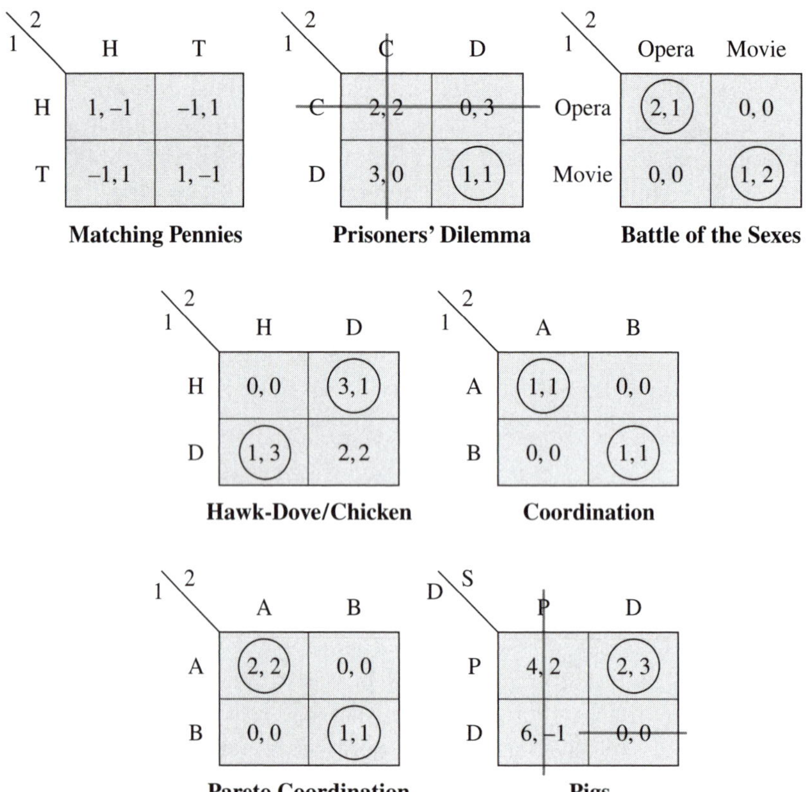

movie. That is, (O, O) and (M, M) are both Nash equilibria of the battle of the sexes. If one's friend selects O, then O is the only best response; if the friend chooses M, one must do the same. Note that, even though it might appear that the payoffs are circled in Figure 9.2, the circles designate the *cells* (strategy profiles). Nash equilibria are strategy profiles, not payoffs. Thus, if you had to report the Nash equilibria of the pigs game, you should write "(P, D)," not "(2, 3)."

A more stringent version of the equilibrium concept is called *strict Nash equilibrium*. A strategy profile s is called a strict Nash equilibrium if and only if $\{s_i\} = BR_i(s_{-i})$ for each player i. In words, s is a strict Nash equilibrium if and only if $\{s\}$ (the set comprising only profile s) is a congruous set of strategies. You should verify that, in the games of Figure 9.2, all of the Nash equilibria are also strict Nash equilibria.

FIGURE 9.3 Determining Nash equilibria.

1 \ 2	X	Y	Z
J	5, 6	3, 7	0, 4
K	8, 3	3, 1	5, 2
L	7, 5	4, 4	5, 6
M	3, 5	7, 5	3, 3

(a)

1 \ 2	X	Y	Z
J	5, 6	3, $\underline{7}$	0, 4
K	$\underline{8}$, $\underline{3}$	3, 1	$\underline{5}$, 2
L	7, 5	4, 4	$\underline{5}$, $\underline{6}$
M	3, $\underline{5}$	$\underline{7}$, $\underline{5}$	3, 3

(b)

In general, Nash equilibria are not difficult to locate in matrix games. Just look for profiles such that each player's strategy is a best response to the strategy of the other. You can check all of the strategy profiles (each cell of the matrix) one at a time. Alternatively, you can find the best responses of player 1 to each of the strategies of player 2 and then find the best responses of player 2 to each of the strategies of player 1. For an illustration of the second methodology, examine the game in Figure 9.3(a). Let us find player 1's best responses to each of player 2's strategies. By scanning player 1's payoffs in the first column of the matrix, you can see that the greatest payoff for player 1 appears in the second row. Therefore, K is player 1's best response to X. In Figure 9.3(b), I have indicated this best response by underlining player 1's payoff of 8 in the (K, X) cell of the matrix. Proceeding to the second column, we find that M is player 1's best response to Y, which is indicated in Figure 9.3(b) by the payoff of 7 that is underlined. Examination of the third column reveals that *both* K and L are best responses to Z; thus, both of the 5s in this column are underlined.

Our analysis of the game in Figure 9.3(a) continues with the evaluation of player 2's best responses. To find player 2's best response to J, we scan across the first row of the matrix and look at player 2's payoff numbers. The largest is 7, which I have underlined in Figure 9.3(b). In other words, Y is player 2's best response to J. Moving ahead to the second through fourth rows of the matrix, we find that X is player 2's best response to K, Z is player 2's best response to L, and both X and Y are best responses to M. These best responses are designated with the appropriately underlined numbers in Figure 9.3(b).

Having located the players' best responses, we can easily find the Nash equilibria of the game in Figure 9.3(a). There is a Nash equilibrium at each cell in which the payoff numbers for *both* players are underlined in Figure 9.3(b).

For example, (K, X) is a Nash equilibrium—K is player 1's best response to X, and X is player 2's best response to K. The game has two other Nash equilibria: (L, Z) and (M, Y). Observe that (K, X) is a strict Nash equilibrium, because K and X are the *only* best responses to one another. Neither (L, Z) nor (M, Y) is a strict Nash equilibrium.

There are a few things you should keep in mind at this point. First, each Nash equilibrium is a rationalizable strategy. (Incidentally, this is not difficult to prove; you might give it a try.) Because of this relation, you can restrict your search for equilibria to rationalizable strategies. Second, as Figure 9.2 indicates, some games have more than one Nash equilibrium. I will address the economic consequences of this later, but for now this should at least be taken as a caveat not to conclude your search for equilibria after finding just one. Third, some games have no equilibrium. Matching pennies is a good example. It is a game of pure conflict, where players can only win or lose. A player who loses under a given strategy profile will want to change her strategy so that she wins and the other loses; but then the other will want to change her strategy, and so forth.

EQUILIBRIUM OF THE PARTNERSHIP GAME

As you have seen, it is not difficult to find Nash equilibria in matrix games. In fact, computing equilibria of games with infinite strategy spaces also is not difficult. One need only compute the best-response mappings for each player and then determine which strategy profiles, if any, satisfy them all simultaneously. This usually amounts to solving a system of equations.

To illustrate the computation of a Nash equilibrium, let us consider the partnership game discussed in Chapter 8. The best-response functions for this game are pictured in Figure 8.2. Player 1's best response function is

$$BR_1(\overline{y}) = 1 + c\overline{y},$$

where $\overline{y}$ is the expected value of player 1's belief about player 2's strategy. Player 2's best-response function is

$$BR_2(\overline{x}) = 1 + c\overline{x},$$

where $\overline{x}$ is the mean of player 2's belief.

A Nash equilibrium for this game is a strategy profile (x^*, y^*) with the property that

$$x^* \in BR_1(y^*) \quad \text{and} \quad y^* \in BR_2(x^*).$$

Because we consider BR_1 and BR_2 to be functions in this example, this property can be expressed as

$$x^* = BR_1(y^*) \quad \text{and} \quad y^* = BR_2(x^*).$$

That is, the point (x^*, y^*) should lie on the best-response functions of both players. There is one such point and it is obviously located where the two best-response functions cross. We can compute this point by solving the following system of equations:

$$x^* = 1 + cy^* \quad \text{and} \quad y^* = 1 + cx^*.$$

Substituting the second equation into the first yields $x^* = 1+c(1+cx^*)$, which simplifies to $x^*(1-c^2) = 1+c$. Noting that $1-c^2 = (1+c)(1-c)$, we therefore have $x^* = 1/(1-c)$. Substituting this equation into the second equation yields $y^* = 1/(1-c)$. Observe that this strategy profile is the same as that computed in Chapter 8. In this game, rationalizability and Nash equilibrium predict the same, single strategy profile.

COORDINATION AND SOCIAL WELFARE

We observed earlier that rationalizability does not necessarily imply coordinated behavior by the players. Nash equilibrium, on the other hand, implies some coordination because it embodies a notion of congruity for a single strategy profile. However, Nash equilibria do not always entail strategies that are preferred by the players as a group. For instance, the only Nash equilibrium of the prisoners' dilemma is inefficient, in that both players would be better off if they played differently. Incentives in the prisoners' dilemma clearly lead to this conclusion. Thus, we note the prisoners' dilemma as an example of individual incentives interfering with the interests of the group.

Sometimes a socially inefficient outcome prevails not because of conflict between individual incentives but because there is more than one way to coordinate. Consider the Pareto coordination game in Figure 9.2. This game has two Nash equilibria on which the players may coordinate. The equilibrium (B, B) is inefficient; both players would rather be in the (A, A) equilibrium than in the (B, B) one. But (B, B) is an equilibrium nonetheless. Given that the other chooses strategy B, each player's only rational move is to select B as well.

Lest you think that inefficient equilibria should be ruled out in practice, consider some examples of inefficient equilibria in the world—which, by the way, are more the rule than the exception. The most well-known historical example is the layout of most computer keyboards and typewriters in the English-speaking world. The standard arrangement of keys (starting with QWERTY in the third row) was devised many years ago by the founder of the typewriter to minimize "jamming" of the keys. On mechanical typewriters, this problem is created when adjacent keys are pushed at nearby points of time, causing the associated arms that strike the ribbon to tangle. The inventor arranged the keys so that those that are likely to be used in close proximity (such as "a" and "n") are not adjacent on the keyboard.

Unfortunately, many people believe the QWERTY keyboard is not the most efficient layout for speedy typing. Furthermore, "jamming" is not a problem with modern typewriters and computers. In the 1930s, August Dvorak and William Dealey, through a careful study of word usage in the English language, devised another keyboard layout that is commonly referred to as the Dvorak keyboard. Some people assert that those who learn to type with this keyboard can do so at rates substantially higher than is possible with QWERTY.[3] We thus must ask the economic question: Why is the QWERTY keyboard still the norm?

The answer is that QWERTY is entrenched and few people have the incentive to switch to a new format, given that most others in the world use QWERTY. In the Pareto coordination game of Figure 9.2, think of player 1 as the typical computer owner and player 2 as the typical computer manufacturer. Strategy A is to buy or produce a Dvorak keyboard, and strategy B is to buy or produce the QWERTY keyboard. If most people in the world are trained to use only the QWERTY design—that is, they have adopted strategy B—it is obviously best for a computer maker to build QWERTY keyboards. Furthermore, because most computer keyboards in existence are QWERTY, the typical youngster will be advised to practice on QWERTY rather than Dvorak, and the next generation of computer buyers will demand QWERTY keyboards. Given that the English-speaking world began typing with QWERTY, the Dvorak keyboard has little chance of widespread success. Besides, computerized speech recognition may make the *keyboard* obsolete, at which time the QWERTY–Dvorak debate will be relegated to history books. I am proud to note that part subs this textbook were gem elated using speech recognition software.

---

[3]A woman named Barbara Blackburn has been cited in the *Guinness Book of World Records* as the fastest typist in the world. She typed at a comfortable 170 words per minute by using the Dvorak system, with a top speed of 212. You can, too, by converting your computer keyboard to the Dvorak layout, which merely requires a minor alteration of software. Of late, some have argued that the advantage of the Dvorak keyboard is actually quite minor, but it still makes for a good example.

The competition between VHS and Betamax video formats during the 1980s is another good example. Many people regarded the Betamax format as superior in the quality of reproduction, yet this format has died out. Its demise may be the simple result of random factors in the early VCR market or shrewd marketing by VHS producers. There was a time in which both formats were popular, although it was obvious that efficiency would be enhanced if everyone used the same format. Gradually, VHS gained a greater market share. At some point, the VHS movement picked up steam and decisively moved toward domination, as new video buyers rationally flocked to the format that appeared to be taking hold. A similar competition is now occurring in the market for high-definition video storage, between Blu-Ray and HD-DVD.

As a final example, consider cigarette lighters in automobiles. Years ago, when there were no cell phones or automobile refrigerator units, auto manufacturers installed round cigarette lighters in dashboards. Many people smoked while driving and they appreciated the little device that, when pressed, allowed the flow of electricity to pass through a resistant wire coil, creating heat sufficient to light a cigarette. Then came the electronic age and the flow of electronic gadgets to market. Manufacturers of cell phones—and other devices—realized that folks would like to use their phones while driving (which is ill advised because cell phone use while driving is on a par with drunkenness in its contribution to traffic accidents and is illegal in some states). These manufacturers discovered that they could power the phones by using an adapter that fits into a cigarette lighter, and so they included the adapter in their designs.

Unfortunately, automobile cigarette lighters do not provide the most secure electrical connection. The flow of electricity is easily disrupted by bumps in the road, as my wife and I discovered on a long camping trip (on which we brought a new portable refrigerator—okay, we should have been "roughing it" anyway). There are better, more reliable ways of getting electricity to devices; more secure plug and socket connections can easily be implemented. But the auto producers and the device manufacturers are now in an inefficiently coordinated outcome, from which neither side has a unilateral incentive to deviate. Automobile companies could, at a reasonable cost, design and install more secure electrical sockets in their cars, but they have no incentive to do so if the device manufacturers are using cigarette lighter adapters. Further, the device manufacturers have no incentive to adopt a different plug if automobile companies do not install the new sockets.

THE THIRD STRATEGIC TENSION

The QWERTY, VHS–Betamax, and auto examples are just a few of the ways in which our world has coordinated inefficiently. It might be helpful for you to look for more examples. On a thematic level, note that inefficient coordination poses a problem *even* if the players have the *same* preferences over outcomes and there is *no* strategic uncertainty—that is, even if the players entirely eschew the first and second strategic tensions.[4] Thus, the specter of inefficient coordination is called the third strategic tension. You might start to contemplate whether specific social or economic institutions can alleviate this tension as they might alleviate strategic uncertainty (the second tension) in the real world. Furthermore, you can ruminate over how institutions help select between equilibria in games such as the battle of the sexes, where players disagree about which equilibrium is preferred.

Because the terms "joint" and "social" come up frequently in the context of optimality and efficiency, some clarification is in order. When using these terms, we should always make clear which set of people we are focusing on. For example, we might say that two firms realize their joint optimal strategy profile; in this case, we are addressing only the firms' welfare, not the utility of other agents in the world (such as the firms' customers). In fact, an outcome that is jointly efficient from the firms' standpoint may be quite inefficient with respect to a larger set of actors. Generally, when I use the terms "joint" and "social," either (1) I will have specified the group or society of interest, or (2) I will be referring to the entire set of players in the game at hand.

ASIDE: EXPERIMENTAL GAME THEORY

At this point in our tour of game theory, it is worthwhile to pause and reflect on the purpose and practicality of the theory. As I have already emphasized (and will continue to emphasize) in this book, game theory helps us to organize our thinking about strategic situations. It provides discipline for our analysis of the relation between the outcome of strategic interaction and our underlying assumptions about technology and behavior. Furthermore, the theory gives us tools for prescribing how people ought to behave—or, at least, what things people ought to consider—in strategic settings.

You might start to ask, however, whether the theory accurately describes and predicts real behavior. The answer is not so straightforward. There are

[4]Recall that the first two tensions are (1) the clash between individual and group incentives, and (2) strategic uncertainty and its implications for joint optimality.

two ways of evaluating whether game theory is successful in this regard. First, you might gather data about how people behave in real strategic situations. For example, you can observe where competing firms locate in a city, how team members interact within a firm, how managers contract with workers, and so forth. Then you can construct game-theoretic models in an attempt to make sense of the data. You can even perform statistical tests of the models. In fact, many empirical economists dedicate themselves to this line of work. These economists are constantly challenged by how to reconcile the complexities of the real world with necessarily abstract and unadorned theoretical models.

The second way of evaluating game theory's predictive power is to bring the real world closer to the simple models. You can, for example, run laboratory experiments in which subjects are asked to play some simple matrix games. In fact, this sort of research—which is called *experimental game theory*—has become a little industry in itself. In many universities throughout the world, experimental economists herd students into laboratories that are filled with computer stations, attracting the students with the prospect of winning significant amounts of money. In comparison with experimental work done by researchers in other disciplines, the economists certainly have gotten one thing right: they pay well. By paying the subjects according to their performance in games, experimenters give them a strong incentive to think about how best to play.[5]

The downside of experimental research is that the games people play in the laboratory are often far removed from the sort of real strategic settings we dearly care about. On the plus side, however, the laboratory affords the experimenter considerable control over the strategic environment. The experimenter can, in essence, put the subjects into any game of interest. There is an important restriction, however: in some settings, the experimenter cannot completely control the subjects' *payoffs*. In particular, the subjects may care about more than just their monetary rewards. People are often motivated in ways the simple theoretical models fail to capture: spite, envy, fairness, schadenfreude, to name a few.

Experiments help us identify people's motivations (and not just when the people are dramatic actors). Experiments also help us determine people's degree of *strategic sophistication,* which concerns whether they are really thinking and behaving in the way that our hyper-rationality-based theories predict. To the extent that behavior deviates from theory in a systematic way, experiments can guide us in improving our models.

---

[5]Experiments in the classroom also are popular, and instructive to boot. If you are enrolled in a game-theory course and your professor does not run any classroom experiments, I dare say that you are being shortchanged.

A good example of high-quality experimental research is the work of Miguel Costa-Gomes, Vincent Crawford, and Bruno Broseta. These economists evaluated strategic sophistication by having subjects play matrix games and by tracking, through a clever computer interface, the way subjects gather and use information. The research attempts to reveal whether subjects are actively thinking about each other's preferences in the way rationalizability and equilibrium theory predict.[6] The researchers find considerable heterogeneity in subjects' strategic thinking, ranging from unsophisticated to moderately sophisticated, "boundedly rational" behavior. Many subjects appeared to follow rules that respect one or two rounds of iterated dominance, leading to equilibrium play in some simple games but deviating systematically from equilibrium in more complex games. This theme is consistent with the greater experimental literature.[7]

GUIDED EXERCISE

Problem: Consider the duopoly game from Exercise 5 of Chapter 8, in which two firms simultaneously and independently select prices that are greater than or equal to zero. Denote firm 1's price as p_1 and firm 2's price as p_2. After the prices are set, consumers demand $10 - p_1 + p_2$ units of the good that firm 1 produces, and they demand $10 - p_2 + p_1$ units of the good that firm 2 produces. Assume that each firm produces at zero cost, so firm i's payoff (profit) is

$$(10 - p_i + p_j)p_i = 10p_i - p_i^2 + p_i p_j,$$

where p_i is firm i's price and p_j is the other firm's price. Find the Nash equilibrium of this game.

Solution: A Nash equilibrium in this game is a profile of prices for the two firms, p_1^* and p_2^*, such that p_1^* is a best response to p_2^* for firm 1 and p_2^* is a best response to p_1^* for firm 2. We begin by calculating the firms' best-response functions. Letting i denote one of the firms, we find i's best-response function

[6] See M. Costa-Gomes, V. Crawford, and B. Broseta, "Cognition and Behavior in Normal-Form Games: An Experimental Study," *Econometrica* 69 (2001):1193–1235.

[7] The behavioral game-theory literature is extensive. If you are interested in the subject, you can browse through some of the books and papers that survey the literature. Two sources are V. Crawford, "Theory and Experiment in the Analysis of Strategic Interaction," in D. Kreps and K. Wallis, eds., *Advances in Economics and Econometrics: Theory and Applications* (Seventh World Congress, vol. I, Econometric Society Monograph No. 27) (Cambridge, UK, and New York: Cambridge University Press, 1997), pp. 206–242; and J. Kagel and A. Roth, eds., *The Handbook of Experimental Economics* (Princeton, NJ: Princeton University Press, 1995).

by taking the derivative of this firm's payoff function with respect to p_i, setting it equal to zero, and solving for p_i. The derivative condition is

$$10 - 2p_i + p_j = 0,$$

which, solving for p_i, yields $p_i = 5 + (p_j/2)$. Thus, firm i's best response, written as a function, is

$$BR_i(p_j) = 5 + \frac{p_j}{2}.$$

Thus, we have the following system of equations:

$$p_1 = 5 + \frac{p_2}{2} \quad \text{and} \quad p_2 = 5 + \frac{p_1}{2},$$

which is solved by $p_1^* = p_2^* = 10$. That is, the Nash equilibrium is (10, 10).

By the way, a shortcut to the solution involves guessing that, because the game is symmetric, the Nash equilibrium might be symmetric as well. That is, $p_1^* = p_2^* = k$ for some number k. Using this to substitute for the prices in the best-response equation $p_i = 5 + (p_j/2)$ (imposing the symmetry), we have $k = 5 + (k/2)$, which simplifies to $k = 10$. This shortcut maneuver is worth trying when examining other symmetric games, but it does not always work because some symmetric games do not have symmetric equilibria.

EXERCISES

1. Consider the normal-form game pictured here.

1 \ 2	a	b	c
w	5, 2	3, 4	8, 4
x	6, 2	2, 3	8, 8
y	1, 1	0, 1	9, 2

(a) What are the Nash equilibria of this game?

(b) Which of these equilibria are efficient?

(c) Is the set $X = \{w, x\} \times \{b, c\}$ weakly congruent?

2. Find the Nash equilibria of and the set of rationalizable strategies for the games in Exercise 1 at the end of Chapter 6.

3. Find the Nash equilibria of the games in Exercise 1 of Chapter 7.

4. Compute the Nash equilibria of the following location game. There are two people who simultaneously select numbers between zero and one. Suppose player 1 chooses s_1 and player 2 chooses s_2. If $s_i < s_j$, then player i gets a payoff of $(s_i + s_j)/2$ and player j obtains $1 - (s_i + s_j)/2$, for $i = 1, 2$. If $s_1 = s_2$, then both players get a payoff of $1/2$.

5. Find the Nash equilibrium of the following normal-form game: $S_1 = [0, 1]$, $S_2 = [0, 1]$, $u_1(s_1, s_2) = 3s_1 - 2s_1s_2 - 2s_1^2$, and $u_2(s_1, s_2) = s_2 + 2s_1s_2 - 2s_2^2$. (The solution is interior, so you can use calculus.)

6. Consider the following normal-form game:

1 \ 2	k	l	m
w	8, 6	6, 8	0, 0
y	6, 8	8, 6	0, 0
z	0, 0	0, 0	5, 5

(a) What are the congruous sets in this game?

(b) Suppose players 1 and 2 meet before playing the game and agree that they will play strategies in some set X. With the assumption that the players will comply with their agreement, what set X do you think they agree to?

(c) Regarding your answer to part (b), does the agreement alleviate strategic uncertainty?

7. Consider the normal-form game pictured here:

1 \ 2	X	Y	Z
A	2, 0	1, 3	5, x
B	5, 4	1, 3	6, 2

All of the payoff numbers are specified, with the exception of that denoted by x. Find a number for x such that the following three statements are *all* true:

(B, X) is a Nash equilibrium, (A, Z) is an efficient strategy profile, and, for the belief $\theta_1 = (\frac{1}{2}, \frac{1}{2})$, Y is a best response for player 2; that is, $Y \in BR_2(\theta_1)$.

8. Consider the nine-region location game presented in Chapter 8, where two vendors simultaneously choose locations and then customers walk to the nearest vendor to purchase a single unit. That is, the strategy space for each player i is $S_i = \{1, 2, 3, 4, 5, 6, 7, 8, 9\}$. Assume that there are ten customers in each region. Suppose that, unlike in the standard model, each customer is only willing to walk up to *two* regions away. For example, customers in region 1 are willing to walk to regions 2 or 3 to purchase a unit, but they will not travel to any higher-numbered regions. Thus, each player's payoff is the number of customers up to two regions away who are closer to this player's location than to the other player's location, with the customers who are indifferent dividing evenly.

(a) Describe the strategies that are eliminated in the first *two* rounds of the rationalizability procedure.

(b) This game has exactly two Nash equilibria. Find them.

9. Consider a two-player game with the following strategy spaces: $S_1 = [0, 5]$ and $S_2 = [0, 5]$. Suppose the players' best-response functions, $s_1 = BR_1(s_2)$ and $s_2 = BR_2(s_1)$, are as pictured here.

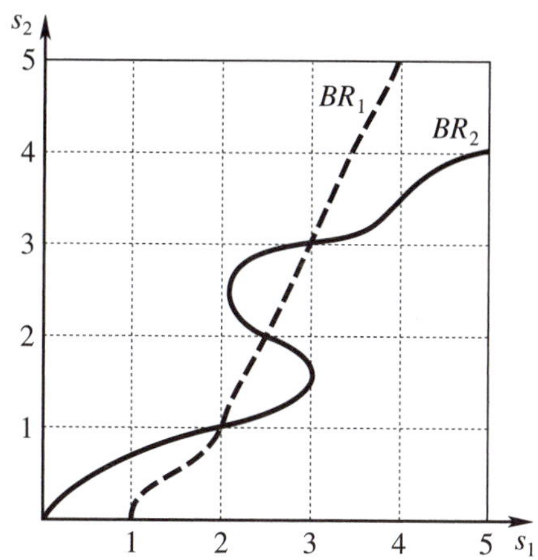

(a) Does this game have any Nash equilibria? If so, what are they?

(b) What is the set of rationalizable strategy profiles for this game?

10. Is the following statement true or false? If it is true, explain why. If it is false, provide a game that illustrates that it is false. "If a Nash equilibrium is not strict, then it is not efficient."

11. This exercise asks you to consider what happens when players choose their actions by a simple rule of thumb instead of by reasoning. Suppose that two players play a specific finite simultaneous-move game many times. The first time the game is played, each player selects a pure strategy at random. If player i has m_i strategies, then she plays each strategy s_i with probability $1/m_i$. At all subsequent times at which the game is played, however, each player i plays a best response to the pure strategy actually chosen by the other player the *previous* time the game was played. If player i has k strategies that are best responses, then she randomizes among them, playing each strategy with probability $1/k$.

(a) Suppose that the game being played is a prisoners' dilemma. Explain what will happen over time.

(b) Next suppose that the game being played is the battle of the sexes. In the long run, as the game is played over and over, does play always settle down to a Nash equilibrium? Explain.

(c) What if, by chance, the players happen to play a strict Nash equilibrium the first time they play the game? What will be played in the future? Explain how the assumption of a strict Nash equilibrium, rather than a nonstrict Nash equilibrium, makes a difference here.

(d) Suppose that, for the game being played, a particular strategy s_i is not rationalizable. Is it possible that this strategy would be played in the long run? Explain carefully.

12. Consider a two-player game and suppose that s^* and t^* are Nash equilibrium strategy profiles in the game. Must it be the case that $\{s_1^*, t_1^*\} \times \{s_2^*, t_2^*\}$ is a weakly congruous strategy set? Explain why or why not.

13. Consider the following n-player game. Simultaneously and independently, the players each select either X, Y, or Z. The payoffs are defined as follows. Each player who selects X obtains a payoff equal to γ, where γ is the number of players who select Z. Each player who selects Y obtains a payoff of 2α, where α is the number of players who select X. Each player who selects Z obtains a payoff of 3β, where β is the number of players who select Y. Note that $\alpha + \beta + \gamma = n$.

(a) Suppose $n = 2$. Represent this game in the normal form by drawing the appropriate matrix.

(b) In the case of $n = 2$, does this game have a Nash equilibrium? If so, describe it.

(c) Suppose $n = 11$. Does this game have a Nash equilibrium? If so, describe an equilibrium and explain *how many* Nash equilibria there are.

OLIGOPOLY, TARIFFS, CRIME, AND VOTING 10

The Nash equilibrium concept is widely applied by practitioners of game theory. Almost every modern social scientist who has dabbled with formal models has called on, or at least argued about, the concept. Nash equilibrium also lies at the heart of more sophisticated concepts of congruity, including most of those discussed in the remainder of this book. This chapter surveys just a few of the classic games to which Nash's concept is routinely applied. The examples will give you some practice in computing equilibria. The examples also demonstrate some of the earliest applications of the theory to the subjects of industrial organization and international relations.

COURNOT DUOPOLY MODEL

In the early 1800s, Augustin Cournot constructed a model of the interaction between two firms, whereby the firms compete by choosing how much to produce.[1] Here is a version of his model. Suppose firms 1 and 2 produce exactly the same good—that is, there is no product differentiation in the market, so consumers do not care from which firm they purchase the good. To be concrete, suppose the product is red brick. Simultaneously and independently, the firms select the number of bricks to produce. Let q_1 denote firm 1's quantity and q_2 denote firm 2's quantity (in thousands). Assume that $q_1, q_2 \geq 0$. The total output in the industry is then $q_1 + q_2$. Assume that all of the brick is sold, but the price that consumers are willing to pay depends on the number of bricks produced.[2] The demand for bricks is given by an inverse relation between quantity and price—an inverse relation is the norm in most markets (when price drops, consumers buy more). Suppose the price is given by the simple function $p = 1000 - q_1 - q_2$. Also suppose each firm must pay a

[1] Cournot's analysis is contained in his book titled *Recherches sur les Principes Mathématiques de la Théorie des Richesses,* published in 1838. A translation is *Researches into the Mathematical Principles of the Theory of Wealth* (New York: Macmillan, 1897).

[2] Indeed, if the price were close enough to zero, I would be motivated to build a brick walkway in my backyard. If the price were close to infinity, I would be motivated to change professions.

production cost of \$100 per thousand bricks. Firms wish to maximize their profits.

To compute the equilibrium of this market game, we start by specifying the normal form. Because each firm selects a quantity, we have $S_1 = [0, \infty)$ and $S_2 = [0, \infty)$. Each firm's payoff is its profit, which is revenue (price times quantity) minus cost. Thus, firm 1's payoff is

$$u_1(q_1, q_2) = (1000 - q_1 - q_2)q_1 - 100q_1$$

and firm 2's payoff is

$$u_2(q_1, q_2) = (1000 - q_1 - q_2)q_2 - 100q_2.$$

Next, we calculate the best-response functions for the firms. Because we will be looking for a Nash equilibrium strategy profile, we can think of each firm's best response as a function of the other firm's quantity (rather than a belief about this quantity). Observe that firm 1's payoff function is a downward parabola, as a function of q_1. To find firm 1's optimal strategy, we use calculus to determine the quantity that maximizes this profit. Taking the partial derivative of u_1 with respect to q_1 and setting this equal to zero yields

$$1000 - 2q_1 - q_2 - 100 = 0.$$

Solving for q_1, we get $q_1 = 450 - q_2/2$. Thus, firm 1's best-response function is $BR_1(q_2) = 450 - q_2/2$. The game is symmetric, so the same analysis reveals firm 2's best-response function to be $BR_2(q_1) = 450 - q_1/2$.

Finally, we determine the quantities that satisfy both best-response functions. That is, we look for q_1^* and q_2^* such that $BR_1(q_2^*) = q_1^*$ and $BR_2(q_1^*) = q_2^*$. You can easily check that these equalities are solved by $q_1^* = 300$ and $q_2^* = 300$. This is the Nash equilibrium strategy profile: each firm produces 300,000 red bricks.[3]

The Nash equilibrium in the Cournot game is inefficient from the firms' point of view. In this sense, the Cournot game is like the prisoners' dilemma. To see this, note that both firms would be better off if they each produced 225,000 bricks—you should convince yourself of this by demonstrating that these quantities actually maximize the sum of the firms' profits. Firms overproduce relative to their joint optimal production levels. Overproduction results

[3] Although the modern rudiments of game theory had not been invented until many years after Cournot's work, his theory predicted the Nash equilibrium outcome as the stable outcome of a dynamic process, whereby firms alternate in playing a best response to each other's strategies. Nash's special contribution (along with those of contemporaries, such as von Neumann) was to build the game theory apparatus and define the equilibrium concept for general games.

because each firm does not value the profit of the other. Consider a marginal increase in the quantity selected by a firm. This increase in quantity expands the firm's sales quantity, but it decreases the market price relative to production (marginal) cost. The firm balances these opposing effects to maximize profit. However, the firm's private costs and benefits of raising quantity do not equal joint costs and benefits. In particular, raising quantity has a detrimental effect on the other firm's payoff through the price change. Because each firm's price effect understates the joint price effect, the firms have the incentive to overproduce relative to their joint optimal levels.

BERTRAND DUOPOLY MODEL

The Cournot model may seem a bit unreasonable because it has firms selecting quantities rather than prices. In reality, firms select both prices and quantities. But consumer demand implies a definite relation between these two variables, so firms can be thought of selecting one first (quantity or price) and then setting the other to whatever the market will bear. It thus makes sense to model firms as selecting quantity.

On the other hand, there are strategic differences between quantity and price selection in a competitive environment. One can generate additional insight by studying models in which firms compete by selecting prices. Consider a variant of the Cournot duopoly model just analyzed. Suppose that the two firms simultaneously and independently set prices and then are forced to produce exactly the number of bricks demanded by customers at these prices. As before, industry demand is given by $p = 1000 - q_1 - q_2$, which can be written as $Q = 1000 - p$, where $Q = q_1 + q_2$. That is, facing price p, consumers demand $1000 - p$ thousand units of brick. Let us assume that consumers purchase brick from the firm that charges the lower price. If the firms set equal prices, then suppose the demand is split evenly—that is, each firm sells $(1000 - p)/2$ thousand units. Assume the cost of production is 100 per thousand bricks, as before. A model like this one was analyzed by Joseph Bertrand in the late 1800s.[4]

To specify the normal form of this game, note that the firms make one decision each (simultaneously and independently). Because each firm selects a

[4]The source is J. Bertrand, "Théorie Mathematique de la Richesse Sociale," *Journal des Savants* 68 (1883): 499–508. Cournot also proposed a price-selection model for the case of "differentiated-product oligopoly," where a firm can charge a higher price than others and still expect some share of the market. Economists generally regard Cournot as the founder of theories of quantity selection and differentiated-product price selection. Bertrand is given credit for the pricing game with homogeneous products. Later, you will see how Bertrand's game has the flavor of a particular auction environment.

price, we have $S_1 = [0, \infty)$ and $S_2 = [0, \infty)$ under the assumption that negative prices are not allowed. As before, each firm's payoff is given by revenue (price times quantity) minus cost. However, in this game, a firm will sell no brick if it charges a price that is higher than the opponent's price. Let p_1 and p_2 be the prices set by the firms; further, given firm i, let j denote the other firm. Firm i's payoff $u_i(p_1, p_2)$ is then equal to

$$(1000 - p_i)p_i - 100(1000 - p_i) = (1000 - p_i)(p_i - 100)$$

if $p_i < p_j$, whereas i's payoff equals 0 if $p_i > p_j$. In the event that $p_i = p_j$, firm i's profit is

$$\frac{(1000 - p_i)p_i}{2} - \frac{100(1000 - p_i)}{2} = \frac{(1000 - p_i)(p_i - 100)}{2}.$$

The Bertrand game requires a different analytical technique from that used for the Cournot game because best-response functions are not well defined in the Bertrand model. To see this, note that, if firm 2's price is 200, then firm 1's optimal response is to select "the largest number that is less than 200." There is no such number, because one can always get closer and closer to 200 (199, 199.99, 199.9999, etc.) from below. However, finding the Nash equilibrium is not difficult. Remember that we are looking for a strategy profile (p_1, p_2) such that p_1 is a best response to p_2 for firm 1 and p_2 is a best response to p_1 for firm 2.

First note that neither firm can be pricing below 100 in an equilibrium, because in this case at least one of the firms earns a negative profit (and each firm can guarantee itself a profit of 0 by pricing at 100). Second, observe that $p_i > p_j \geq 100$ cannot be the case for $i = 1$ or $i = 2$. Here, if $p_j > 100$, then firm i can raise its profit (from 0) by changing its price to be between 100 and p_j. Further, if $p_j = 100$, then firm j can increase its profit by raising its price while still undercutting the price charged by firm i. Third, note that $p_i = p_j > 100$ is not possible in equilibrium, because then each firm gets half of the market demand but could grab all of the quantity demanded by dropping its price a tiny amount. These facts imply that the *only* possible equilibrium prices are $p_1 = p_2 = 100$; that is, price equals marginal cost. We verify that this is an equilibrium strategy profile by noting that neither firm can gain by raising or lowering its price.

Interestingly, the equilibrium of the Bertrand game yields zero profit for the firms, whereas the equilibrium of the Cournot game gives the firms positive profits. In addition, prices are lower and quantities are higher in the Bertrand case. For intuition, note that, in the Bertrand model, firms always have the incentive to undercut each other's prices as long as price exceeds the marginal cost of production, because a firm can grab the entire market by doing so. But in the Cournot model, firms have to raise output more than just a little to grab significant market share. In addition, large quantity increases cause large price decreases, which have a negative effect on profit. There is thus a sense in which price-setting environments are more competitive than quantity-setting environments in markets with homogeneous products.

TARIFF SETTING BY TWO COUNTRIES

National governments can influence international trade (trade between consumers and firms of different countries) by imposing barriers that restrict trade. The most common of these barriers are taxes on the importation of foreign commodities, commonly referred to as *tariffs*. A large country (or union of countries) usually benefits by setting a small import tariff, assuming that other countries do not raise their tariffs, too. Consider, for example, the European Union (EU) as an importer of bananas. Because the EU is a large economy in regard to its share of world trade, an increase in the EU's banana tariff causes the world's quantity demand for bananas to decrease and the international price of bananas to fall. Simultaneously, the tariff drives up the price of bananas in the EU. Thus, the tariff creates a wedge between the international price of bananas and the price of bananas in the EU. When this wedge is large enough, the tariff revenue may be larger than the loss in consumer welfare incurred by Europeans due to the higher prices for bananas in the EU. Similar reasoning holds for the United States as an importer of European cheese.

Thus, the United States and the EU have unilateral incentives to impose tariffs (the United States sets a tariff on the importation of cheese, and the EU sets a tariff on bananas). Unfortunately, the United States and the EU are both worse off when tariffs are uniformly high, relative to uniformly low tariffs. Thus, the tariff-setting game is a form of the prisoners' dilemma. The two economies would benefit by cooperating to keep tariffs low, instead of narrowly pursuing their individual interests. In other words, they would benefit by finding a way to enforce free trade.

A game-theoretic model can be used to illustrate the strategic aspects of tariffs. Suppose there are two countries that are labeled 1 and 2. Let x_i be the tariff level of country i (in percent), for $i = 1, 2$. If country i picks x_i and the other country (j) selects x_j, then country i gets a payoff of $2000 + 60x_i + x_i x_j - x_i^2 - 90x_j$ (measured in billions of dollars). Assume that x_1 and x_2 must be between 0 and 100 and that the countries set tariff levels simultaneously and independently.[5] Exercise 3 at the end of this chapter asks you to compute the Nash equilibrium of this game.

A MODEL OF CRIME AND POLICE

Game theory and the Nash equilibrium concept can be used to study the interaction between criminals and law-enforcement agencies. Gary Becker, a Nobel Prize-winning economist, led the way on this kind of research and showed that economic analysis is extremely useful in this policy arena. According to Becker's theory, "The optimal amount of enforcement is shown to depend on, among other things, the cost of catching and convicting offenders, the nature of punishments—for example, whether they are fines or prison terms—and the response of offenders to changes in enforcement."[6] Becker also argued that, with the optimal enforcement system, crime does occur.

Here is a game that illustrates how the government balances the social cost of crime with law-enforcement costs and how criminals balance the value of illegal activity with the probability of arrest. The game has two players: a criminal (C) and the government (G). The government selects a level of law enforcement, which is a number $x \geq 0$. The criminal selects a level of crime, $y \geq 0$. These choices are made simultaneously and independently. The government's payoff is given by $u_G = -xc^4 - y^2/x$ with the interpretation that $-y^2/x$ is the negative effect of crime on society (moderated by law enforcement) and c^4 is the cost of law enforcement, per unit of enforcement. The number c is a positive constant. The criminal's payoff is given by $u_C = y^{1/2}/(1 + xy)$, with the interpretation that $y^{1/2}$ is the value of criminal activity when the criminal is not caught, whereas $1/(1 + xy)$ is the probability that the criminal evades capture. Exercise 4 of this chapter asks you to compute the Nash equilibrium of this game.

THE MEDIAN VOTER THEOREM

Consider an expanded version of the location-choice model from Chapter 8, where two political candidates (players 1 and 2) decide where to locate on the political spectrum. Suppose the policy space is given by the interval [0, 1], with the location 0 denoting extreme liberal and 1 denoting extreme conservative. Citizens (the voters) are distributed across the political spectrum, not

[5]You can read more about tariff games in J. McMillan, *Game Theory in International Economics* (New York: Harwood Academic Publishers, 1986).

[6]G. Becker, "Crime and Punishment: An Economic Approach," *Journal of Political Economy* 76(2): 169–217, at p. 170.

necessarily uniformly as was the case in the basic location game described in Chapter 8. Each citizen has an ideal point (a favorite policy) on the interval $[0, 1]$. Let function F describe the distribution of the citizens. Specifically, for any location $x \in [0, 1]$, $F(x)$ is the fraction of the citizens whose ideal point is less than or equal to x. Assume that F is a continuous function, with $F(0) = 0$ and $F(1) = 1$. Technically, this means that there is a "continuum" of citizens—an infinite number, smoothly distributed across the political spectrum.

Voting takes place after the candidates simultaneously and independently select their policy locations. Each citizen is assumed to vote for the candidate who is located closest to the citizen's ideal point. A candidate wins by obtaining more votes than the other candidate does. Assume that if the candidates select exactly the same policy position, then the voters split evenly between them and they tie in the election, with the eventual winner determined by an unmodeled Supreme Court decision. Each candidate wants to win the election, so let us assume that a candidate obtains a payoff of 1 if he wins, 0 if he loses, and $1/2$ if there is a tie.[7]

Let us analyze the game by looking for a Nash equilibrium. Suppose that player 1 selects location s_1 and player 2 selects s_2. Guided by the intuition developed from the location model in Chapter 8, we can guess that the players will choose the same policy location in equilibrium—that is, $s_1 = s_2$. To see why this is so, consider the case in which $s_1 < s_2$. Could this arise in equilibrium? In fact, no. If the players were to tie with these strategies, then an individual player would fare strictly better—that is, he would get more votes and win—by moving closer to the position of the other player. Similarly, if one of the players were to win with strategy profile (s_1, s_2), then the other player could improve from a loss to a tie by matching the first player's policy position. Thus, in equilibrium it must be that $s_1 = s_2 = x$, for some number $x \in [0, 1]$.

The question then becomes: What is the equilibrium policy location x that both players choose? Our intuition from Chapter 8 is that the players will locate in the center, but the definition of "center" surely depends on the distribution of voters. Let x^* be the number satisfying $F(x^*) = 1/2$. Then x^* is the *median voter*—that is, the voter who is more conservative than exactly half of the citizens and who is more liberal than exactly half of the citizens.

The Nash equilibrium of the location is game is for both players to select x^*, which induces a tie in the election. To see this, note what would happen

[7] Recall that this is not exactly the same as wanting to maximize the number of votes; nothing changes if we assume vote maximization in the example studied here, however.

if candidate i were to deviate from locating at x^*. Then candidate i would obtain strictly fewer votes than would candidate j and, as a result, candidate i would lose the election. Furthermore, at no other number x is there a Nash equilibrium. This is because, if there were, there would be strictly more voters on the side of x containing x^* than there are on the other side and so, from such a position, a candidate would be able to guarantee victory by moving toward x^*. That the equilibrium occurs at x^* is Anthony Downs's celebrated **median voter theorem**, which is a cornerstone of political science theory.[8]

STRATEGIC VOTING

An underlying assumption of the analysis of candidates' policy choices in the previous section is that each citizen votes for the candidate whose policy is closest to the citizen's ideal point. Is it, however, safe to say that voters behave in this way? Would a sophisticated citizen always vote for her favorite candidate, even if this candidate were sure to lose the election? A story and numerical example will shed light on this issue.

On October 7, 2003, in an historic recall election, Gray Davis was removed from office as the governor of California. Popular confidence in Davis had been eroded in the preceding years by revelations of a severe imbalance in the state's budget, a crisis in the electricity market, declining trust in government officials in general, and a growing sense that Davis lacked the gumption to lead. In fact, Davis's approval ratings had been falling well before his reelection in November 2002, but his shrewd and disciplined political machine was able to orchestrate his reelection anyway.

The recall election was highly unusual; many people considered its rules to be ill suited to modern political decision making. On the same ballot, voters were asked whether to remove Davis from office and then, conditional on the recall passing, they were asked to select among the—get this—135 registered replacement candidates. Among the field of replacement candidates were dozens of people who entered the race for publicity or just on a whim. Candidates included a star of pornographic movies, a former child star from a television series, and many others with colorful histories. The election was won by Arnold Schwarzenegger, a movie star and former bodybuilder, who

[8]The location-choice model was developed for analysis of oligopoly by H. Hotelling, "Stability in Competition," *The Economic Journal* 39 (1929): 41-57, and for analysis of politics by A. Downs, *An Economic Theory of Democracy* (New York: Harper & Row, 1957). The median voter theorem extends to sequential candidate choices.

had little political experience but excellent name recognition and was himself a savvy enough politician to later get reelected.[9]

Gray Davis's journey through reelection and disgraceful fall from the governorship provides two examples of strategic behavior in elections. The first took place in the regular election of 2002, in which Davis was campaigning for reelection. Facing no serious opposition for nomination in the Democratic primary, Davis's campaign allocated part of its war chest to manipulate the contest between Bill Simon and Los Angeles Mayor Richard Riordan, who were sparring in the Republican primary for the right to oppose Davis in the general election.

Riordan, a moderate Republican, was widely viewed as a greater threat to Davis than was the right-wing Simon. In other words, Riordan was closer to Davis on the political spectrum and thus, as we know from the basic location model, in a more competitive stance against Davis than was Simon. In fact, polls suggested that Riordan could have beaten Davis in the general election. So Davis did something consistent with our location model: he funded advertisements that questioned Riordan's credentials (even from the right) and thus helped Simon to defeat Riordan in the primary. Davis then defeated Simon in the general election.[10]

The second noteworthy example concerns Davis's recall election, in which the leading Republican candidates to replace him were Schwarzenegger and State Senator Tom McClintock. It was another example of a moderate (Schwarzenegger) versus a right-wing (McClintock) Republican. Conservative and die-hard Republican voters favored McClintock. However, much of the Republican party feared that a Republican vote split between Schwarzenegger and McClintock would propel Lieutenant Governor Cruz Bustamante, the focal liberal candidate, into the governorship. As the recall election neared, the state Republican Party's board of directors took the unprecedented step of endorsing one of the Republican candidates, Schwarzenegger, who was viewed as more moderate than Bustamante. In essence, the Republican leadership asked conservative Republicans to vote for their second-favorite candidate as their only hope of securing a Republican governor.[11]

[9]The recall election, and events leading to it and from it, have been extensively recorded and discussed in the media. A brief newspaper account of the election is J. Marelius, "Schwarzenegger Wins: Decisive Support for Dramatic Change Ends Historic Race," *San Diego Union-Tribune*, October 8, 2003.

[10]A summary appears in J. Marelius, "Davis Weighs In with Anti-Riordan Ad," *San Diego Union-Tribune*, January 26, 2002; and J. Marelius, "Simon Storms Past Riordan: Rookie Politician to Face Incumbent Davis," *San Diego Union-Tribune*, March 6, 2002.

[11]See J. Marelius and P. J. LaVelle, "State GOP Taps Schwarzenegger: Party Chairman Calls on McClintock Backers Not to 'Waste' Votes," *San Diego Union-Tribune*, September 30, 2003.

FIGURE 10.1
Voter preferences.

	Most preferred	Second choice	Least preferred
Player L	Bustamante	Schwarzenegger	McClintock
Player M	Schwarzenegger	Bustamante	McClintock
Player C	McClintock	Schwarzenegger	Bustamante

Here is a simple voting game that illustrates the story. Suppose that Davis is recalled and that there are three candidates to replace him: Bustamante, Schwarzenegger, and McClintock. The candidates have no actions in the game. Suppose there are three voting blocks: liberals (L), moderates (M), and conservatives (C). For simplicity, assume that each voting block behaves as a single player, so we have a three-player game. Simultaneously and independently, players L, M, and C select single candidates. The candidate who obtains the most votes wins the election.

Suppose that the voting blocks contain roughly the same number of voters, but that the liberal voting block is slightly bigger than the other two. Thus, if L votes for Bustamante, M votes for Schwarzenegger, and C votes for McClintock, then Bustamante wins. If two or three of the players vote for the same candidate, then this candidate wins.

The general preferences of the voters are shown in Figure 10.1. For example, L favors Bustamante over Schwarzenegger, and Schwarzenegger over McClintock. Suppose that each voter cares both about whom she votes for and about who wins the election. Specifically, assume that a voter's payoff is the sum of an amount determined directly by her voting action and an amount determined by who wins. On the first component, the voter gets 2 if she votes for her most preferred candidate, 1 if she votes for her second choice, and 0 if she votes for her least preferred candidate. On the second component, the voter gets 4 if her most preferred candidate wins, 2 if her second choice wins, and 0 if her least preferred candidate wins the election. For example, if player M votes for Schwarzenegger yet Bustamante wins, then player M gets a payoff of 4 (that is, 2 from voting for her favorite candidate, plus 2 for her second favorite winning the election).

In the voting game, it is *not* rational for each player to vote for his most preferred candidate. In particular, the strategy profile (Bustamante, Schwarzenegger, McClintock) is not a Nash equilibrium of the game. From this profile, which implies victory for Bustamante, player C can strictly gain by deviating to vote for Schwarzenegger. The switch to Schwarzenegger changes the outcome of the race (now won by Schwarzenegger), and player C gains more from the outcome change (from 0 to 2) than she loses by voting for her second choice (from 2 to 1). Thus, this example shows that it is not always rational for citizens to vote "truthfully," as we assumed in the basic election model.

The example also substantiates the logic behind the Republican Party's endorsement of Schwarzenegger at the eleventh hour in the recall election. The endorsement helped convince conservatives to vote for Schwarzenegger if they didn't want to see their least favorite candidate elected. Note that the strategy profile (Bustamante, Schwarzenegger, Schwarzenegger) is a Nash equilibrium of the game. With this strategy profile, player M is voting for and getting her most preferred candidate and therefore has no incentive to deviate. Player L cannot change the outcome of the election by switching his vote, so he rationally selects Bustamante. Finally, if player C switched to Bustamante, then Bustamante would be elected and player C would get zero; alternatively, if player C switched to McClintock, then Bustamante would win and player C would obtain a payoff of just 2. By voting for Schwarzenegger, player C obtains 3, and this is the best player C can do.

GUIDED EXERCISE

Problem: Consider a market with ten firms. Simultaneously and independently, the firms choose between locating downtown and locating in the suburbs. The profit of each firm is influenced by the number of other firms that locate in the same area. Specifically, the profit of a firm that locates downtown is given by $5n - n^2 + 50$, where n denotes the number of firms that locate downtown. Similarly, the profit of a firm that locates in the suburbs is given by $48 - m$, where m denotes the number of firms that locate in the suburbs. In equilibrium, how many firms locate in each region and what is the profit of each?

Solution: Note that there is a negative congestion effect for the suburban region in that a firm's value of locating there decreases in the number of other firms that locate in the same region. The same it true for the downtown region when the number of firms locating there exceeds three. Thus, intuition suggests that equilibrium will not feature the vast majority of firms congregating in one or the other region, but instead will have the firms dividing between the regions.

Another way to think about this is that, in equilibrium, the value of locating downtown will be roughly the same as the value of locating in the suburbs. If, for instance, the value of locating in the suburbs were much higher than that of locating downtown, then a firm that was planning to locate downtown would strictly gain by switching to the strategy of locating in the suburbs. Let us determine the number of firms in each location that would be required to equate the values of the two locations. Noting that $m = 10 - n$, equating the values implies

$$5n - n^2 + 50 = 48 - (10 - n),$$

which simplifies to

$$(n - 6)(n + 2) = 0.$$

The solution $n = -2$ is not meaningful. The equilibrium therefore has $n = 6$ and $m = 4$. That is, six firms locate downtown and four locate in the suburbs. Each firm earns a profit of 44. You should verify that no firm would gain by switching locations unilaterally.

EXERCISES

1. Consider a more general Cournot model than the one presented in this chapter. Suppose there are n firms. The firms simultaneously and independently select quantities to bring to the market. Firm i's quantity is denoted q_i, which is constrained to be greater than or equal to zero. All of the units of the good are sold, but the prevailing market price depends on the total quantity in the industry, which is $Q = \sum_{i=1}^{n} q_i$. Suppose the price is given by $p = a - bQ$ and suppose each firm produces with marginal cost c. There is no fixed cost for the firms. Assume $a > c > 0$ and $b > 0$. Note that firm i's profit is given by $u_i = p(Q)q_i - cq_i = (a - bQ)q_i - cq_i$. Defining Q_{-i} as the sum of the quantities produced by all firms except firm i, we have $u_i = (a - bq_i - bQ_{-i})q_i - cq_i$. Each firm maximizes its own profit.

(a) Represent this game in the normal form by describing the strategy spaces and payoff functions.

(b) Find firm i's best-response function as a function of Q_{-i}. Graph this function.

(c) Compute the Nash equilibrium of this game. Report the equilibrium quantities, price, and total output. (Hint: Summing the best-response functions over the different players will help.)

(d) Show that, for the Cournot duopoly game ($n = 2$), the set of rationalizable strategies coincides with the Nash equilibrium.

2. Consider a more general Bertrand model than the one presented in this chapter. Suppose there are n firms that simultaneously and independently select their prices, $p_1, p_2, \ldots, p_n$, in a market. These prices are greater than or equal to zero. The lowest price offered in the market is defined as

$\underline{p} = \min\{p_1, p_2, \ldots, p_n\}$. Consumers observe these prices and purchase only from the firm (or firms) charging $\underline{p}$, according to the demand curve $Q = a - \underline{p}$. That is, the firm with the lowest price gets all of the sales. If the lowest price is offered by more than one firm, then these firms equally share the quantity demanded. Assume that firms must supply the quantities demanded of them and that production takes place at a cost of c per unit. That is, a firm producing q_i units pays a cost cq_i. Assume $a > c > 0$.

(a) Represent this game in the normal form by describing the strategy spaces and payoff (profit) functions.

(b) Find the Nash equilibrium of this market game.

(c) Is the notion of a best response well defined for every belief that a firm could hold? Explain.

3. Consider the tariff game described in this chapter.

(a) Find the best response functions for the countries.

(b) Compute the Nash equilibrium.

(c) Show that the countries would be better off if they made a binding agreement to set *lower* tariffs (than in equilibrium). You do not need to speculate how such an agreement could be enforced.

(d) Using the graph of the best-response functions, determine the set of rationalizable strategies in the tariff game.

4. Consider the game between a criminal and the government described in this chapter.

(a) Write the first-order conditions that define the players' best-response functions and solve them to find the best-response functions. Graph the best-response functions.

(b) Compute the Nash equilibrium of this game.

(c) Explain how the equilibrium levels of crime and enforcement change as c increases.

5. In the years 2000 and 2001, the bubble burst for many Internet and computer firms. As they closed shop, some of the firms had to liquidate sizable assets, such as inventories of products. Suppose eToys is going out of business and the company seeks a buyer for a truckload of Elmo dolls in its warehouse. Imagine that eToys holds an auction on eBay to sell the dolls and that two retailers (players 1 and 2) will bid for them. The rules of the auction are: the retailers simultaneously and independently submit sealed bids and then eToys gives the merchandise to the highest bidder, who must pay his bid. It is common knowledge that the retailer who obtains the load of dolls can resell the load for a total of \$15,000. Thus, if player i wins the auction with bid b_i, then player i's payoff is $\$15{,}000 - b_i$. The losing retailer gets a payoff of \$0. If the retailers make the same bids ($b_1 = b_2$), then eToys declares each player the winner with probability 1/2, in which case player i obtains an expected payoff of $(1/2)(15{,}000 - b_i)$. What will be the winning bid in the Nash equilibrium of this auction game? If you can, describe the equilibrium strategies and briefly explain why this is an equilibrium. (Hint: This is similar to the Bertrand game.)

6. Imagine that a zealous prosecutor (P) has accused a defendant (D) of committing a crime. Suppose that the trial involves evidence production by both parties and that, by producing evidence, a litigant increases the probability of winning the trial. Specifically, suppose that the probability that the defendant wins is given by $e_{\mathrm{D}}/(e_{\mathrm{D}} + e_{\mathrm{P}})$, where e_{D} is the expenditure on evidence production by the defendant and e_{P} is the expenditure on evidence production by the prosecutor. Assume that e_{D} and e_{P} are greater than or equal to 0. The defendant must pay 8 if he is found guilty, whereas he pays 0 if he is found innocent. The prosecutor receives 8 if she wins and 0 if she loses the case.

(a) Represent this game in normal form.

(b) Write the first-order condition and derive the best-response function for each player.

(c) Find the Nash equilibrium of this game. What is the probability that the defendant wins in equilibrium.

(d) Is this outcome efficient? Why?

7. Consider an asymmetric Cournot duopoly game, where the two firms have different costs of production. Firm 1 selects quantity q_1 at a production cost of $2q_1$. Firm 2 selects quantity q_2 and pays the production cost $4q_2$. The market price is given by $p = 12 - q_1 - q_2$. Thus, the payoff functions are $u_1(q_1, q_2) = (12 - q_1 - q_2)q_1 - 2q_1$ and $u_2(q_1, q_2) = (12 - q_1 - q_2)q_2 - 4q_2$. Calculate the firms' best-response functions $BR_1(q_2)$ and $BR_2(q_1)$, and find the Nash equilibrium of this game.

8. Recall the candidate location game discussed in this chapter, whose analysis led to the median voter theorem. Consider a variant of the game in which some of the voters have to be motivated to vote. In particular, suppose that the policy space [0, 1] is divided into three regions: $[0, \frac{1}{2} - \alpha]$, $(\frac{1}{2} - \alpha, \frac{1}{2} + \alpha)$, and $[\frac{1}{2} + \alpha, 1]$, where α is a fixed parameter that is smaller than 1/2. Moderate citizens, whose ideal points are in the interval $(\frac{1}{2} - \alpha, \frac{1}{2} + \alpha)$, always go to the polls and vote as described in the basic model; that is, each moderate citizen votes for the closest candidate. Liberals, whose ideal points are in the interval $[0, \frac{1}{2} - \alpha]$, go to the polls only if there is a liberal candidate—that is, someone who locates left of the point $\frac{1}{2} - \alpha$. Conditional on voting, each liberal votes for the closest candidate. Likewise, conservatives, whose ideal points are in the interval $[\frac{1}{2} + \alpha, 1]$, will vote only if there is a conservative candidate—someone who locates to the right of $\frac{1}{2} + \alpha$.

For example, suppose one of the candidates locates at $1/2$ and the other candidate locates at 1. Then the candidate at 1 received votes from the citizens between $3/4$ and 1, whereas the candidate at $1/2$ receives votes from only the citizens between $\frac{1}{2} - \alpha$ and $3/4$ because the liberal citizens are not motivated to vote. Assume that the citizens are uniformly distributed over the policy space. Thus, if F describes the distribution then $F(x) = x$ for every $x \in [0, 1]$.

(a) For what values of α is there an equilibrium of the candidate location-choice game in which both candidates locate at $1/2$ (in the center, at the median voter)?

(b) Show that, for some values of α, there is an equilibrium in which one of the candidates locates at $\frac{1}{2} - \alpha$ and the other locates at $\frac{1}{2} + \alpha$. For what values of α does such an equilibrium exist? Comment on the reason why the median-voter result fails.

9. Consider the strategic voting example discussed at the end of this chapter, where we saw that the strategy profile (Bustamante, Schwarzenegger, Schwarzenegger) is a Nash equilibrium of the game. Show that (Bustamante, Schwarzenegger, Schwarzenegger) is, in fact, the only rationalizable strategy profile. Do this by first considering the dominated strategies of player L.

10. Consider a game that has a continuum of players. In particular, the players are uniformly distributed on the interval $[0, 1]$. (See Appendix A for a definition of uniform distribution.) Each $x \in [0, 1]$ represents an individual player; that is, we can identify a player by her location on the interval $[0, 1]$. In the game, the players simultaneously and independently select either F or G. The story is that each player is choosing a type of music software to buy, where F and G are the competing brands. The players have different values of the two brands; they also have a preference for buying what other people are buying (either because they want to be in fashion or they find it easier to exchange music with others who use the same software). The following payoff function represents these preferences. If player x selects G, then her payoff is the constant g. If player x selects F, then her payoff is $2m - cx$, where c is a constant and m is the fraction of players who select F. Note that m is between 0 and 1.

(a) Consider the case in which $g = 1$ and $c = 0$. What are the rationalizable strategies for the players? Is there a symmetric Nash equilibrium, in which all of the players play the same strategy? If so, describe such an equilibrium.

(b) Next, consider the case in which $g = 1$ and $c = 2$. Calculate the rationalizable strategy profiles and show your steps. (Hint: Let $\overline{m}$ denote an upper bound on the fraction of players who rationally select F. Use this variable in your analysis.)

(c) Describe the rationalizable strategy profiles for the case in which $g = -1$ and $c = 4$. (Hint: Let $\overline{m}$ denote an upper bound on the fraction of players who rationally select F and let $\underline{m}$ denote a lower bound on the fraction of players who rationally select F.)

11 MIXED-STRATEGY NASH EQUILIBRIUM

Recall that some games do not have a Nash equilibrium. Consider, for example, the matching pennies game shown in Figure 9.2. In this game, no strategy profile is stable because each one has a "winner" and a "loser"—a status that is flipped if either player alters his or her strategy. In matching pennies, players actually have an interest in deceiving each other. Suppose that you and I are playing the game and I, as player 1, have privately decided to select strategy H. This would be rational if I thought that you would select H as well, but your selection of H relies on a belief that I am likely to play T. In other words, I would like you to believe that I will choose T while I, in fact, plan to select H. Coordination of beliefs and behavior is, in this example, seemingly at odds with best-response behavior.

Yet, let us think a bit further. When players match wits in the matching pennies game, they might reasonably conclude that there is no way for a rational player to outwit another rational player. Perhaps the best that I can do is to *randomize* between H and T, so that you cannot take advantage of me. In fact, players' beliefs and rational behavior can be coordinated if they are both mixing appropriately. Notice that if each player randomizes with probability $1/2$ on both strategies, then neither player has a strict incentive to play H or T; in fact, all strategies—H, T, and every mixed strategy—are best responses to the opponent randomizing with equal probability. To see this, observe that if you play H and T with equal probabilities, then I will get an expected payoff of zero—that is, $(1/2)(1) + (1/2)(-1)$—regardless of whether I choose H or T. Furthermore, if I mix between H and T, I will still expect zero.

The point here is that if we extend the notions of best-response and equilibrium to the consideration of mixed strategies, we find that the mixed-strategy profile $((1/2, 1/2), (1/2, 1/2))$, where both players randomize equally between their pure strategies, has the Nash equilibrium property in the matching pennies game. With this mixed-strategy profile, each player is best responding to the other. We then can say that $((1/2, 1/2), (1/2, 1/2))$ is a *mixed-strategy Nash equilibrium* of the matching pennies game.

Formally, then, the Nash equilibrium concept extends to mixed strategies. For general games, the definition of a mixed-strategy Nash equilibrium is a

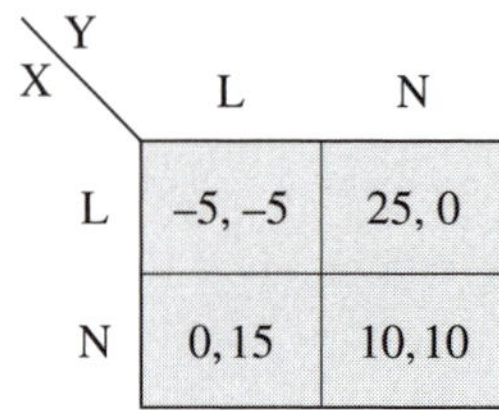

FIGURE 11.1
A lobbying game.

mixed-strategy profile having the property that no player could increase his or her payoff by switching to any other strategy, given the other player's strategy:

> Consider a strategy profile $\sigma = (\sigma_1, \sigma_2, \ldots, \sigma_n)$, where $\sigma_i \in \Delta S_i$ for each player i. Profile σ is a **mixed-strategy Nash equilibrium** if and only if $u_i(\sigma_i, \sigma_{-i}) \geq u_i(s_i', \sigma_{-i})$ for each $s_i' \in S_i$ and each player i. That is, σ_i is a best response to σ_{-i} for every player i.

For a mixed strategy to be a best response (as required in the definition), *it must put positive probability only on pure strategies that are best responses.* This demonstrates how to calculate a mixed-strategy Nash equilibrium.[1]

For an example, consider a lobbying game between two firms. Each firm may lobby the government in hopes of persuading the government to make a decision that is favorable to the firm. The two firms, X and Y, independently and simultaneously decide whether to lobby (L) or not (N). Lobbying entails a cost of 15. Not lobbying costs nothing. If both firms lobby or neither firm lobbies then the government takes a neutral decision, which yields 10 to both firms. (A firm's payoff is this value minus the lobbying cost if it lobbied.) If firm Y lobbies and firm X does not lobby, then the government makes a decision that favors firm Y, yielding zero to firm X and 30 to firm Y. (Thus, firm Y's payoff in this case is $30 - 15 = 15$.) If firm X lobbies and firm Y does not lobby, then the government makes a decision that favors firm X, yielding 40 (with a payoff to firm X of $40 - 15 = 25$) to firm X and zero to firm Y. The normal form of this game is pictured in Figure 11.1.

You can quickly verify that there are two pure-strategy Nash equilibria for this game: (N, L) and (L, N). Check this by evaluating each cell of the matrix and asking whether at least one of the players would want to make a unilateral deviation. In addition to these pure-strategy equilibria, there is also a mixed-strategy equilibrium. To find it, recall what must hold in a mixed-strategy equilibrium: a player must achieve a best response by selecting a mixed strategy. For example, let's guess that firm X mixed between L and N. If this strategy

[1] Note that, in the definition, σ_i is compared against each pure strategy s_i' rather than against all mixed strategies. This is sufficient because, if another mixed strategy were to deliver a higher payoff than does σ_i, then such would be the case for some pure strategy s_i'. Thus, checking the pure-strategy deviations is enough.

is optimal for firm X (in response to the other firm's strategy), then it must be that the expected payoff from playing L equals the expected payoff from playing N; otherwise, firm X would strictly prefer to pick either L or N.

But how can firm X's strategies L and N yield the same expected payoff? It must be that firm Y's behavior generates this expectation (because if firm Y played a pure strategy, then X would strictly prefer one of its strategies over the other). Let q denote the probability that firm Y plays L; that is, $(q, 1 - q)$ is firm Y's mixed strategy. Against this mixed strategy, firm X expects to earn $q(-5) + (1 - q)(25) = 25 - 30q$ by choosing L and $q(0) + (1 - q)(10) = 10 - 10q$ by choosing N. For firm X to be willing to randomize, it must be that $25 - 30q = 10 - 10q$. This simplifies to $q = 3/4$. In other words, firm X can randomize in playing a best response if firm Y's strategy is $(3/4, 1/4)$.

Let us move on to firm Y's incentives and let p be the probability that firm X plays L. Then, if firm Y selects L, its expected payoff is

$$p(-5) + (1 - p)(15) = 15 - 20p.$$

By choosing N, firm Y obtains $10 - 10p$. Firm Y is indifferent between its two strategies (and therefore willing to randomize) if $15 - 20p = 10 - 10p$, which simplifies to $p = 1/2$.

The mixed-strategy profile $((1/2, 1/2), (3/4, 1/4))$ is a mixed-strategy Nash equilibrium. Given firm Y's mixed strategy, firm X's mixed strategy is a best response—in fact, *every* strategy is a best response for firm X. Likewise, given firm X's strategy, firm Y's prescribed strategy is a best response. Note that constructing a mixed-strategy equilibrium entails an interesting new twist: we look for a mixed strategy for one player that makes the *other* player indifferent between her pure strategies. This is the best method of calculating mixed-strategy equilibria.

You should try your hand at computing mixed-strategy Nash equilibria. To guide you in this regard, know that—like pure-strategy equilibria—mixed-strategy equilibria never use dominated strategies. Examine the games in Figure 9.2 again. Convince yourself of the mixed equilibrium notion by demonstrating that $((1/2, 1/2), (1/2, 1/2))$ is the mixed-strategy equilibrium of both the matching pennies and coordination games. Players select one strategy with probability $1/3$ in the mixed equilibria of the battle of the sexes and Pareto coordination games. Also find the mixed-strategy equilibrium of the hawk–dove game.

For another example, take the tennis-service game of Chapter 7's Guided Exercise, whose payoff matrix is reproduced in Figure 11.2 on the next page. Recall that each player's strategy F is removed in the iterated-dominance procedure, so the set of rationalizable strategies for each player is $\{C, B\}$.

FIGURE 11.2
A tennis-service game.

1 \ 2	F	C	B
F	0, 5	2, 3	2, 3
C	2, 3	0, 5	3, 2
B	5, 0	3, 2	2, 3

The game has no Nash equilibrium in pure strategies. In any mixed-strategy equilibrium, the players will put positive probability on only rationalizable strategies. Thus, we know a mixed-strategy equilibrium will specify a strategy $(0, p, 1-p)$ for player 1 and a strategy $(0, q, 1-q)$ for player 2. In this strategy profile, p is the probability that player 1 selects C, and $1-p$ is the probability that he selects B; likewise, q is the probability that player 2 selects C, and $1-q$ is the probability that she selects B.

To calculate the mixed-strategy equilibrium in the tennis example, observe that against player 2's mixed strategy, player 1 would get an expected payoff of

$$q \cdot 0 + (1-q) \cdot 3 = 3 - 3q$$

if he selects C; whereas by choosing B, he would expect

$$q \cdot 3 + (1-q) \cdot 2 = 2 + q.$$

In order for player 1 to be indifferent between C and B, which is required to motivate him to randomize, it must be that player 2's probability q solves

$$3 - 3q = 2 + q,$$

implying $q = 1/4$. Turning to player 2's incentives, note that she would get

$$p \cdot 5 + (1-p) \cdot 2 = 2 + 3p$$

by choosing C and

$$p \cdot 2 + (1-p) \cdot 3 = 3 - p$$

by choosing B. Equating these and solving for p yields $p = 1/4$. Thus, the mixed-strategy equilibrium of the tennis-service example is $(1/4, 3/4)$ for player 1 and $(1/4, 3/4)$ for player 2.

The following summarizes the steps required to calculate mixed-strategy Nash equilibria for simple two-player games.

Procedure for finding mixed-strategy equilibria:

1. Calculate the set of rationalizable strategies by performing the iterated-dominance procedure.
2. Restricting attention to rationalizable strategies, write equations for each player to characterize mixing probabilities that make the other player indifferent between the relevant pure strategies.
3. Solve these equations to determine equilibrium mixing probabilities.

If each player has exactly two rationalizable strategies, this procedure is quite straightforward. If a player has more than two rationalizable strategies, then there are several cases to consider; the various cases amount to trying different combinations of pure strategies over which the players may randomize. For example, suppose that A, B, and C are all rationalizable for a particular player. Then, in a mixed-strategy equilibrium, it may be that this player mixes between A and B (putting zero probability on C), mixes between A and C (putting zero probability on B), mixes between B and C (putting zero probability on A), or mixes between A, B, and C. There are also cases in which only one of the players mixes.

Note that every pure-strategy equilibrium can also be considered a mixed-strategy equilibrium—where all probability is put on one pure strategy. All of the games analyzed thus far have at least one equilibrium (in pure or mixed strategies). In fact, this is a general theorem.[2]

> **Result:** Every finite game (having a finite number of players and a finite strategy space) has at least one Nash equilibrium (in pure or mixed strategies).

For more on this theorem, see Appendix B. The result is quite useful, because it guarantees that the Nash equilibrium concept provides a prediction for every finite game. You should now ask, "Is the prediction reasonable?" What about examples such as the lobbying game, where there are both pure-strategy equilibria and nontrivial mixed-strategy equilibria? Is it more reasonable to expect the mixed equilibrium to occur in some cases?

---

[2]John Nash presented this theorem in "Non-Cooperative Games," *Annals of Mathematics* 51(1951):286–295.

GUIDED EXERCISE

Problem: Consider the following n-player game. Simultaneously, each of the n players chooses between X and Y. The payoff of player i is 1 if he/she selects Y. In the event that player i selects X, his payoff is 2 if no other player chooses X, and his payoff is 0 if at least one other player chooses X as well. In this game, there is a mixed-strategy Nash equilibrium in which each player selects Y with probability α. (The probability α is the same for all players.) Calculate α.

Solution: In the mixed-strategy equilibrium, a given player i must be indifferent between selecting X and Y. Suppose that each of the other $n - 1$ players selects Y with probability α. Note that, because these players act independently, the probability that they all select Y is α^{n-1}. It is in this event that player i would obtain a payoff of 2 by choosing X; otherwise, the choice of X would yield 0. Thus, player i's expected payoff of selecting X is

$$\alpha^{n-1} \cdot 2 + \left(1 - \alpha^{n-1}\right) \cdot 0.$$

Equating player i's expected payoffs of X and Y, we obtain

$$2\alpha^{n-1} = 1,$$

which simplifies to

$$\alpha = \left[\frac{1}{2}\right]^{\left(\frac{1}{n-1}\right)}.$$

Thus, there is a mixed-strategy Nash equilibrium in which each player selects Y with this probability.

EXERCISES

1. Consider the following normal-form game.

1 \ 2	Q	W	Z
X	1, 7	1, 5	3, 4
Y	2, 3	0, 4	0, 6

   (a) Determine the set of rationalizable strategies for this game.

(b) The game has only one Nash equilibrium, and it is a mixed-strategy Nash equilibrium. Compute this equilibrium.

2. Suppose you know the following about a particular two-player game: $S_1 = \{A, B, C\}$, $S_2 = \{X, Y, Z\}$, $u_1(A, X) = 6$, $u_1(A, Y) = 0$, and $u_1(A, Z) = 0$. In addition, suppose you know that the game has a mixed-strategy Nash equilibrium in which (a) the players select each of their strategies with positive probability, (b) player 1's expected payoff in equilibrium is 4, and (c) player 2's expected payoff in equilibrium is 6. Do you have enough information to calculate the probability that player 2 selects X in equilibrium? If so, what is this probability?

3. Consider another version of the lobbying game introduced in this chapter. Suppose the payoffs are the same as presented earlier, except in the case in which firm X lobbies and firm Y does not lobby. In this case, suppose the government's decision yields x to firm X and zero to firm Y. Assume that $x > 25$. The normal form of this game is pictured here.

X \ Y	L	N
L	–5, –5	$x - 15$, 0
N	0, 15	10, 10

(a) Designate the (pure-strategy) Nash equilibria of this game (if it has any).

(b) Compute the mixed-strategy Nash equilibrium of the game.

(c) Given the mixed-strategy equilibrium computed in part (b), what is the probability that the government makes a decision that favors firm X? (It is the probability that (L, N) occurs.)

(d) As x rises, does the probability that the government makes a decision favoring firm X rise or fall? Is this good from an economic standpoint?

4. Compute the mixed-strategy equilibria of the following games.

1 \ 2	A	B
A	2, 4	0, 0
B	1, 6	3, 7

1 \ 2	L	M	R
U	8, 3	3, 5	6, 3
C	3, 3	5, 5	4, 8
D	5, 2	3, 7	4, 9

5. This exercise explores how, in a mixed-strategy equilibrium, players must put positive probability only on best responses. Consider the game in the following figure.

1 \ 2	L	M	R
U	x, x	$x, 0$	$x, 0$
C	$0, x$	2, 0	0, 2
D	$0, x$	0, 2	2, 0

Compute the pure-strategy and mixed-strategy Nash equilibria for this game, and note how they depend on x. In particular, what is the difference between $x > 1$ and $x < 1$?

6. Determine all of the Nash equilibria (pure-strategy and mixed-strategy equilibria) of the following games.

(a)

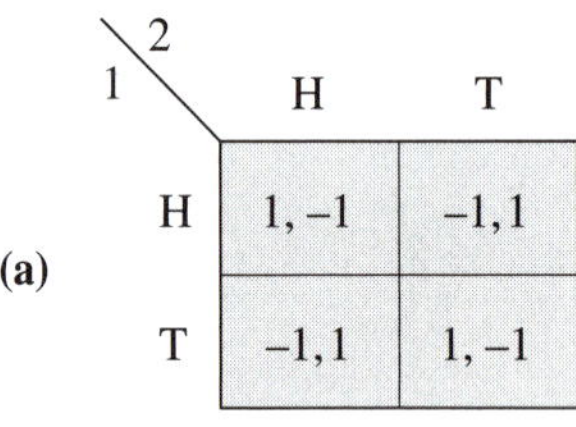

1 \ 2	H	T
H	1, –1	–1, 1
T	–1, 1	1, –1

(b)

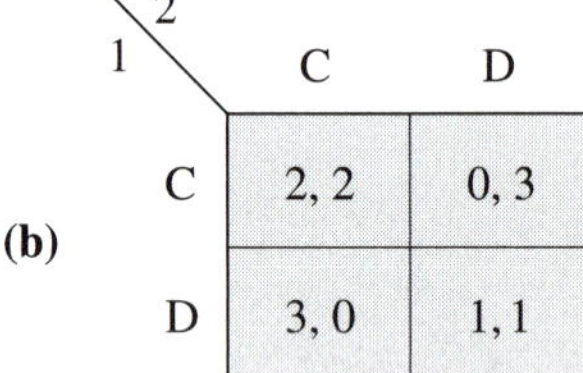

1 \ 2	C	D
C	2, 2	0, 3
D	3, 0	1, 1

(c)

1 \ 2	H	D
H	2, 2	3, 1
D	3, 1	2, 2

(d)

1 \ 2	A	B
A	1, 4	2, 0
B	0, 8	3, 9

(e)

1 \ 2	A	B
A	2, 2	0, 0
B	0, 0	3, 4

(f)

1 \ 2	L	M	R
U	8, 1	0, 2	4, 3
C	3, 1	4, 4	0, 0
D	5, 0	3, 3	1, 4

7. Compute the mixed-strategy Nash equilibria of the following games. (First convert the games into the normal form.)

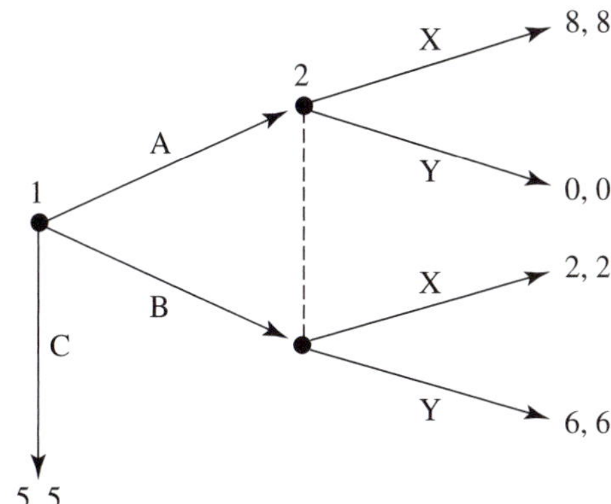

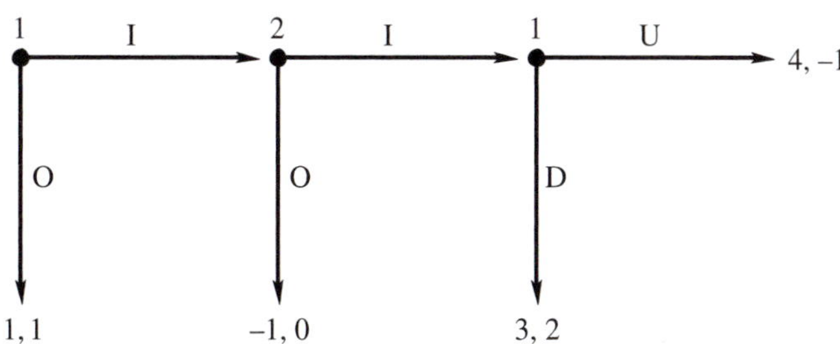

8. Consider the following social problem.[3] A pedestrian is hit by a car and lies injured on the road. There are n people in the vicinity of the accident. The injured pedestrian requires immediate medical attention, which will be forthcoming if at least one of the n people calls for help. Simultaneously and independently, each of the n bystanders decides whether or not to call for help (by dialing 911 on a cell phone or pay phone). Each bystander obtains v units of utility if someone (anyone) calls for help. Those who call for help pay a personal cost of c. That is, if person i calls for help, then he obtains the payoff $v - c$; if person i does not call but at least one other person calls, then person i gets v; finally, if none of the n people calls for help, then person i obtains zero. Assume $v > c$.

(a) Find the symmetric Nash equilibrium of this n-player normal-form game. (Hint: The equilibrium is in mixed strategies. In your analysis, let p be the probability that a person does not call for help.)

[3] The social situation described here was featured in A. M. Rosenthal, *Thirty-Eight Witnesses* (New York: McGraw-Hill, 1964) and in J. Darley and B. Latané, "Bystander Intervention in Emergencies: Diffusion of Responsibility," *Journal of Personality and Social Psychology* 8(1968):377–383.

(b) Compute the probability that at least one person calls for help in equilibrium. (This is the probability that the injured pedestrian gets medical attention.) Note how this depends on n. Is this a perverse or intuitive result?

9. Prove that every 2×2 game has a Nash equilibrium (in either pure or mixed strategies). Do this by considering the following general game and breaking the analysis into two categories: (a) one of the pure-strategy profiles is a Nash equilibrium, and (b) none of the pure-strategy profiles is a Nash equilibrium.

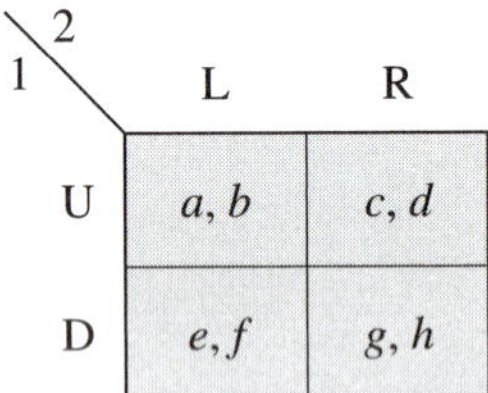

1 \ 2	L	R
U	a, b	c, d
D	e, f	g, h

10. Does the rock–paper–scissors game have any pure-strategy Nash equilibria? Find and report all of the mixed-strategy Nash equilibria of this game. (Take an educated guess to find one, verify it mathematically, and then search for others.) If you forget the representation of this game, refresh your memory by looking at Exercise 4 of Chapter 2 and Exercise 5 of Chapter 6.

11. The famous British spy 001 has to choose one of four routes, a, b, c, or d (listed in order of speed in good conditions) to ski down a mountain. Fast routes are more likely to be struck by an avalanche. At the same time, the notorious rival spy 002 has to choose whether to use (y) or not to use (x) his valuable explosive device to cause an avalanche. The payoffs of this game are represented here.

001 \ 002	x	y
a	12, 0	0, 6
b	11, 1	1, 5
c	10, 2	4, 2
d	9, 3	6, 0

(a) Let $\theta_2(\text{x})$ denote the probability that 001 believes 002 selects x. Explain what 001 should do if $\theta_2(\text{x}) > 2/3$, if $\theta_2(\text{x}) < 2/3$, and if $\theta_2(\text{x}) = 2/3$.

(b) Imagine that you are Mr. Cue, the erudite technical advisor to British military intelligence. Are there any routes you would advise 001 definitely **not** to take? Explain your answer.

(c) A viewer of this epic drama is trying to determine what will happen. Find a Nash equilibrium in which one player plays a pure strategy s_i and the other player plays a mixed strategy σ_j. Find a different mixed-strategy equilibrium in which this same pure strategy s_i is assigned zero probability. Are there any other equilibria?

12. Consider a game with n players. Simultaneously and independently, the players choose between X and Y. That is, the strategy space for each player i is $S_i = \{X, Y\}$. The payoff of each player who selects X is $2m_x - m_x^2 + 3$, where m_x is the number of players who choose X. The payoff of each player who selects Y is $4 - m_y$, where m_y is the number of players who choose Y. Note that $m_x + m_y = n$.

(a) For the case of $n = 2$, represent this game in the normal form and find the pure-strategy Nash equilibria (if any).

(b) Suppose that $n = 3$. How many Nash equilibria does this game have? (Note: you are looking for pure-strategy equilibria here.) If your answer is more than zero, describe a Nash equilibrium.

(c) Continue to assume that $n = 3$. Determine whether this game has a symmetric mixed-strategy Nash equilibrium in which each player selects X with probability p. If you can find such an equilibrium, what is p?

12 STRICTLY COMPETITIVE GAMES AND SECURITY STRATEGIES

The concepts introduced in this chapter are of peripheral importance in the grand scheme of things, but they are well worth digesting for several reasons. First, they are low fat. Second, working with them will enhance your understanding of rationality and strategy in general. Third, they will be used to produce some useful results later in this tour of game theory. The first concept is the definition of a particular class of games:

> A **two-player, strictly competitive game** is a two-player game with the property that, for every two strategy profiles $s, s' \in S$, $u_1(s) > u_1(s')$ if and only if $u_2(s) < u_2(s')$.

Thus, in a strictly competitive game, the two players have exactly opposite rankings over the outcomes. In comparisons of various strategy profiles, wherever one player's payoff increases, the other player's payoff decreases. Strictly competitive games offer no room for joint gain or compromise.

Matching pennies (see Figure 3.4) is a good example of a two-player, strictly competitive game. There are only two payoff vectors in matching pennies, one that favors player 1 and one that favors player 2. The payoff vectors describe who wins the game. When you think about it, lots of games have outcomes that are limited to: (a) player 1 wins and player 2 loses, (b) player 2 wins and player 1 loses, and (c) the players tie (draw). Because players prefer winning to tying and tying to losing, *every* two-player game with outcomes described in this way is a strictly competitive game. Such games include chess, checkers, tennis, football (with each team considered to be one player), and most other sports and leisure games. Another example of a strictly competitive game appears in Figure 12.1. Note that the outcomes of this game cannot be put in terms of a simple "winner" and "loser." Incidentally, the matching pennies game is a special type of strictly competitive game called *zero-sum,* in which the players' payoffs always sum to zero.[1] The strictly competitive game in Figure 12.1 is not a zero-sum game.

[1] A great deal of the early game-theory literature concerned zero-sum games, as analyzed in J. von Neumann and O. Morgenstern, *Theory of Games and Economic Behavior* (Princeton, NJ: Princeton University Press, 1944).

FIGURE 12.1
A two-player, strictly competitive game.

1 \ 2	X	Y
A	3, 2	0, 4
B	6, 1	1, 3

Next consider the concept of a *security strategy,* which is based on evaluating "worst case scenarios." In any game, the *worst* payoff that player i can get when he plays strategy s_i is defined by

$$w_i(s_i) \equiv \min_{s_j \in S_j} u_i(s_i, s_j).$$

Here we look across player j's strategies to find the one that gives player i the lowest payoff given that player i plays s_i. If player i selects strategy s_i, then he is guaranteed to get a payoff of *at least* $w_i(s_i)$. A security strategy gives player i the best of the worst cases.

A strategy $\underline{s}_i \in S_i$ for player i is called a **security strategy** if $\underline{s}_i$ solves $\max_{s_i \in S_i} w_i(s_i)$. Player i's **security payoff level** is $\max_{s_i \in S_i} w_i(s_i)$.

Note that the security payoff level can also be written as

$$\max_{s_i \in S_i} \min_{s_j \in S_j} u_i(s_i, s_j).$$

The exercises at the end of this chapter will help you explore the relation between security strategies, best response, and equilibrium.[2]

There is no general relation between security strategies and equilibrium strategies. But an important exception exists for strictly competitive games.

Result: If a two-player game is strictly competitive and has a Nash equilibrium $s^* \in S$, then s_1^* is a security strategy for player 1 and s_2^* is a security strategy for player 2.[3]

In other words, if s_i is a Nash-equilibrium strategy for player i in a strictly competitive game, then s_i guarantees player i at least her security payoff level,

[2]Note that my definition of security strategies focuses on pure strategies. One can also define security strategies using mixed strategies and, in fact, the mixed-strategy definition is more proper for many applications. With mixed strategies, player i's security payoff level is

$$\max_{\sigma_i \in \Delta S_i} \min_{s_j \in S_j} u_i(\sigma_i, s_j)$$

and player i's security strategies solve the first maximization problem.

[3]J. von Neumann, "Zur Theorie der Gesellschaftsspiele," *Mathematishce Annalen* 100(1928):295–320.

regardless of what the other player does. In fact, we can elaborate. By virtue of s^* being a Nash equilibrium, s_j^* is a best response to s_i^*. Thus, any deviation from s_j^* would weakly lower player j's payoff and, because the game is strictly competitive, it would raise player i's payoff. We therefore know that with the equilibrium profile s^*, player i receives the lowest payoff conditional on selecting s_i^*. This means that player i's security payoff level is $u_i(s^*)$. To summarize, by choosing s_i^*, player i guarantees her equilibrium payoff, whether or not player j selects his equilibrium strategy.

The result is not difficult to prove mathematically, so I include the proof for those who want to see it. The proof is a bit technical, so stay focused. Let us prove the result for player 1's strategy; the same arguments work for player 2's strategy as well. Because s^* is a Nash equilibrium, player 2 cannot increase her payoff by unilaterally deviating from s_2^*, which means that

$$u_2(s_1^*, s_2) \leq u_2(s^*)$$

for every strategy s_2. But, because the game is strictly competitive, we can put this inequality in terms of player 1's payoff by reversing the direction of inequality. That is, we know that

$$u_1(s_1^*, s_2) \geq u_1(s^*)$$

for every strategy s_2. In words, at strategy profile s^*, player 1 gets the lowest possible payoff given that he plays s_1^*. Thus, $u_1(s^*) = w_1(s_1^*)$. We're almost done. Because s^* is a Nash equilibrium, player 1 cannot gain by deviating from s_1^*, which means that

$$u_1(s^*) \geq u_1(s_1, s_2^*)$$

for every strategy s_1. In addition, we know that $u_1(s_1, s_2^*)$ is not less than $w_1(s_1)$, because the latter is the worst-case payoff when player 1 selects s_1. We conclude that $w_1(s_1^*) = u_1(s^*) \geq w_1(s_1)$ for every strategy s_1, which means s_1^* is a security strategy.

GUIDED EXERCISE

Problem: Are all security strategies rationalizable? Prove this or find a counterexample.

Solution: In fact, not all security strategies are rationalizable, as the following game demonstrates.

1 \ 2	X	Y
A	3, 5	−1, 1
B	2, 6	1, 2

In this game, the only rationalizable strategy profile is (A, X), yet B is player 1's security strategy.

EXERCISES

1. Determine which of the following games are strictly competitive.

1 \ 2	X	Y	Z
A	2, 9	6, 5	7, 4
B	5, 6	8, 2	3, 8
C	9, 1	4, 7	7, 3

(a)

1 \ 2	X	Y	Z
A	8, 1	7, 2	3, 6
B	9, 0	2, 8	4, 5
C	7, 2	8, 1	6, 4

(b)

1 \ 2	X	Y
A	1, 1	2, 0
B	0, 2	1, 1

(c)

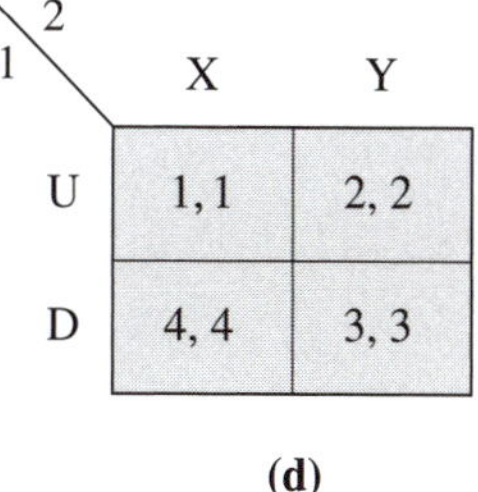

1 \ 2	X	Y
U	1, 1	2, 2
D	4, 4	3, 3

(d)

2. Find the players' security strategies for the games pictured in Exercise 1.

3. Give some examples of games people play for entertainment that are strictly competitive.

4. For a two-player game, two pure-strategy Nash equilibria (s_1, s_2) and (t_1, t_2) are called *equivalent* if $u_i(s) = u_i(t)$ for $i = 1, 2$; they are called *interchangeable*

if (s_1, t_2) and (t_1, s_2) also are Nash equilibria. Mathematically prove that any two pure-strategy Nash equilibria of a two-player, strictly competitive game are both equivalent and interchangeable.

13 CONTRACT, LAW, AND ENFORCEMENT IN STATIC SETTINGS

As indicated in the discussion of the Nash equilibrium concept, many institutions help align beliefs and behavior to achieve congruous outcomes. In particular, contracting institutions—such as the legal system—govern a wide variety of relationships in our society. Most economic relationships feature some degree of contracting. For example, firms and workers generally negotiate and agree on wages, job assignments, and other aspects of the employment relation. Homeowners write contracts with building contractors. Firms contract with their suppliers as well as with their customers. Nations contract with one another on the subject of terms of trade and security interests.

Contracting does more than alleviate strategic uncertainty. In fact, it also helps resolve the other two tensions of strategic situations discussed in preceding chapters. Specifically, deliberate contracting gives players a way of avoiding inefficient coordination. Further, to the extent that the participation of third parties changes the "game" to be played by the players, contracts can also mitigate conflicts between joint and individual incentives—that is, contracts may help align incentives.

In this chapter, I explain how to use game theory to define and study contracting, I describe the various ways in which contracts can be enforced, and I present some examples of how legal institutions facilitate cooperation. I begin with some definitions.[1] First, and most important, is the concept of a contract. The U.S. legal system generally regards a contract as a "promise or a set of promises for the breach of which the law gives a remedy, or the performance of which the law in some way recognizes as a duty."[2] Although it is a useful working definition in the legal world, this definition is too much a product of the development of formal legal institutions to serve us well at the general conceptual level. I use the following, more comprehensive, definition.

> A **contract** is an agreement about behavior that is intended to be enforced.

[1] The concepts noted here are covered more deeply in J. Watson, "Contract and Game Theory: Basic Concepts for Settings with Finite Horizons," University of California, San Diego, working paper, 2005.

[2] Section 1, *Restatement (Second) of Contracts*.

As the definition suggests, we ought to differentiate between contracting and enforcing contracts.

I call interaction between two or more economic agents a *contractual relationship* if the parties, with some deliberation, work together to set the terms of their relationship. By "work together," I mean that the parties negotiate and agree on a course of action. Some aspects of their agreement may include the interaction of third parties, such as judges or arbitrators. Often, we wish to study the involvement of these third parties without actually modeling their preferences, in which case we will call the third parties "external players." (Players whose preferences we address, such as the players in all the games that we have analyzed to this point, are considered "internal players.") An external player may be considered to take actions on the basis of information about the contractual relationship that is **verifiable** to this external party.

We can roughly divide any contractual relationship into two phases: a *contracting phase,* in which players set the terms of their contract, and an *implementation phase,* in which the contract is carried out and enforced. I do not explicitly model the contracting phase in this chapter. Instead, I focus on simple static games that highlight the nature of contract and methods of enforcement.[3]

There are three methods of contract enforcement that are worth isolating with game theory. First, a contract is said to be **self-enforced** if the players have the individual incentives to abide by the terms of the contract. A contract is said to be **externally enforced** if the players are motivated to behave in accordance with a contract by the actions of an external player, such as a judge or arbitrator. Finally, a contract is said to be **automatically enforced** if implementing the contract is instantaneous with the agreement itself. People generally rely on a combination of these three methods of enforcement.

We will study automatically enforced contracts later, through the analysis of joint decision problems in Chapter 20. To illustrate the other two enforcement methods and to address some of the deeper issues regarding contracting institutions, let us consider a specific example. Suppose Jessica and Mark, a happily married couple, wish to remodel their home. Remodeling requires the efforts of both an architect and a building contractor. The architect's job is to make precise measurements of the house, determine the work to be done, and decide what materials and techniques should be used in the process. Suppose Mark himself is an architect, so he plays the architect's role in the remodeling project. The duty of the building contractor is to handle the physical construction, with the assistance of workers whom he or she hires. Let us name the

---

[3]In the next part of the book (in particular, in Chapters 18–21), I analyze the contracting phase as a bargaining problem.

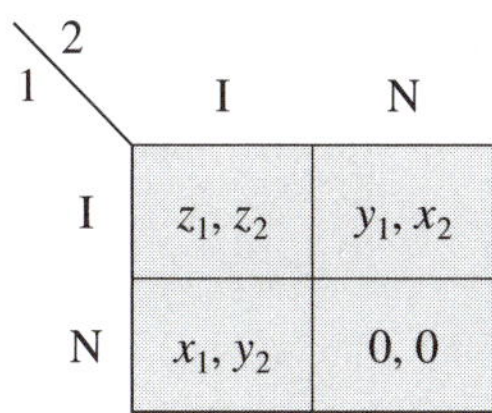

FIGURE 13.1
Technology of the contractual relationship.

building contractor Denny. Jessica, an attorney, will use her legal expertise to design a contract between Denny and the couple.

I begin by reviewing the **technology of the relationship,** which is the game-theoretic representation of productive interaction in the contractual relationship. Production (the remodeling job) entails the inputs of Mark and Denny, whose interaction is represented by the normal-form game pictured in Figure 13.1. I call it the *underlying game*. Mark is player 1 and Denny is player 2. Simultaneously and independently, they each choose between investing effort in the project (I) and not investing (N). For Mark, the investment of effort entails architectural work that may be individually costly, but it makes Denny's job much easier and increases the value of the project. Denny's investment refers to effort expended that increases the quality of the remodeling work.

The payoff numbers are in thousands of dollars. Each player wants to maximize the number of dollars that he receives. Let us focus on the case in which

$$z_1 + z_2 > x_1 + y_2, \quad z_1 + z_2 > x_2 + y_1, \quad \text{and} \quad z_1 + z_2 > 0.$$

In words, this means that the sum of the players' monetary payoffs—the *joint value* of their relationship—is highest when (I, I) is played. In fact, if the players have the ability to *transfer* money between themselves (that is, give money to each other), then (I, I) is the *only* efficient outcome of the underlying game. This is because both players can always be made better off when they share a larger joint value than when they share a smaller joint value.[4] For the contracting problems studied in this chapter, I make the reasonable assumption that the parties can make monetary transfers. Thus, given that (I, I) is the best outcome, our fundamental question is: Can the players enforce a contract specifying that (I, I) will be played?

[4]For example, suppose $z_1 = z_2 = 5$, $x_1 = 6$, and $y_2 = 2$, and so the joint value of (I, I) is 10 and the joint value of (N, I) is 8. Without transfers, player 1 prefers (N, I) to (I, I). However, if player 2 were to transfer 2 units to player 1 when (I, I) is played, then both players would prefer (I, I) to (N, I). In Chapter 18, I elaborate on this point about transfers and efficiency.

FIGURE 13.2
Induced game.

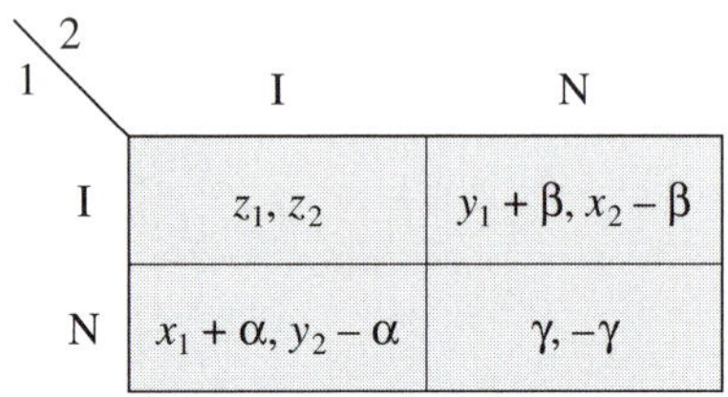

	I	N
I	z_1, z_2	$y_1 + \beta, x_2 - \beta$
N	$x_1 + \alpha, y_2 - \alpha$	$\gamma, -\gamma$

The players can easily implement (I, I) if this profile is a Nash equilibrium of the underlying game. In other words, an agreement to play (I, I) is a self-enforced contract only if (I, I) is a Nash equilibrium, which is the case if

$$z_1 \geq x_1 \quad \text{and} \quad z_2 \geq x_2.$$

However, if either of these inequalities fails to hold, then the players cannot rely on self-enforcement to support the outcome (I, I). They need the participation of a third party, whose actions may serve to change the nature of the game between Mark and Denny. A court is such a third party.

The court enters the picture as follows: After Mark and Denny choose whether to invest—so that the outcome of the underlying game is realized—the court may step in and compel a transfer of money between them. The court will not impose any transfer if (I, I) is played, because in this case the players have cooperated as desired. But, if one or both players has deviated by selecting N, then the court takes remedial action. Let m denote the transfer required by the court; assume it is a transfer from Denny to Mark. Thus, Mark's payoff increases by m while Denny's payoff decreases by m. If $m < 0$, then it means that Mark actually gives Denny money.

You might imagine that, in some scenarios, the court's intervention does not lead to a transfer between the players but rather leads to an outcome whereby both players lose money. For example, court costs impose a transfer from both players, at least jointly. For now, we shall leave aside such costs and concentrate on transfers from one player to another.

The court-imposed transfer m may depend on the outcome of the underlying game. For instance, the transfer may be different if (I, N) occurred than it would be if (N, I) occurred. In terms of the underlying game, then, there are possibly four different transfers, one associated with each cell of the matrix. Assuming the players know what the transfers will be, the court's intervention *changes the game* played between Mark and Denny. Instead of the underlying game, they actually play the game depicted in Figure 13.2. This game adds the transfers to the underlying game, implying the players' true payoffs. Note that no transfer is made when (I, I) occurs, as heretofore assumed. Transfers in the other cells are denoted α, β, and γ. In each case, these numbers represent a transfer from Denny (player 2) to Mark (player 1). The game in Figure 13.2 is called the **induced game**.

The induced game captures the two most important types of contract enforcement: self-enforcement and external enforcement. External enforcement is associated with how the actions of the external party (the court) change the game to be played by the contracting agents—that is, how the formal contract between Mark and Denny influences the behavior of the court and, in turn, their own payoffs in the induced game. Technically, external enforcement transforms the underlying game into the induced game. Self-enforcement relates to the fact that, in the end, the players can only sustain a Nash equilibrium (where they coordinate on a weakly congruous outcome). Let us presume that the players will jointly select the *best* Nash equilibrium in the induced game.

The key to effective external enforcement is the extent to which the court-imposed monetary transfers can be chosen arbitrarily. In regard to Figure 13.2, what determines the transfers α, β, and γ? Can the players select them before the game is played? If so, to what extent are the players constrained? The answers to these questions lie in a deeper understanding of the external enforcement institution. To this end, it is worthwhile to think about two contractual settings. In the first setting, the court allows people to write contracts as they see fit and the court enforces contracts verbatim. In the second setting, the court puts tighter constraints on the set of feasible contracts.

COMPLETE CONTRACTING IN DISCRETIONARY ENVIRONMENTS

Return your thoughts to the example of a remodeling project. Suppose the court allows the players to write a **complete contract,** which specifies a transfer for *each* of the outcomes of the underlying game. That is, the players can sign a document assigning values to α, β, and γ. After the underlying game is played, the court reviews this document and observes the outcome of the game. The court then enforces the agreement between the players by compelling them to make the appropriate transfer.[5]

Enter Jessica and her legal sensibilities. She notices that an appropriately designed contract can easily induce the efficient outcome (I, I). This is accomplished by specifying externally enforced transfers sufficient to make (I, I) a Nash equilibrium in the induced game. Any α and β will do, as long as

$$z_1 \geq x_1 + \alpha \quad \text{and} \quad z_2 \geq x_2 - \beta.$$

[5]The contracting process (how players establish a contract) is addressed in Chapters 18 and 20 in Part III of this book. For now, it is enough to focus on maximizing the joint value of the contractual relationship by an appropriately specified externally enforced contract and coordination by the players on the best self-enforced component (Nash equilibrium).

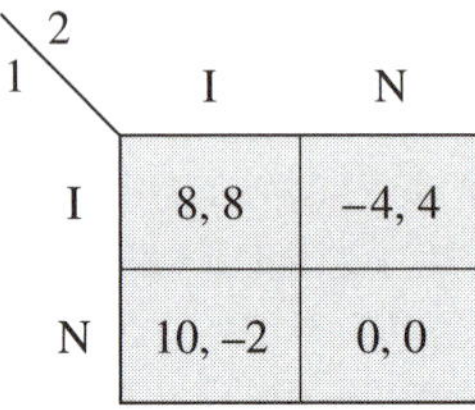

1 \ 2	I	N
I	8, 8	−4, 4
N	10, −2	0, 0

FIGURE 13.3
Specific example.

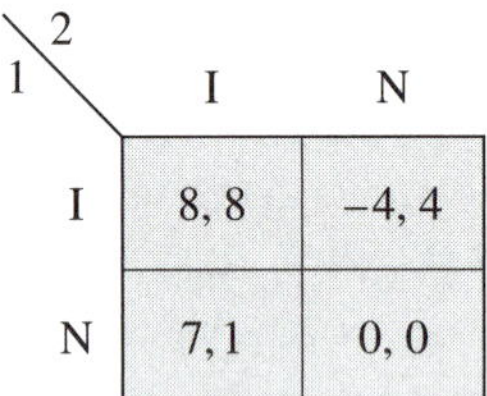

1 \ 2	I	N
I	8, 8	−4, 4
N	7, 1	0, 0

FIGURE 13.4
Induced game for the specific example.

For example, consider the technology represented in Figure 13.3. Jessica might recommend that Mark and Denny sign a document specifying $\alpha = -3$, $\beta = 0$, and $\gamma = 0$. In words, this contract instructs the court to enforce a transfer of 3 from Mark to Denny in the event that Mark does not invest while Denny does invest. This externally enforced component implies the induced game pictured in Figure 13.4. Because (I, I) is a Nash equilibrium in the induced game, the agreement to play (I, I) is self-enforced.

The lesson from the preceding example is that, under certain conditions, complete contracts allow the players to induce *any* profile in the underlying game, regardless of the technology of the relationship—that is, regardless of the payoffs in the underlying game. The essential condition is that the court must be able to differentiate between all of the different outcomes of the underlying game. For example, the court must be able to verify whether each player selected I or N in the game. Then the court knows when to impose transfer α, when to impose β, and so on. We call this informational condition **full verifiability**. To summarize:

> **Result:** With full verifiability, there is an enforced contract yielding the efficient outcome (that which maximizes the players' total payoff).

In legal terms, full verifiability means that sufficient evidence exists to prove to the court exactly what the outcome of the underlying game was.[6]

[6] Although the model does not address how the evidence is produced and by whom, it is reasonable at this level of analysis to suppose that the evidence will make its way to the court.

FIGURE 13.5
Induced game with limited verifiability.

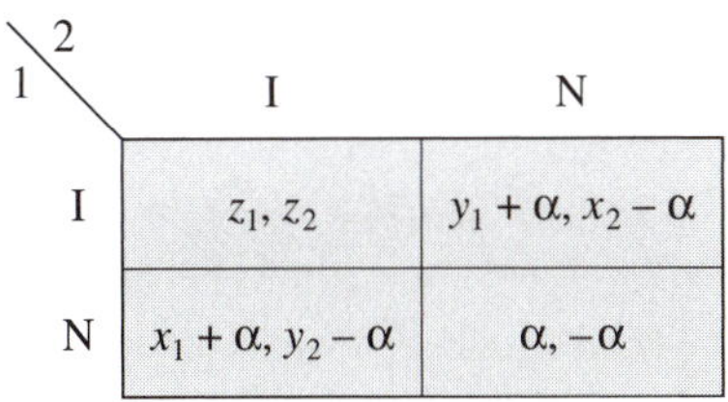

1 \ 2	I	N
I	z_1, z_2	$y_1 + \alpha, x_2 - \alpha$
N	$x_1 + \alpha, y_2 - \alpha$	$\alpha, -\alpha$

Unfortunately, full verifiability is more often the exception than the rule. Usually there is not sufficient evidence to completely reveal the outcome of the underlying game. For example, the court may have no way of knowing whether Mark or Denny individually invested in the remodeling project. The court may be able to confirm only whether the end result is good or poor. A good job indicates that both Mark and Denny invested as promised; a bad job reveals that one or both players did not invest.

To be more specific, suppose the remodeling job entails the design and construction of cabinetry. The court can observe only whether, in the end, the cabinets fit together seamlessly. If they do, then the court can conclude that both Mark and Denny invested. If the cabinets do not fit together, then the court knows that at least one of the players failed to invest. However, the court does not know if the faulty outcome was the result of Mark's failure to supply the correct measurements and instruction or the result of Denny's failure to take care during construction. This is an example of *limited verifiability,* defined as a setting in which the court cannot perfectly verify the players' productive actions.

In the limited-verifiability setting, Jessica observes that it is impossible to specify different externally enforced transfers for each cell of the game matrix. Because the court cannot distinguish between (I, N), (N, I), and (N, N), the contract must specify the same transfer for each of these outcomes. In legal terms, the players sign a document directing the court to impose transfers on the basis of what the court verifies about the relationship. Because the court can verify only whether the cabinets fit, the transfer can be conditioned only on whether this is the case. In game-theory terms, the externally enforced component of the contract consists of a single number α, yielding the induced game pictured in Figure 13.5.

It can be difficult, even impossible, to support (I, I) with limited verifiability. To see this, note that, although raising α does reduce player 2's incentive to play N, it *increases* player 1's incentive to select N. In order for (I, I) to be a Nash equilibrium of the induced game, α must be selected to balance the incentives of the players. Specifically, we need

$$z_1 \geq x_1 + \alpha \quad \text{and} \quad z_2 \geq x_2 - \alpha.$$

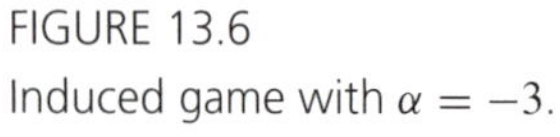

FIGURE 13.6
Induced game with $\alpha = -3$.

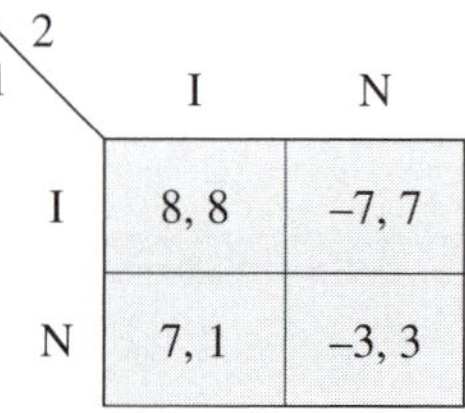

1 \ 2	I	N
I	8, 8	−7, 7
N	7, 1	−3, 3

FIGURE 13.7
Another specific example.

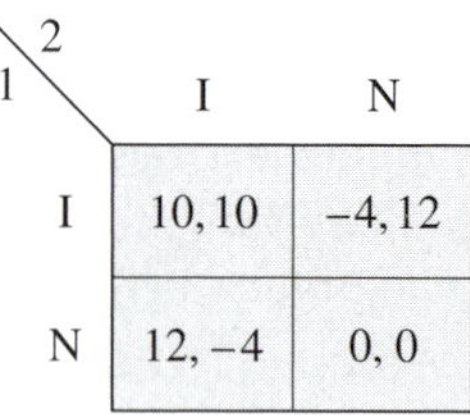

1 \ 2	I	N
I	10, 10	−4, 12
N	12, −4	0, 0

Rearranging terms simplifies this to

$$x_2 - z_2 \leq \alpha \leq z_1 - x_1.$$

We conclude that there is an α that satisfies these inequalities if and only if $z_1 + z_2 \geq x_1 + x_2$.

As an example, take the underlying game shown in Figure 13.3. Setting $\alpha = -3$ yields the induced game in Figure 13.6, for which (I, I) is a Nash equilibrium. Next consider the underlying game shown in Figure 13.7. Under limited verifiability, (I, I) cannot be enforced. (Check that there is no α that makes this a Nash equilibrium of the induced game.) However, as with all technologies, (I, I) can be supported with full verifiability. Thus, limited verifiability puts real constraints on players' ability to achieve efficient outcomes. Contractual imperfections, such as limited verifiability, lead to the three strategic tensions identified earlier in this book.[7]

CONTRACTING WITH COURT-IMPOSED BREACH REMEDIES

Courts do not always enforce what players write into their contracts. In fact, in the United States, courts are more likely to impose transfers on the basis of certain legal principles, rather than be dictated to by the contract document.

[7] In addition, the three tensions discussed earlier operate on the level of contracting. For example, players 1 and 2 may contract separately from players 3 and 4, yet these two contractual relationships may affect one another in prisoners' dilemma fashion. Players 1 and 2 may have the narrow incentive to select a contract of form "D" and players 3 and 4 may have the same incentive, yet all four would prefer a grand contract specifying "C" in each bilateral relationship.

Further, contracts in the real world tend not to be complete specifications of transfers as a function of verifiable information. There are several reasons why players write such "incomplete" contracts. For example, it may be expensive or time consuming to list all of the contingencies that may arise in a contractual relationship. More importantly, players may count on the court to "complete" a contract (fill in the gaps) during litigation, trusting that the court will be able to determine the appropriate remedial action. Because courts are active with regard to assessing relationships and imposing transfers, the players need not write a detailed contract up front.

An externally enforced contract is often merely a statement of how the players intend to behave in their relationship. For example, Mark and Denny may agree to play (I, I) in the game. That is, Mark promises to supply Denny with accurate measurements and thorough instructions, whereas Denny promises to exercise care in the building process. The players may never even mention the possibility that one or both of them may deviate from this plan, much less what the legal remedy should be in such a case. In this sense, the contract is *incomplete*. The players' contract provides no guidance to the court for the eventuality in which one of the players fails to invest.

In this section, I sketch three legal principles that guide damage awards in U.S. commercial cases.[8] The principles apply in the setting of incomplete contracts, where either the players do not specify a transfer for each contingency or the court ignores the players' statement of transfers. For simplicity, I assume that the externally enforced component of a contract is simply a joint statement by the players describing the actions that they intend to take in the underlying game. For the examples, we are interested in the contract specifying (I, I). A deviation from this profile by one or both players is called a **breach**. If one player commits a breach, then this player is called the *defendant* and the other player is called the *plaintiff*. On verifying a breach, the court imposes a **breach remedy,** which implies a transfer between the players. For now, assume that the court can fully verify the outcome of the underlying game—that is, that there is full verifiability. With reference to Figure 13.2, the court's breach remedy defines α, β, and γ.

Under the legal principle of **expectation damages**, the court forces the defendant to transfer to the plaintiff the sum of money needed to give the plaintiff the payoff that he or she would have received had the contract been fulfilled. In terms of the underlying game in Figure 13.1, player 1's expectation is z_1 and player 2's expectation is z_2. Thus, if Mark (player 1) breaches, then he

[8]The standard breach remedies analyzed here were studied by S. Shavell, "Damage Measures for Breach of Contract," *Bell Journal of Economics* 11(1980):466–490. For an overview of contract law, you can browse through a law school textbook such as R. Barnett, *Contracts: Cases and Doctrine,* 3rd ed. (New York: Aspen Law & Business, 2003).

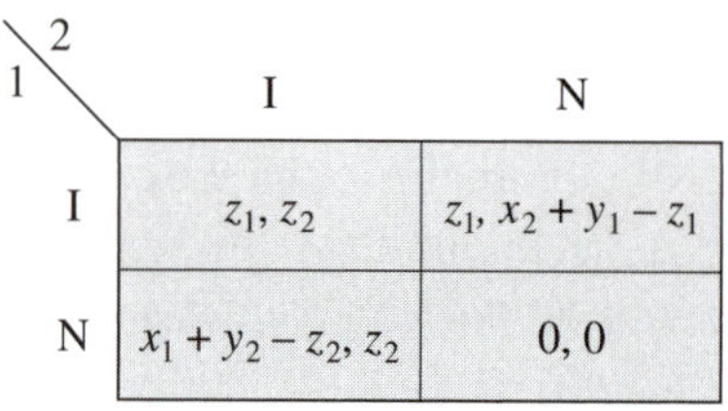

1 \ 2	I	N
I	z_1, z_2	$z_1, x_2 + y_1 - z_1$
N	$x_1 + y_2 - z_2, z_2$	0, 0

FIGURE 13.8
Induced game under expectation damages.

is forced to pay Denny an amount to make Denny's payoff z_2. This implies that $\alpha = y_2 - z_2$. Likewise, the transfer imposed if Denny breaches is $\beta = z_1 - y_1$. For simplicity, we can let $\gamma = 0$. Under expectation damages, the induced game is that depicted in Figure 13.8.

In the induced game, (I, I) is a Nash equilibrium if and only if

$$z_1 \geq x_1 + y_2 - z_2 \quad \text{and} \quad z_2 \geq x_2 + y_1 - z_1.$$

Rearranging these inequalities yields

$$z_1 + z_2 \geq x_1 + y_2 \quad \text{and} \quad z_1 + z_2 \geq x_2 + y_1,$$

which should be familiar to you—they are the inequalities that define conditions under which (I, I) is efficient. Thus:

> **Result:** Under expectation damages with the underlying game of Figure 13.1, (I, I) is enforceable if and only if (I, I) is efficient.

This result is quite strong, because it means that expectation damages alleviate the tension between individual and joint incentives. In fact, even if the players do not know the values x_i, y_i, and z_i at the time of contracting, these damages encourage the efficient outcome. Specifically, if the agents formally agree to play (I, I) and then observe the values x_i, y_i, and z_i before deciding whether to invest, then they will have the incentive to breach only if (I, I) turns out to be inefficient—in which case breach is an efficiency-enhancing choice.

Expectation damages are the ideal court remedy, both in theory and in practice. Courts generally impose expectation awards whenever possible. But expectation requires a great deal of information—more information than exists in many economic settings. First, the court must be able to confirm which of the two players has breached the contract (full verifiability). Second, the players must know the payoff parameters before deciding whether to invest. Third, and most critically, the court must be able to determine the payoffs in the breach and nonbreach outcomes of the underlying game (in particular, z_i and y_i).

The parameter y_i is sometimes easily verified by courts. For example, y_i may include expenses borne by the defendant, for which the defendant can produce receipts. That is, Denny may be able to prove how much he spent on materials and assistants. On the other hand, a court may find it very difficult to determine z_i, in particular following breach. If the cabinetry is not built properly, how can the court accurately estimate the value that would have resulted if the job had been completed with care? An even better example arises in a joint venture on the development of a new product. If contractual breach destroyed the chance that the product could be brought to market, then one can hardly expect the court to accurately estimate the value that the product would have generated.

The second breach remedy applies well to cases in which z_i cannot be observed. Under the principle of **reliance damages**, the court imposes a transfer that returns the plaintiff to the state in which he or she would have been but for the contract. That is, the court determines the payoff that the plaintiff would have received had no contract been written, and then the court forces a transfer sufficient to give the plaintiff this total payoff. The idea behind this legal principle is that the plaintiff should be compensated for the initial investment that he makes in the contractual relationship—an investment that, in a contractual commitment, places one party in the position of relying on the other to fulfill the terms of the contract. For example, when Mark makes detailed drawings of his house, he relies on Denny to follow the drawings with care during construction. A player is due compensation for his investment if the other player does not honor the contract.

In our model, we can capture the "no contract" value in two ways. First, we might view this value as a particular value w_i that player i would have received had the underlying game not been played. Second, we might suppose that the underlying game is played with or without a contract, in which case the absence of a contract implies the play of a Nash equilibrium in the underlying game. Our choice between these alternatives should be sensitive to the actual case at hand. For simplicity, let us use the second modeling approach. Assuming that $y_i \leq 0$ for $i = 1, 2$, (N, N) is a Nash equilibrium of the underlying game; this is our prediction for the no-contract outcome.

With reference to Figures 13.1 and 13.2, reliance damages imply $\alpha = y_2$ and $\beta = -y_1$, and so the induced game is the one shown in Figure 13.9 on the next page. Note that (I, I) is a Nash equilibrium in the induced game if and only if

$$z_1 \geq x_1 + y_2 \quad \text{and} \quad z_2 \geq x_2 + y_1.$$

Unless the defendant's damage y_i is sufficiently large, reliance awards may not support the efficient outcome.

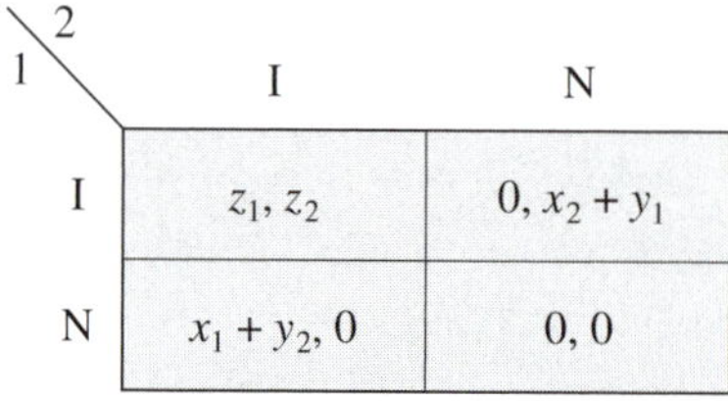

FIGURE 13.9
Induced game under reliance damages.

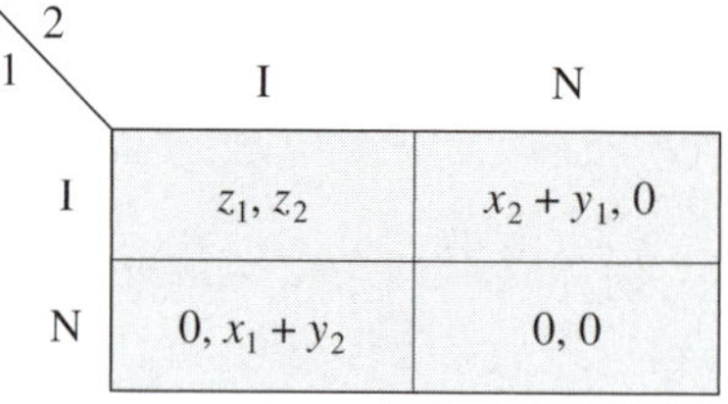

FIGURE 13.10
Induced game under restitution damages.

The third common legal principle on breach refers to **restitution damages**, which seek to cancel any unjust enrichment that the defendant obtained by breaching the contract, relative to the no-contract state. The best way to think about this principle is to imagine the following comparison that a mischievous-minded Denny might make. On one hand, he could enter into a contract with Mark with the thought of breaching it. For example, Denny may value Mark's detailed drawings independently of the project at hand (he can present them to a future customer perhaps). Denny may be able to take advantage of Mark in other ways as well, say, by raiding Jessica and Mark's refrigerator and watching television while they are out of the house. On the other hand, Denny can obtain zero by not having a contract with Mark. The breach value x_2 represents the unjust enrichment (relative to the no-contract outcome yielding him zero) that an opportunistic Denny can obtain from the contract.

Restitution damages imply $\alpha = -x_1$ and $\beta = x_2$, which yields the induced game pictured in Figure 13.10. Note that (I, I) is a Nash equilibrium in the induced game if and only if $z_1 \geq 0$ and $z_2 \geq 0$. Restitution awards do not imply the efficient outcome in two cases: (1) when one value of z_i is negative, and (2) when both values of z_i are positive yet (I, I) is not efficient.

As Jessica readily points out to Mark and Denny, expectation damages are best; however, such damages require the court to know the value of the hypothetical performance outcome in the event that the contract is actually breached. When z_i cannot be estimated with precision, the court may adopt a reliance or restitution test. Reliance requires the court to know the plaintiff's expenditures related to the contract, as well as the value of the plaintiff's foregone opportunities. Restitution requires knowledge of the defendant's unjust enrichment.

Complicating factors often make it difficult to estimate the payoff parameters. For example, evaluating the plaintiff's foregone opportunities can be a slippery exercise. If Mark and Denny had not entered into a contract, would Mark have benefited from searching for another building contractor? How much effort should Mark have expended in looking for an alternative contractor? How much effort should Mark expend to minimize damage to his house following a breach by Denny? The answers to these questions and more can be provided—or are at least addressed—by your local attorney and economist.

GUIDED EXERCISE

Problem: Consider a contractual relationship, where the technology of the relationship is given by the following matrix:

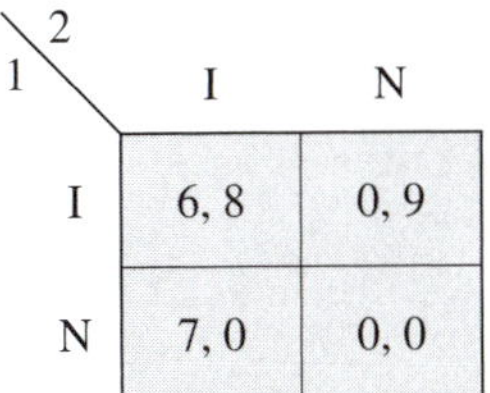

1 \ 2	I	N
I	6, 8	0, 9
N	7, 0	0, 0

(a) Suppose that the players write a contract whereby they agree to play (I, I). Draw the matrix that describes the induced game under the assumption that the court enforces *expectation damages*.

(b) Suppose that the players can write a complete contract, but that they are in a setting of *limited verifiability*, where the court can verify only whether N was played by one or both of the players (but the court cannot verify *which* of the players chose N). Is there a contract that induces play of (I, I)?

Solution:

(a) Recall that, with expectation damages, the defendant must pay the plaintiff the amount that would make the plaintiff's payoff the same as it would be if the contract were fulfilled. For instance, if player 1 breaches the contract, leading to profile (N, I), then player 1 is compelled to transfer 8 to player 2. The induced game is therefore the following:

1 \ 2	I	N
I	6, 8	6, 3
N	–1, 8	0, 0

(b) No, (I, I) cannot be induced. To prevent player 1 from deviating, the court must require player 1 to transfer at least 1 unit to player 2 if cell (N, I) is reached in the underlying game. On the other hand, to prevent player 2 from deviating, the court must require player 2 to transfer at least 1 unit to player 1 if cell (I, N) is reached in the underlying game. Limited verifiability requires the transfers specified for cells (N, I) and (I, N) to be the same, which contradicts these two conditions.

EXERCISES

1. Consider a contractual setting in which the technology of the relationship is given by the following underlying game:

1 \ 2	I	N
I	5, 5	–1, 1
N	7, –1	0, 0

Suppose an external enforcer will compel transfer α from player 2 to player 1 if (N, I) is played, transfer β from player 2 to player 1 if (I, N) is played, and transfer γ from player 2 to player 1 if (N, N) is played. The players wish to support the investment outcome (I, I).

(a) Suppose there is limited verifiability, so that $\alpha = \beta = \gamma$ is required. Assume that this number is set by the players' contract. Write the matrix representing the induced game and determine whether (I, I) can be enforced. Explain your answer.

(b) Suppose there is full verifiability, but that α, β, and γ represent reliance damages imposed by the court. Write the matrix representing the induced game and determine whether (I, I) can be enforced. Explain your answer.

2. Consider a contractual setting in which the technology of the relationship is given by the following partnership game:

1 \ 2	I	N
I	4, 4	−4, 9
N	2, −4	0, 0

Suppose the players contract in a setting of court-imposed breach remedies. The players can write a formal contract specifying the strategy profile they intend to play; the court observes their behavior in the underlying game and, if one or both of them cheated, imposes a breach transfer. The players wish to support the investment outcome (I, I).

(a) Write the matrix representing the induced game under the assumption that the court imposes *expectation damages*. Can a contract specifying (I, I) be enforced? Explain your answer.

(b) Write the matrix representing the induced game under the assumption that the court imposes *restitution damages*. Can a contract specifying (I, I) be enforced?

(c) Write the matrix representing the induced game under the assumption that the court imposes *reliance damages*. Can a contract specifying (I, I) be enforced with reliance transfers? Explain your answer.

(d) Suppose litigation is costly. When a contract is breached, each player has to pay a court fee of c in addition to the *reliance* transfer imposed by the court. What is the induced game in this case?

(e) Under what condition on c can (I, I) be enforced with reliance transfers and court costs?

(f) Continue to assume the setting of part (d). Suppose the court intervenes after a breach only if the plaintiff brings suit. For what values of c does the plaintiff have the incentive to sue?

(g) How does your answer to part (e) change if the court forces the losing party to pay all court costs?

3. Consider a setting of complete contracting in a discretionary environment, where the court will impose transfers as specified by the players. For each of the following two underlying games, how much should the players be jointly willing to pay to transform the setting from one of *limited verifiability*—where the court cannot distinguish between (I, N), (N, I), and (N, N) but knows whether (I, I) was played—to one of *full verifiability*—where the court knows exactly what the players did? To answer this question, you must determine the outcomes in the two different information settings.

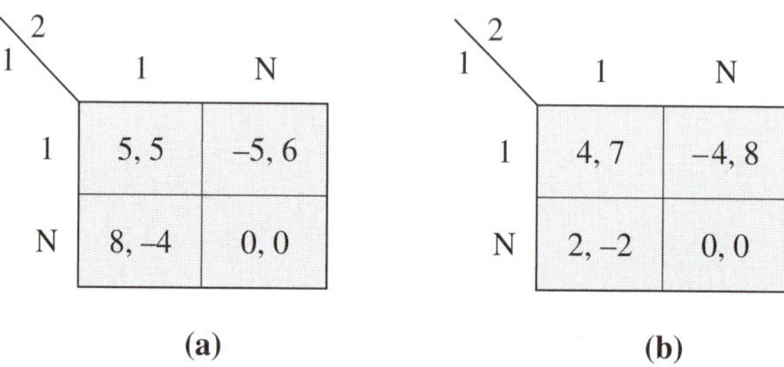

4. Reconsider the pedestrian-injury example of Exercise 8 in Chapter 11. Suppose you are a prominent legal expert. The government has asked for your advice on a proposed law specifying punishment of bystanders who fail to call for help. If the law is enforced, then with some probability a "do nothing" bystander will be forced to pay a fine.

(a) In words, explain how this law changes the game played between the bystanders. If you can, sketch and solve a version of the model incorporating the fine and the probability that a bystander is prosecuted.

(b) Compare the following two rules: (1) a bystander can be prosecuted if he or she fails to call, regardless of whether another person calls; and (2) a bystander can be prosecuted for not calling only if no one else calls (in which case the pedestrian's injury is not treated, resulting in lifelong health problems).

(c) What kind of bystander law would you recommend? Explain your criterion (efficiency?).

5. Which is more critical to effective external enforcement—breach or verifiability? Explain.

6. Which remedy is more likely to achieve efficiency—expectation damages or restitution damages? Explain.

7. The technologies of interaction for two different contractual relationships are given by matrices A and B.

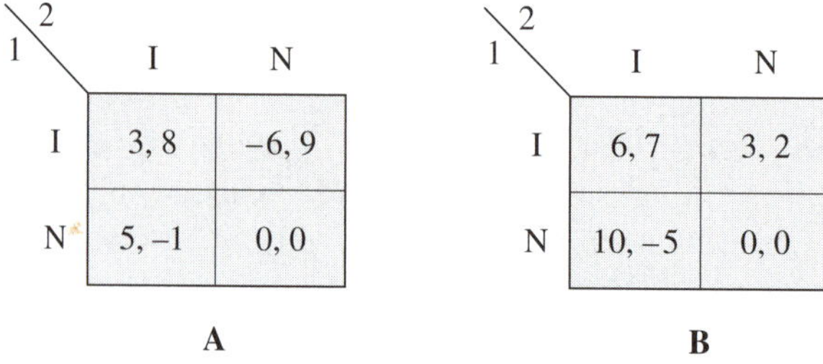

Answer the following questions for both technology A and technology B.

(a) Suppose the players can obtain externally enforced transfers by writing a contract. Under the assumption that the court can verify exactly the outcome of production, what value can the players obtain and what contract is written? Describe both the externally and self-enforced components.

(b) Repeat part (a) under the assumption that the court can only verify whether (I, I) is played—that is, it cannot distinguish between (I, N), (N, I), and (N, N).

(c) Explain or calculate the outcome for the setting of court-imposed breach remedies. Separately study expectation, reliance, and restitution. Assume (0, 0) is the benchmark, nonrelationship payoff for reliance and restitution.

8. Suppose a manager (player 1) and a worker (player 2) have a contractual relationship with the following technology of interaction. Simultaneously and independently, the two parties each select either low (L) or high (H) effort. A party that selects high effort suffers a disutility. The worker's disutility of high effort (measured in dollars) is 2, whereas the manager's disutility of high effort is 3. The effort choices yield revenue to the manager, as follows: If both choose L, then the revenue is zero. If one party chooses L while the other selects H, then the revenue is 3. Finally, if both parties select H, then the revenue is 7. The worker's payoff is zero minus his disutility of effort (zero if he exerted low effort), while the manager's payoff is her revenue minus her disutility of effort.

(a) Draw the normal-form matrix that represents the underlying game.

(b) If there is no external enforcement, what would be the outcome of this contractual relationship?

(c) Suppose there is external enforcement of transfers, but that there is limited verifiability in that the court can only observe the revenue generated by the parties. Explain how this determines which cells of the matrix the court can distinguish between.

(d) Continuing to assume limited verifiability, what is the best outcome that the parties can achieve and what contract will they write to achieve it? (Assume maximization of joint value.)

(e) Now assume that there is full verifiability. What outcome will be reached and what contract will be written?

9. Suppose that Shtinki Corporation operates a chemical plant, which is located on the Hudson River. Downstream from the chemical plant are a group of fisheries. The Shtinki plant emits some byproducts that pollute the river, causing harm to the fisheries. The profit Shtinki obtains from operating the chemical plant is a positive number X. The harm inflicted on the fisheries due to water pollution is measured to be Y in terms of lost profits. If the Shtinki plant is shut down, then Shtinki loses X while the fisheries gain Y. Suppose that the fisheries collectively sue Shtinki Corporation. It is easily verified in court that Shtinki's plant pollutes the river. However, the values X and Y cannot be verified by the court, although they are commonly known to the litigants.

Suppose that the court requires the fisheries' attorney (player 1) and the Shtinki attorney (player 2) to play the following litigation game. Player 1 is supposed to announce a number y, which the court interprets as the fisheries' claim about their profit loss Y. Player 2 is to announce a number x, which the court interprets as a claim about X. The announcements are made simultaneously and independently. Then the court uses *Posner's nuisance rule* to make its decision.[9] According to the rule, if $y > x$, then Shtinki must shut down its chemical plant. If $x \geq y$, then the court allows Shtinki to operate the plant, but the court also requires Shtinki to pay the fisheries the amount y. Note that the court cannot force the attorneys to tell the truth. Assume the attorneys want to maximize the profits of their clients.

(a) Represent this game in the normal form by describing the strategy spaces and payoff functions.

(b) For the case in which $X > Y$, compute the Nash equilibria of the litigation game.

(c) For the case in which $X < Y$, compute the Nash equilibria of the litigation game.

(d) Is the court's rule efficient?

10. Discuss a real-world example of a contractual situation with limited verifiability. How do the parties deal with this contractual imperfection?

---

[9] See R. Posner, *Economic Analysis of Law,* 5th ed. (Boston: Little, Brown, 1997). The exercise here is from I. Kim and J. Kim, "Efficiency of Posner's Nuisance Rule: A Reconsideration," *Journal of Institutional and Theoretical Economics* 160(2004):327–333.

Analyzing Behavior in Dynamic Settings

Rationalizability and Nash equilibrium are based on the normal-form representation of a game. As demonstrated in Part II of this book, these concepts generate useful and precise conclusions regarding how people behave in many applied settings. In this part, I focus on extensive-form games and I show how features of the extensive form suggest *refined* versions of the rationalizability and Nash equilibrium concepts.[1] I also embark on a detailed analysis of several classes of games that command the interest of economists and researchers in other disciplines: models of competition between firms, employment relations, negotiation, reputation, and parlor games. The analysis pays close attention to the information that players have at different points in a game. In addition, it considers how players should look ahead to evaluate how others will respond to their actions. It is an important lesson to think about the consequences of your actions before making a choice.

[1]Note that there is no difficulty in applying the standard forms of rationalizability and Nash equilibrium to games in the extensive form, because every extensive form generates a single normal form (to which we apply the concepts). But, as you will see, some straightforward refinements present themselves vividly when we consider the extensive form.

14 DETAILS OF THE EXTENSIVE FORM

Before proceeding to the analysis of extensive-form games, I will elaborate on the technical aspects of the extensive-form representation.[1] As I described in Chapter 2, an extensive-form game is defined by a tree that has been properly constructed and labeled. Those who love to flaunt technical jargon will appreciate knowing that, in formal mathematical terms, a tree is a "directed graph." I do not think there is any reason to flaunt technical jargon.

Recall that trees consist of *nodes* connected by *branches*. Each branch is an arrow, pointing from one node to another. Thus, you can start from a given node and trace through the tree by following arrows. Nodes that can be reached in this manner are called *successors* of the node at which you start. Branches from a given node point to its *immediate successors*. Analogously, by tracing backward through the tree from a given node, you can define a node's *predecessors* and *immediate predecessor*. In Figure 14.1 on the next page, nodes b and c are successors of node a. Node b is an immediate successor of node a and the immediate predecessor of node c. Obviously, a given node x is a successor of node y if and only if y is a predecessor of x. Also note that, for nodes x, y, and z, if x is a predecessor of y and y is a predecessor of z, then it must be that x is a predecessor of z. That is, trees have a "transitive precedence relation."

A tree starts with the initial node and ends at terminal nodes (from which no arrows extend). We would like the initial node to clearly designate the "beginning" of the tree, and so we impose the following rule on trees:

> **Tree Rule 1** *Every node is a successor of the initial node, and the initial node is the only one with this property.*

A *path* through the tree is a sequence of nodes that (1) starts with the initial node, (2) ends with a terminal node, and (3) has the property that successive nodes in the sequence are immediate successors of each other. Each path is one way of tracing through the tree by following branches. For example, in Figure 14.1, one path starts at the initial node, goes to node a and then node b,

[1] If you want to get really technical, read the formal definition of an extensive-form game in D. M. Kreps and R. Wilson, "Sequential Equilibria," *Econometrica* 50(1982):863–894.

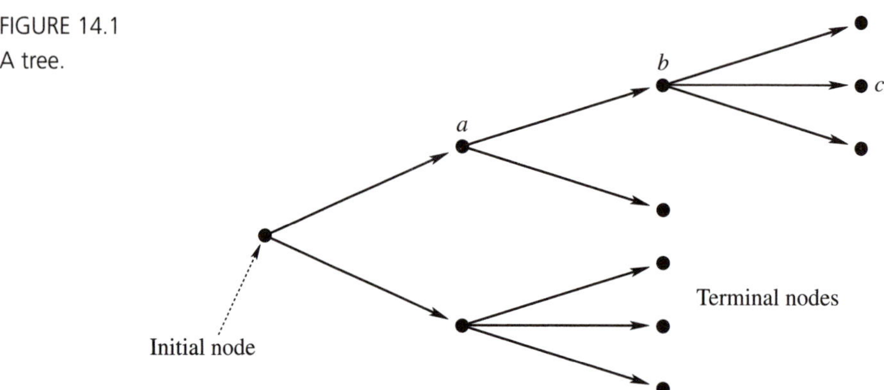

FIGURE 14.1
A tree.

and ends at node c. We require that terminal nodes exist and that each terminal node completely and uniquely describes a path through the tree. Thus, no two paths should cross. The following rule captures this requirement.

> **Tree Rule 2** *Each node except the initial node has exactly one immediate predecessor. The initial node has no predecessors.*

To check your understanding of paths and precedence, I recommend drawing some trees to verify that paths do not cross as long as Tree Rule 2 holds.[2]

Remember that branches represent actions that the players can take at decision nodes in the game. We label each branch in a tree with the name of the action that it represents.

> **Tree Rule 3** *Multiple branches extending from the same node have different action labels.*

Information sets represent places where players have to make decisions. Formally, an information set is a *set of nodes between which a player cannot distinguish when making a decision.* The concept of an information set adds the following two rules for trees.

> **Tree Rule 4** *Each information set contains decision nodes for only one of the players.*

> **Tree Rule 5** *All nodes in a given information set must have the same number of immediate successors and they must have the same set of action labels on the branches leading to these successors.*

---

[2]"Family trees" do not satisfy Tree Rule 2, because humans reproduce sexually. In a family tree, one normally associates a node with a single individual, and so everyone has two immediate predecessors (a mother and a father). Of course, a family tree is not a game. But families play games and you know "the family that plays together stays together."

FIGURE 14.2 Crazy information sets.

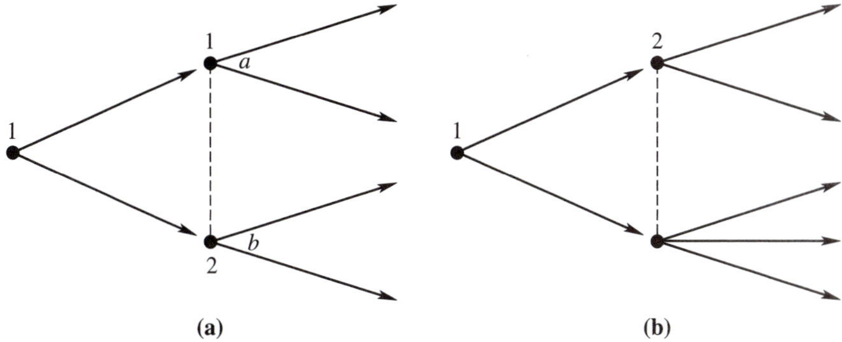

Regarding Tree Rule 4, it does not make sense to have, say, an information set containing nodes *a* and *b* if *a* is a decision node for player 1 and *b* is a decision node for player 2, as shown in Figure 14.2(a). It would imply that, at some point in the game, the players do not know who is to make a decision. The intuition behind Tree Rule 5 is that, if a player has a different number of available actions at nodes *a* and *b*, then he must be able to distinguish between these nodes (that is, they are not in the same information set). A tree that violates Tree Rule 5 is pictured in Figure 14.2(b).

In addition to the tree rules, it is generally reasonable to assume that players remember their own past actions as well as any other events that they have observed. A game satisfying this assumption is said to exhibit *perfect recall.* Figure 14.3 depicts a setting of *im*perfect recall. In this example, player 1 first chooses between U and D and then chooses between X and Y. That nodes *a* and *b* are contained in the same information set means that player 1 has forgotten whether he chose U or D when he makes the second choice. Although there are real-world examples of imperfect recall (Where did I park my car?), game theory focuses on models with perfect recall. Some game theorists have studied imperfect recall, but they seem to have misplaced their notes.

FIGURE 14.3
Imperfect recall.

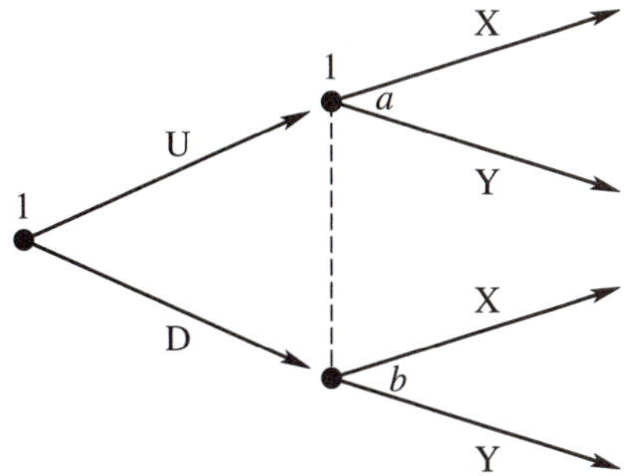

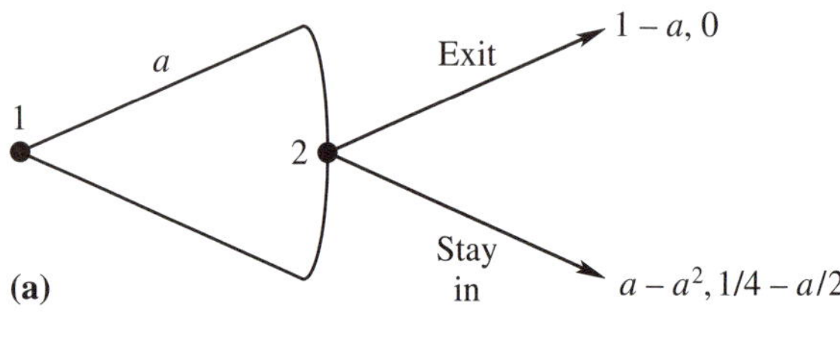

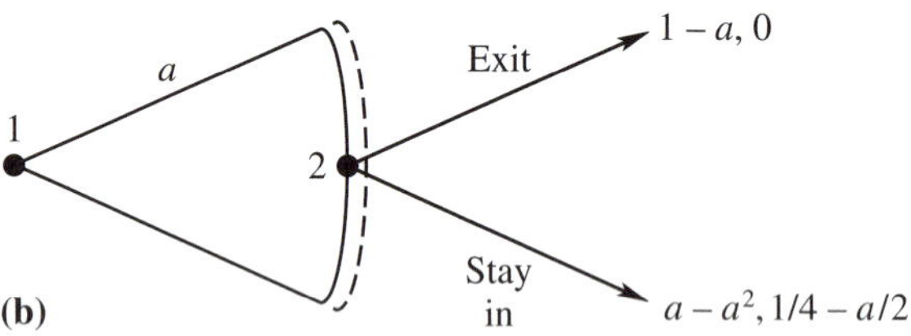

FIGURE 14.4
Advertising/exit.

In summary, an extensive-form game consists of a properly labeled tree obeying rules 1 through 5. If every information set of the extensive form is a single node—if there are no dashed lines in the picture—then we say that the game is of *perfect information;* otherwise, the game is of *imperfect information.* In a game of imperfect information, there is at least one contingency in which the player on the move does not know exactly where he is in the tree.

One issue that has not yet been addressed herein is how to model random events not under the control of the players, such as whether it rains when the Super Bowl is played or whether there is a drought in a particular growing season. Such random events ("moves of nature") require analysis that is best put off until later, which in this book means discussion in Part IV.

On the subject of drawing game trees, I have one more point to make. As you have seen, in some games a player can choose from an infinite number of actions. For example, consider a simple market game between two firms in which firm 1 first decides how much to spend on advertising and then firm 2, after observing firm 1's choice, decides whether to exit or stay in the market. Suppose that firm 1 may choose any advertising level between zero and one (million dollars). This firm may choose advertising levels of .5, .3871, and so forth. Because there are an infinite number of possible actions for firm 1, we cannot draw a branch for each action. One way of representing firm 1's potential choices, though, is to draw two branches from firm 1's decision node, designating advertising levels of zero and one, as in Figure 14.4(a). These branches are connected by an arc, which shows that firm 1 can select any number between zero and one. We label this graphical configuration with a

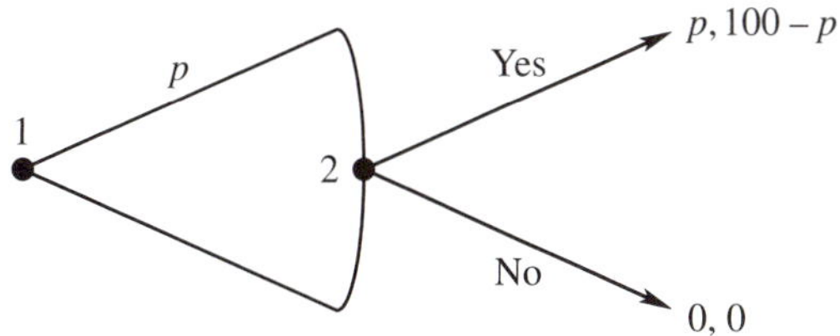

FIGURE 14.5
Ultimatum bargaining.

variable (in this case, the letter a), which stands for firm 1's action. In the interior of the arc, we draw a generic node that continues the tree. Note that, in the game illustrated in Figure 14.4(a), the payoffs of the players depend on a. In another version of the game, player 2 does not observe the advertising level of firm 1, as depicted in Figure 14.4(b).

I conclude this chapter with a simple model of bargaining, which has played a significant role in economic analysis. Economists spend a lot of time studying how people trade goods. In some markets, goods are traded through posted prices and with very little negotiation. In other markets, buyers and sellers bargain over the price at which they will trade. Negotiation is a part of buying homes, automobiles, souvenirs in the shops of tourist destinations, merchandise at swap meets, and in many other markets. As a rule, regardless of what is for sale, one can always bargain over the price. Because bargaining is such a widespread means of establishing the terms of trade, economists, business people, and everyday consumers seek to understand how negotiation takes place and how to master its art.

Imagine that person 1 wishes to sell a painting to person 2. Person 1, the seller, has no interest in the painting; it is worth nothing to her. On the other hand, the painting is worth \$100 to the buyer. Thus, if the painting is sold for any price between 0 and \$100, both parties will be better off than if the painting is not sold. There are many procedures by which bargaining over the price can take place. I will later try to help you develop an understanding through some simple models. For now, consider the simplest model of negotiation, called *ultimatum bargaining*.

Suppose that, if the parties do not reach an agreement over the price quickly, then the opportunity for trade disappears. In fact, there is only time for one take-it-or-leave-it offer by one of the parties. For the sake of argument, assume that the seller can make the offer. She states a price at which she is willing to sell the painting, and the buyer then either accepts or rejects it. If the buyer rejects the price, then he and the seller must part without making a trade. If the buyer accepts the price, then they trade the painting at this price. This simple game is represented in Figure 14.5. Note the payoffs. If the painting is not traded, both parties obtain nothing. If the painting is traded, then the seller obtains the price and the buyer gets the value of the trade to her (the difference between what the painting is worth to her and the price).

GUIDED EXERCISE

Problem: Provide an example to show that more than one strategy profile can correspond to a single path through an extensive-form tree.

Solution: In the following extensive-form game, the strategy profiles (A, ac) and (A, ad) induce the same path.

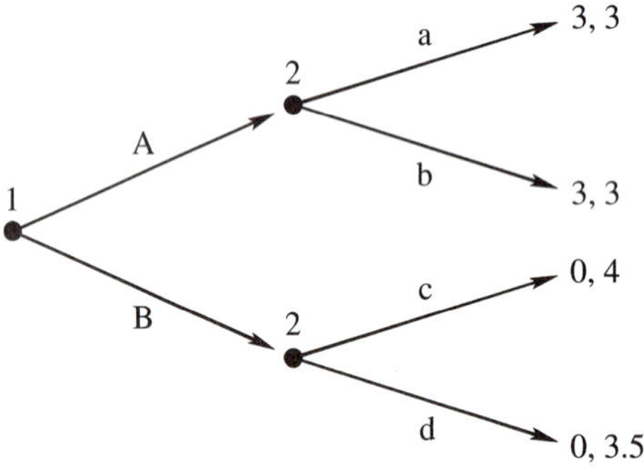

EXERCISES

1. Do normal-form games exhibit perfect or imperfect information? Explain.

2. Given an extensive-form game, prove that each pure-strategy profile induces a unique path through the tree.

3. Consider the following two-player game: First, player 1 selects a number x, which must be greater than or equal to zero. Player 2 observes x. Then, simultaneously and independently, player 1 selects a number y_1 and player 2 selects a number y_2, at which point the game ends. Player 1's payoff is $u_1 = y_1 y_2 + x y_1 - y_1^2 - (x^3/3)$ and player 2's payoff is $u_2 = -(y_1 - y_2)^2$. Represent this game in the extensive form.

4. Draw the extensive form of the following game, a version of the *Cournot duopoly game* (analyzed in Chapter 10). Two firms compete in a homogeneous good market, where the firms produce exactly the same good. The firms simultaneously and independently select quantities to produce. The quantity selected by firm i is denoted q_i and must be greater than or equal to zero, for $i = 1, 2$.

The market price is given by $p = 2 - q_1 - q_2$. For simplicity, assume that the cost to firm i of producing any quantity is zero. Further, assume that each firm's payoff is defined as its profit. That is, firm i's payoff is $(2 - q_i - q_j)q_i$, where j denotes firm i's opponent in the game.

5. Consider a variation of the Cournot duopoly game in which the firms move sequentially rather than simultaneously. Suppose that firm 1 selects its quantity first. After observing firm 1's selection, firm 2 chooses its quantity. This is called the *von Stackelberg duopoly* model (referred to previously in Exercise 6 of Chapter 3). Draw the extensive form of this game.

15 BACKWARD INDUCTION AND SUBGAME PERFECTION

To this point, our analysis of rational behavior has centered on the normal form—where we examine strategies and payoff functions. Although the normal-form concepts are valid for *every* game (because an extensive form can be translated into the normal form), they are most convincing for games in which all of the players' decisions are made simultaneously and independently. I want you to now consider games with interesting dynamic properties, where the extensive form more precisely captures the order of moves and the information structure.

To see that there are some interesting issues regarding comparison of the normal and extensive forms, study the game pictured in Figure 15.1. There are two firms in this market game: a potential entrant (player 1) and an incumbent (player 2). The entrant decides whether or not to enter the incumbent's industry. If the entrant stays out, then the incumbent gets a large profit and the entrant gets zero. If the potential entrant does enter the industry, then the incumbent must choose whether or not to engage in a price war. If the firm triggers a price war, then both firms suffer. If the incumbent accommodates the entrant, then they both obtain modest profits.

Pictured in Figure 15.1 are both the extensive- and the normal-form representations of this market game. A quick look at the normal form will reveal that the game has two Nash equilibria (in pure strategies): (I, A) and (O, P). The former equilibrium should not strike you as controversial. The latter equi-

FIGURE 15.1 Entry and predation.

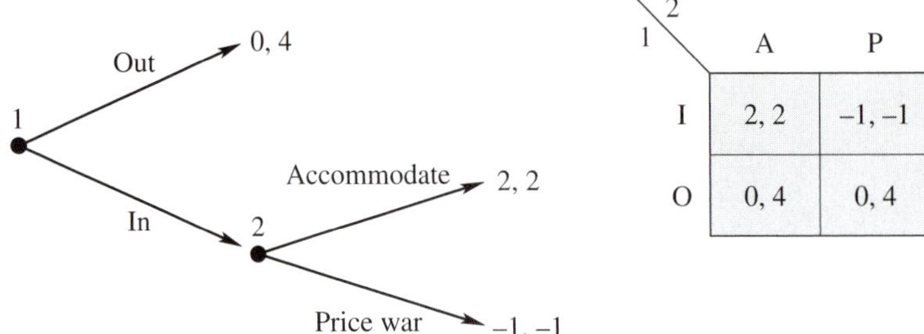

1 \ 2	A	P
I	2, 2	–1, –1
O	0, 4	0, 4

librium calls for an explanation. With the extensive form in mind, equilibrium (O, P) has the following interpretation. Suppose the players do not select their strategies in real time (as the game progresses), but rather choose their strategies in advance of playing the game. Perhaps they write their strategies on slips of paper, as instructions to their managers regarding how to behave. The entrant prefers not to enter if she expects the incumbent to initiate a price war. That is, she responds to the threat that player 2 will trigger a price war in the event that player 1 enters the industry. For his part, the incumbent has no incentive to deviate from the price-war strategy when he is convinced that the entrant will stay out. Note that the incumbent's thought process here consists of an *ex ante* evaluation; that is, the incumbent is willing to *plan* on playing the price-war strategy when he believes that his decision node will not be reached.

Is the (O, P) equilibrium really plausible? Many theorists think not, because games are rarely played by scribbling strategies on paper and ignoring the real-time dimension. In reality, player 2's threat to initiate a price war is incredible. Even if player 2 plans on adopting this strategy, would he actually carry it out in the event that his decision node is reached? He would probably step back for a moment and say, "Whoa, Penelope! I didn't think this information set would be reached. But given that this decision really counts now, perhaps a price war is not such a good idea after all." The incumbent will therefore accommodate the entrant. Furthermore, the entrant ought to think ahead to this eventuality. Even though the incumbent may "talk tough," claiming that he will initiate a price war if entry occurs, the entrant understands that such behavior would not be rational when the decision has to be made. Thus, the potential entrant can enter the industry with confidence, knowing that the incumbent will capitulate.[1] There are many examples of games such as these, where an equilibrium specifies an action that is irrational, conditional on a particular information set being reached. Not all of the examples concern incredible threats. You might flip through preceding chapters and search for such games.

SEQUENTIAL RATIONALITY AND BACKWARD INDUCTION

As the example in Figure 15.1 indicates, more should be assumed about players than that they select best responses *ex ante* (from the point of view of the

[1] Children sometimes play games like these with their parents. Parents may threaten to withhold a trip to the ice-cream parlor, unless a child eats his vegetables. But the parents may have an independent interest in going to the ice-cream parlor and, if they go, then they will have to take the child with them, lest he be left at home alone. They can punish the child by skipping sundaes, but they would also be punishing themselves in the process. A shrewd child knows when his parents do not have the incentive to carry out punishments; such children grow up to be economists or attorneys. They also feed peas to their dogs from time to time.

beginning of the game). Players ought to demonstrate rationality whenever they are called on to make decisions. This is called *sequential rationality*.

> **Sequential rationality:** An optimal strategy for a player should maximize his or her expected payoff, conditional on every information set at which this player has the move. That is, player i's strategy should specify an optimal action from each of player i's information sets, even those that player i does not believe (*ex ante*) will be reached in the game.

If sequential rationality is common knowledge between the players (at every information set), then each player will "look ahead" to consider what players will do in the future in response to her move at a particular information set.

To determine the strategies that are consistent with the common knowledge of sequential rationality, we evaluate whether each strategy is *conditionally dominated*. A strategy s_i for player i is **conditionally dominated** if, contingent on reaching some information set of player i, there is another strategy σ_i that strictly dominates it. One can remove conditionally dominated strategies by using an iterated procedure that is roughly similar to the procedure for normal-form games. However, a complete description of the procedure requires more analysis than is appropriate herein.[2] It is more worthwhile to focus on a simpler version of dominance known as *backward induction*.

> **Backward induction:** This is the process of analyzing a game from back to front (from information sets at the end of the tree to informations sets at the beginning). At each information set, one strikes from consideration actions that are dominated, given the terminal nodes that can be reached.

To determine the optimal action at an information set, one must look ahead in the game. Thus, the easiest information sets at which to evaluate rationality are those at the end of the tree—that is, those whose nodes are immediate predecessors of terminal nodes. When the optimal actions have been determined for these information sets, one can move to immediate predecessors and continue the process until the initial node is reached.

Backward induction can be tricky for games of imperfect information, where it is best to use the full-blown conditional-dominance concept that is not introduced here. (Such games are addressed later in this chapter in the context of equilibrium.) For now, consider the class of extensive-form games with perfect information. Each game in this class can be easily analyzed using backward induction. If there are no ties in the payoffs—where two or more

[2]Conditional dominance is defined and studied in M. Shimoji and J. Watson, "Conditional Dominance, Rationalizability, and Game Forms," *Journal of Economic Theory* 83(1998):161–195.

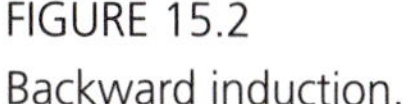

FIGURE 15.2
Backward induction.

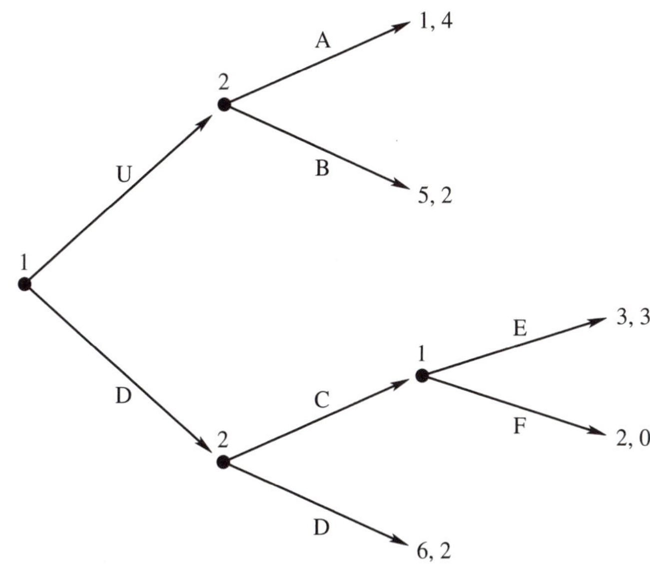

terminal nodes lead to the same payoff for one of the players—then backward induction completely solves the game. That is, backward induction identifies a single rational strategy profile for the players.

For a demonstration of backward induction, examine the game in Figure 15.2. There are two nodes at which player 1 makes a decision. At the second node, player 1 decides between E and F. On reaching this node, player 1's only rational choice is E. We can therefore cross out F as a possibility. Player 2 knows this and, therefore, in her lower decision node she ought to select C (which she knows will eventually yield the payoff 3 for her). We thus cross out action D for player 2. Furthermore, A is the best choice at player 2's upper decision node, so we cross out B. Finally, we can evaluate the initial node, where player 1 has the choice between U and D. He knows that, if he chooses U, then player 2 will select A and he will get a payoff of 1. If he chooses D, then player 2 will select C, after which player 1 will select E, yielding a payoff of 3. Player 1's optimal action at the initial node is therefore D; action U should be crossed out.

Using the process of backward induction in this example, we have identified a single sequentially rational strategy profile: (DE, AC). In fact, it is also easy to check that (DE, AC) is a Nash equilibrium. Strategy DE is a best response for player 1 to player 2's strategy AC; further, AC is a best response to DE for player 2. In general, backward-induction outcomes are Nash equilibria. To see this, just notice that the property of sequential rationality implies that no player has an incentive to deviate from his prescribed action at any information set. Further, because every finite game with perfect information can be solved by backward induction, every such game has a Nash equilibrium.

> **Result:** Every finite game with perfect information has a pure-strategy Nash equilibrium. Backward induction identifies an equilibrium.[3]

If there are ties in the payoffs, then there may be more than one such equilibrium and there may be more than one sequentially rational strategy profile.

SUBGAME PERFECTION

The concept of backward induction can be expanded to cover general extensive-form games. One way of doing this is to think of a sequential version of rationalizability, where players must play best responses to their beliefs at all information sets. Such a notion has been developed and is related to the conditional-dominance concept, but, as with conditional dominance, it is a bit too technical to present here.[4] Instead, I focus on equilibrium and define a refinement of Nash's concept that incorporates sequential rationality. Versions of this kind of refinement are identified by the term "perfection."

The concept of *subgame perfection* starts with a description of a subgame.

> Given an extensive-form game, a node x in the tree is said to *initiate a subgame* if neither x nor any of its successors are in an information set that contains nodes that are not successors of x. A **subgame** is the tree structure defined by such a node x and its successors.

This definition may seem complicated on first sight, but it is not difficult to understand. Take any node x in an extensive form. Examine the collection of nodes given by x and its successors. If there is a node y that is not a successor of x but is connected to x or one of its successors by a dashed line, then x does not initiate a subgame. In other words, once players are inside a subgame, it is common knowledge between them that they are inside it. Subgames are self-contained extensive forms—meaningful trees on their own. Subgames that start from nodes other than the initial node are called *proper subgames*. Observe that, in a game of perfect information, *every* node initiates a subgame.

[3]H. W. Kuhn, "Extensive Games and the Problem of Information," in *Contributions to the Theory of Games,* vol. II (*Annals of Mathematics Studies*, 28), ed. H. W. Kuhn and A. W. Tucker (Princeton, NJ: Princeton University Press, 1953), pp. 193–216.

[4]Extensive-form rationalizability was introduced in D. Pearce, "Rationalizable Strategic Behavior and the Problem of Perfection," *Econometrica* 52(1984):1029–1050. It was clarified by P. Battigalli, "On Rationalizability in Extensive Games," *Journal of Economic Theory* 74(1997):40–61. The relation to conditional dominance is reported in M. Shimoji and J. Watson, "Conditional Dominance, Rationalizability, and Game Forms," *Journal of Economic Theory* 83(1998):161–195.

FIGURE 15.3
Subgames.

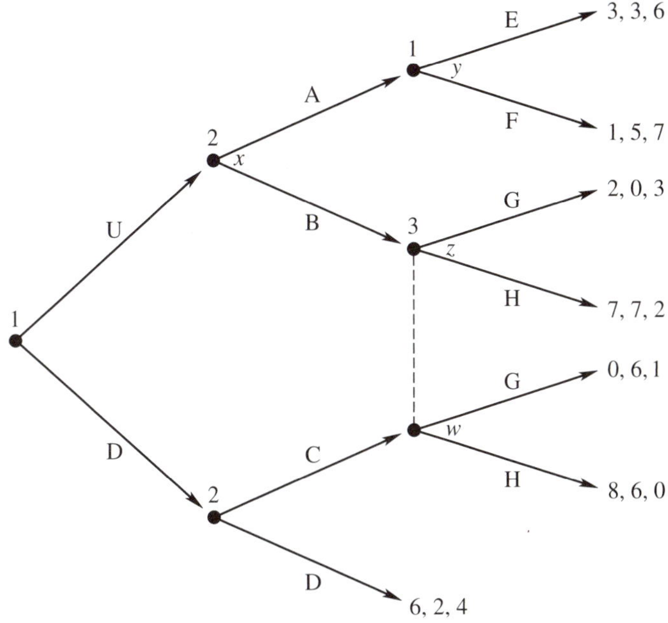

To test your comprehension, examine the game in Figure 15.3. Note that this game is a three-player game. The initial node initiates a subgame, because the entire game is itself a subgame. Node y also initiates a subgame, because y is not tied to any other nodes by a dashed line and all of y's successors are terminal nodes. Node x, on the other hand, does not initiate a subgame, because one of its successors (node z) is in the same information set as a node that is not one of its successors (node w). In other words, starting from node x, there is a contingency in which player 3 does not know whether he is at a successor of x. By the same reasoning, neither z nor w initiates a subgame. Thus, this extensive form has two subgames: (1) the entire game, and (2) the proper subgame starting at node y.

Here is the version of Nash equilibrium that incorporates sequential rationality by evaluating subgames individually.

> A strategy profile is called a **subgame perfect Nash equilibrium** if it specifies a Nash equilibrium in every subgame of the original game.[5]

[5]In 1994, Reinhard Selten was awarded the Nobel Prize in economics for his work on this and another "perfection" refinement of Nash equilibrium. Corecipients of the prize were John Nash and John Harsanyi. Harsanyi studied games with incomplete information, which is the topic of Part IV of this book. Selten's contributions originated with "Spieltheoretische Behandlung eines Oligopolmodells mit Nachfragetragheit," *Zeitschrift fur die gesamte Staatswissenschaft* 121(1965):301–324, 667–689, and "Reexamination of the Perfectness Concept for Equilibrium Points in Extensive Games," *International Journal of Game Theory* 4(1975):25–55.

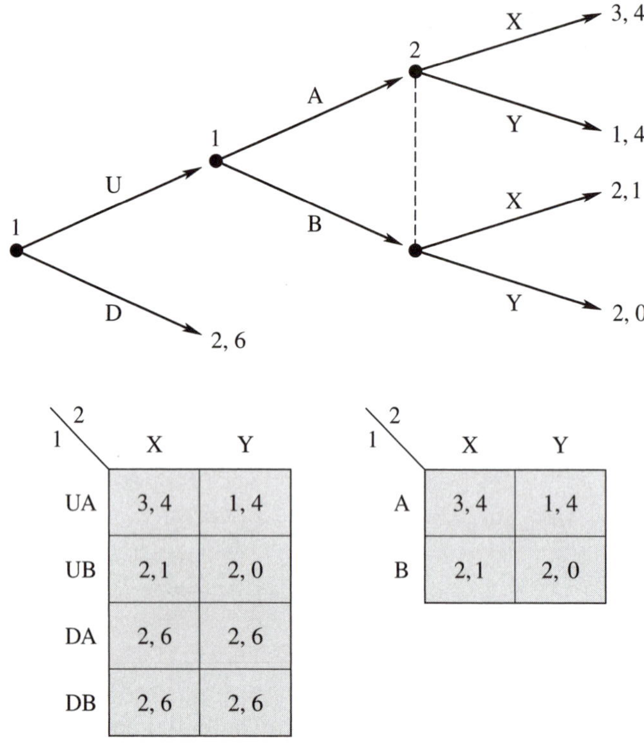

1 \ 2	X	Y
UA	3, 4	1, 4
UB	2, 1	2, 0
DA	2, 6	2, 6
DB	2, 6	2, 6

1 \ 2	X	Y
A	3, 4	1, 4
B	2, 1	2, 0

FIGURE 15.4
Subgame perfection.

The basic idea behind subgame perfection is that a solution concept should be consistent with its own application from anywhere in the game where it can be applied. Because Nash equilibrium can be applied to well-defined extensive-form games, subgame perfect Nash equilibrium requires the Nash condition to hold on all subgames.

To understand how to work with the concept of subgame perfection, consider the game pictured in Figure 15.4. Note first that this game has one proper subgame, which starts at the node reached when player 1 plays U at the initial node. Thus, there are two subgames to evaluate: the proper subgame as well as the entire game. Because strategy profiles tell us what the players do at every information set, each strategy profile specifies behavior in the proper subgame even if this subgame would not be reached. For example, consider strategy profile (DA, X). If this profile is played, then play never enters the proper subgame. But (DA, X) does include a specification of what the players *would* do conditional on reaching the proper subgame; in particular, it prescribes action A for player 1 and action X for player 2.

Continuing with the example, observe that two normal-form games appear next to the extensive form in Figure 15.4. The first is the normal-form version of the entire extensive form, whereas the second is the normal-form version of the proper subgame. The former reveals the Nash equilibria of the entire game; we examine the latter to find the Nash equilibria of the proper subgame.

In the entire game, (UA, X), (DA, Y), and (DB, Y) are the Nash equilibria. But not all of these strategy profiles are subgame perfect. To see this, observe that the proper subgame has only one Nash equilibrium, (A, X). The strategies (DA, Y) and (DB, Y) are not subgame perfect, because they do not specify Nash equilibria in the proper subgame. In particular, (DA, Y) stipulates that (A, Y) be played in the proper subgame, yet (A, Y) is not an equilibrium. Furthermore, (DB, Y) specifies that (B, Y) be played in the subgame, yet (B, Y) is not an equilibrium. Thus, there is a single subgame perfect Nash equilibrium in this example: (UA, X).

The subgame perfection concept requires weak congruity at every subgame, meaning that, if any particular subgame is reached, then we can expect the players to follow through with the prescription of the strategy. This is an important consideration, whether or not a given equilibrium actually reaches the subgame in question. The example in Figure 15.4 illustrates this point vividly. Nash equilibrium (DA, Y) entails a kind of threat that the players jointly make to give player 1 the incentive to select D at his first information set. But the threat is not legitimate. In the proper subgame, (A, Y) is about as stable as potassium in water. Player 1 has the strict incentive to deviate from this prescription.[6]

Several facts are worth noting at this point. First, and most obvious, recognize that a subgame perfect Nash equilibrium is a Nash equilibrium because such a profile must specify a Nash equilibrium in every subgame, one of which is the entire game. We thus speak of subgame perfection as a *refinement* of Nash equilibrium. Second, for games of perfect information, backward induction yields subgame perfect equilibria. Third, on the matter of finding subgame perfect equilibria, the procedure just described is useful for many finite games. One should examine the matrices corresponding to all of the subgames and locate Nash equilibria. For infinite games and for mixed equilibria, it is often easier to start with the subgames that are toward the end of the extensive form, with the hope that these subgames have unique Nash equilibria. Then one can work backward by embedding these equilibrium outcomes into the "larger"

[6]That player 2 is indifferent between (A, X) and (A, Y) in the proper subgame implies a very different conclusion using the notion of conditional dominance. In the proper subgame, although player 1's action B is stricken, neither of player 2's actions are dominated. Thus, player 1 could rationally play DA with the belief that player 2 selects strategy Y. The difference between subgame perfect Nash equilibrium and conditional dominance is therefore due to the fact that profile (A, Y) is rationalizable, yet it is not a Nash equilibrium, in the proper subgame.

subgames. This procedure is illustrated in the following guided exercise and, more generally, in Chapter 16.

GUIDED EXERCISE

Problem: Consider the ultimatum bargaining game described at the end of Chapter 14. In this game, the players negotiate over the price of a painting that player 1 can sell to player 2. Player 1 proposes a price p to player 2. Then, after observing player 1's offer, player 2 decides whether to accept it (yes) or reject it (no). If player 2 accepts the proposal, then player 1 obtains p and player 2 obtains $100 - p$. If player 2 rejects the proposal, then each player gets zero. Verify that there is a subgame perfect Nash equilibrium in which player 1 offers $p^* = 100$ and player 2 has the strategy of accepting any offer $p \leq 100$ and rejecting any offer $p > 100$.

Solution: The exercise here is to verify that a particular strategy profile is a subgame perfect equilibrium. We do *not* have to determine *all* of the subgame perfect equilibria. Note that there is an infinite number of information sets for player 2, one for each possible offer of player 1. Each one of these information sets consists of a single node and initiates a simple subgame in which player 2 says "yes" or "no" and the game ends. The suggested strategy for player 2, which accepts (says "yes") if and only if $p \leq 100$, implies an equilibrium in every such subgame. To see this, realize that player 2 obtains a payoff of $100 - p$ if he accepts, whereas he receives 0 if he rejects, so accepting is a best response in every subgame with $p \leq 100$; rejecting the offer is a best response in every subgame with $p \geq 100$. For $p = 100$, both accepting and rejecting are best responses; thus, player 2 is indifferent between accepting and rejecting when $p = 100$. We are assuming that, in this case of indifference, player 2 accepts.

Having verified that player 2 is using a sequentially rational strategy, which implies equilibrium in every proper subgame, let us turn to the entire ultimatum game (from the initial node) and player 1's incentives. Observe that the offer $p^* = 100$ is a best response for player 1 to the strategy of player 2. If player 1 offers this price then, given player 2's strategy, player 2 will accept it and player 1's payoff would be 100. On the other hand, if player 1 were to offer a higher price, then player 2 would reject it and player 1 would get a payoff of zero. Furthermore, if player 1 were to offer a lower price $p < 100$, then player 2 would accept it and player 1 would obtain a payoff of only p. Also note that, in the entire game, player 2's strategy is a best response to player 1's strategy of offering $p^* = 100$, because player 2 accepts this offer.

We have thus established that the suggested strategy profile induces a Nash equilibrium in all of the subgames, including the entire game, and so we have a subgame perfect Nash equilibrium. You will find further analysis of the ultimatum bargaining game in Chapter 19, where it is shown that the equilibrium we have just identified is, in fact, the unique subgame perfect equilibrium.

EXERCISES

1. Consider the following extensive-form games.

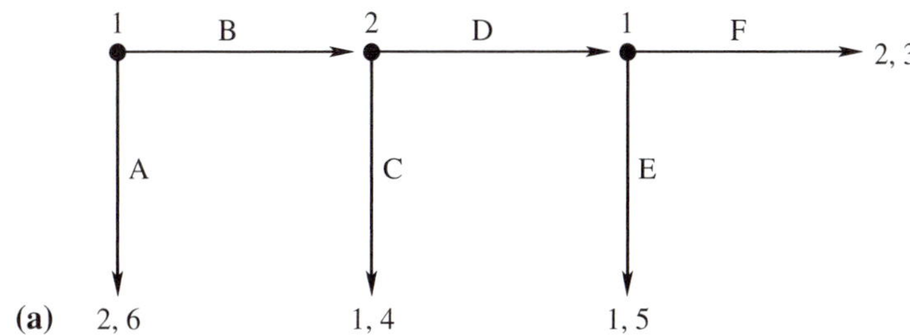

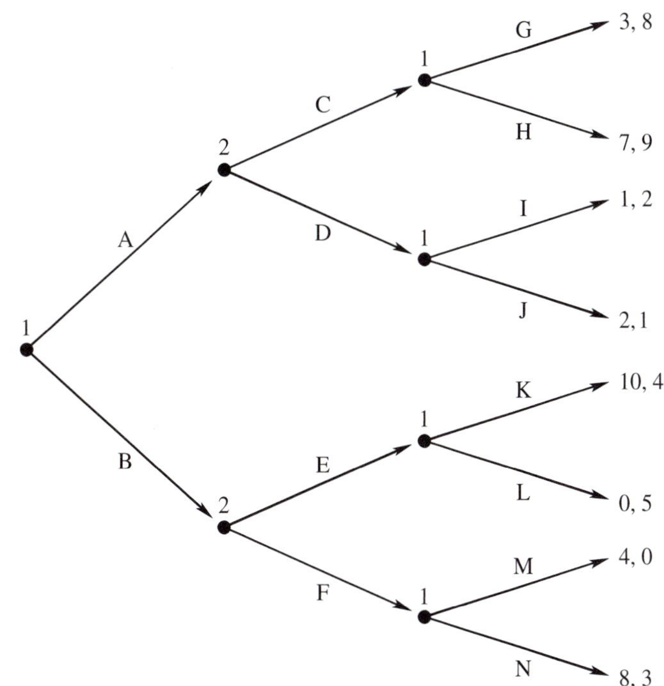

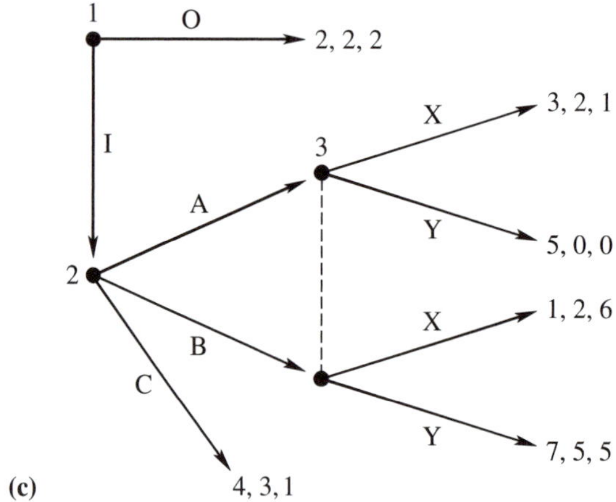

(c)

Solve the games by using backward induction.

2. Compute the Nash equilibria and subgame perfect equilibria for the following games. Do so by writing the normal-form matrices for each game and its subgames. Which Nash equilibria are not subgame perfect?

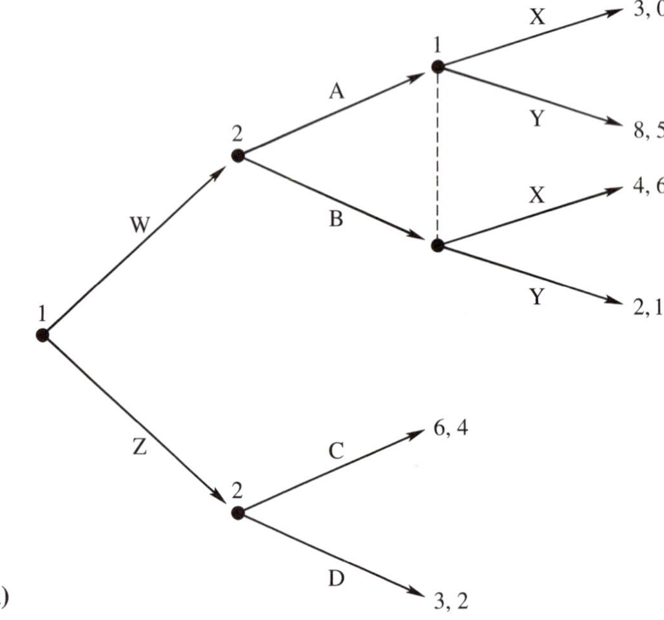

(a)

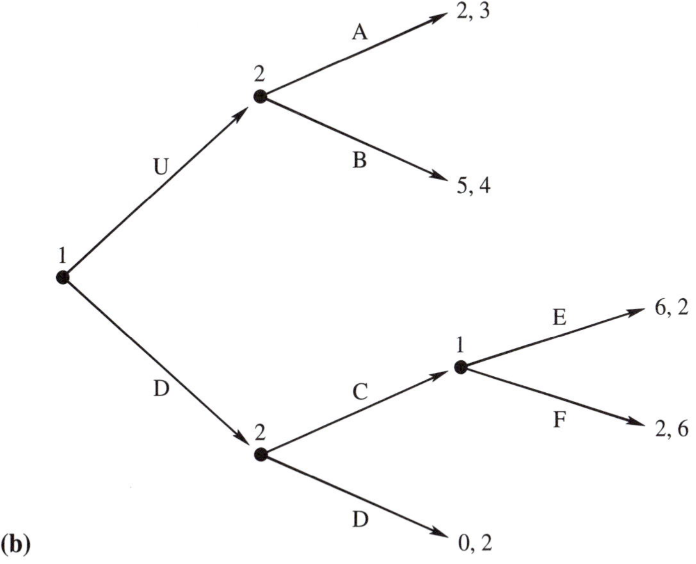

3. Consider the following game.

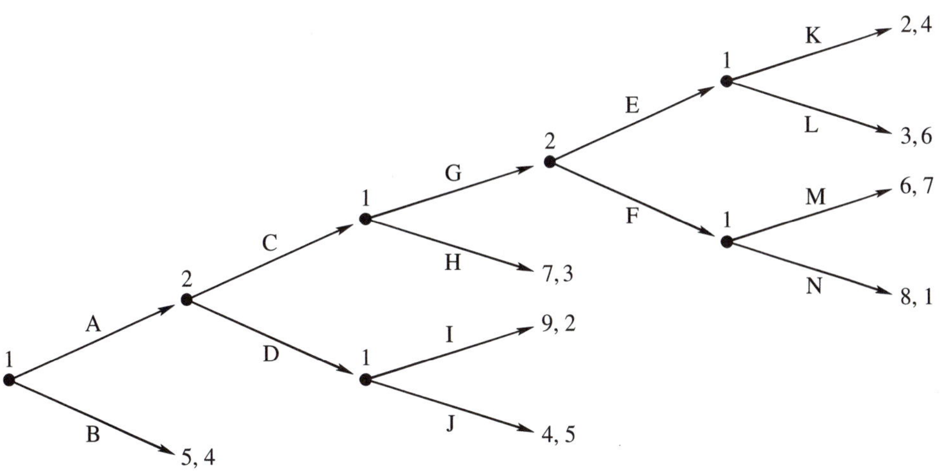

(a) Solve the game by backward induction and report the strategy profile that results.

(b) How many proper subgames does this game have?

4. Calculate and report the subgame perfect Nash equilibrium of the game described in Exercise 3 in Chapter 14.

5. In the Envelope Game, there are two players and two envelopes. One of the envelopes is marked "player 1," and the other is marked "player 2." At the beginning of the game, each envelope contains one dollar. Player 1 is given the choice between stopping the game and continuing. If he chooses to stop, then each player receives the money in his own envelope and the game ends. If player 1 chooses to continue, then a dollar is removed from his envelope and two dollars are added to player 2's envelope. Then player 2 must choose between stopping the game and continuing. If he stops, then the game ends and each player keeps the money in his own envelope. If player 2 elects to continue, then a dollar is removed from his envelope and two dollars are added to player 1's envelope. Play continues like this, alternating between the players, until either one of them decides to stop or k rounds of play have elapsed. If neither player chooses to stop by the end of the kth round, then both players obtain zero. Assume players want to maximize the amount of money they earn.

(a) Draw this game's extensive-form tree for $k = 5$.

(b) Use backward induction to find the subgame perfect equilibrium.

(c) Describe the backward induction outcome of this game for any finite integer k.

6. Consider a variation of the television station broadcast game of Exercise 4 in Chapter 7. Suppose the stations interact sequentially. First, MBC chooses between 6:00 and 7:00. Then, after observing MBC's choice, RBC decides between 6:00 and 7:00. Finally, after observing the behavior of both MBC and RBC, CBC chooses either 6:00 or 7:00. Payoffs are as given before. Draw the extensive form of this sequential game and compute the subgame perfect equilibrium. Is the outcome different from that in the simultaneous-play game? Explain.

7. Consider a game in which player 1 first selects between I and O. If player 1 selects O, then the game ends with the payoff vector $(x, 1)$ (x for player 1), where x is some positive number. If player 1 selects I, then this selection is revealed to player 2 and then the players play the battle-of-the-sexes game in which they simultaneously and independently choose between A and B. If they coordinate on A, then the payoff vector is $(3, 1)$. If they coordinate on B, then the payoff vector is $(1, 3)$. If they fail to coordinate, then the payoff vector is $(0, 0)$.

(a) Represent this game in the extensive and normal forms.

(b) Find the pure-strategy Nash equilibria of this game.

(c) Calculate the mixed-strategy Nash equilibria and note how they depend on x.

(d) Represent the proper subgame in the normal form and find its equilibria.

(e) What are the pure-strategy subgame perfect equilibria of the game? Can you find any Nash equilibria that are not subgame perfect?

(f) What are the mixed-strategy subgame perfect equilibria of the game?

8. Imagine a game in which players 1 and 2 simultaneously and independently select A or B. If they both select A, then the game ends and the payoff vector is $(5, 5)$. If they both select B, then the game ends with the payoff vector $(-1, -1)$. If one of the players chooses A while the other selects B, then the game continues and the players are required to simultaneously and independently select positive numbers. After these decisions, the game ends and each player receives the payoff $(x_1 + x_2)/(1 + x_1 + x_2)$, where x_1 is the positive number chosen by player 1 and x_2 is the positive number chosen by player 2.

(a) Describe the strategy spaces of the players.

(b) Compute the Nash equilibria of this game.

(c) Determine the subgame perfect equilibria.

16 TOPICS IN INDUSTRIAL ORGANIZATION

One of the first areas of economic research to be transformed by the game-theory revolution is the field of industrial organization—the study of market structure, firm behavior, and overall industry performance. Game theory provides a useful methodology for studying issues of market dynamics, such as how firms compete or collude over time and how a firm can attain a monopoly and exercise monopoly power. In this chapter, I sketch just a few examples of how game theory is used to study industrial organization.

ADVERTISING AND COMPETITION

To generate profit, firms must do more than just produce goods or services. They must also *market* their products to consumers or other firms. For example, when computer software company Columbus Research designs a new software package to handle inventory control, it cannot expect all of its potential customers to beat down the door in search of the software. Most potential customers will probably not even know about the software when it is first available. By advertising, Columbus Research can make customers aware of its new product, as well as tout its advantages over competing products of other firms.

Advertisements come in many forms and have different effects on demand and welfare. Some advertisements function to announce the availability of a new product; for example, a toothpaste producer may run a television advertisement to let potential customers know that it has added to its product line a new toothpaste that cleans teeth and freshens breath. An advertisement that highlights a product's advantages ("It freshens breath twice as well as did our older formula") is called a *positive advertisement*. A *negative advertisement* highlights the disadvantages of competing products ("The other major brand leaves your teeth feeling gritty"). Politicians often use negative advertisements, leaving voters gritting their teeth. An extreme form of negative advertisement attempts to make people feel bad unless they purchase a particular product. Sadly, this form is effective with the young ("You are ugly unless you wear our CoolDudeRapper tennis shoes") as well as the old ("If you are bald, then you are ugly, so you better try our hair-growth formula").

Firms advertise to increase the demand for their products. Sometimes the increased demand is achieved at the expense of competing firms, as is often the case with advertisements that accentuate the disadvantages of the competing products. For example, firm A may point out that firm B's product is prone to failure, causing the demand for A's product to increase and the demand for B's product to decrease. In other cases, advertisements by one firm can increase demand for *all* firms in the industry. For example, when a major producer of clothing (such as Levi's) advertises its line of jeans, it may increase consumers' general interest in jeans and boost the demand of all producers. Here is a simple model of strategic interaction when advertising enhances industry demand.

Consider an elaboration of the Cournot duopoly game in which firm 1 engages in advertising before the firms compete in the market. Firm 1 selects an advertising level a, which is a number greater than or equal to zero. Advertising has a positive effect on the demand for the good sold in the industry, enhancing the price that the consumers are willing to pay for the output of both firms. In particular, the market price is $p = a - q_1 - q_2$, where q_1 is the output of firm 1 and q_2 is the output of firm 2. After firm 1 selects a, it is observed by the other firm. Then the two simultaneously and independently select their production levels. Assume for simplicity that the firms produce at zero cost. However, firm 1 must pay an advertising cost of $2a^3/81$.

This game has an infinite number of proper subgames—namely, the Cournot duopoly game that is played after firm 1 chooses a (for each of the possible advertisement levels). Each subgame has an infinite action space, making it difficult to begin the analysis by searching for Nash equilibria of the full game. Because we know that the Cournot game has a unique Nash equilibrium and because this equilibrium must be a part of any subgame perfect equilibrium, it is easier to begin by identifying the equilibrium of each of the proper subgames. That is, we should study the game by working backward.

Suppose a subgame is reached following advertisement level a selected by firm 1. To find the Nash equilibrium of the subgame, compute the best-response functions of the players. Player 1's profit is

$$(a - q_1 - q_2)q_1 - \frac{2a^3}{81}.$$

Noting that a is a constant (having already been chosen by firm 1), we take the derivative of this payoff function with respect to q_1. To find this firm's best response to q_2, we set this derivative equal to zero and solve for q_1, yielding

$$q_1^* = BR_1(q_2) = \frac{a - q_2}{2}.$$

Likewise, firm 2's best-response function is

$$BR_2(q_1) = \frac{a - q_1}{2}.$$

Solving this system of equalities to find where they are both satisfied, we obtain $q_1 = q_2 = a/3$. The equilibrium price is $p = a/3$. Plugging these values into the firms' profit functions, we see that firm 1's profit as a function of a is

$$z_1(a) = \frac{a^2}{9} - \frac{2a^3}{81}.$$

Next, evaluate firm 1's choice of an advertisement level at the beginning of the game. Looking ahead, firm 1 knows that, by advertising at level a, it induces a subgame whose equilibrium yields a profit of $z_1(a)$. Firm 1 maximizes this payoff. The profit-maximizing choice of advertising level is found by taking the derivative of $z_1(a)$ and setting it equal to zero. That is, the optimal a satisfies

$$\frac{2a}{9} - \frac{6a^2}{81} = 0.$$

Solving for a, we obtain $a^* = 3$, which completes the analysis.

The strategy profile identified is $a^* = 3$, $q_1(a) = a/3$, and $q_2(a) = a/3$. Note that q_1 and q_2 are reported as functions of a. This is critically important because a strategy specifies what a player does at every information set at which he or she takes an action. There are an infinite number of information sets in this game. In particular, each player has a distinct information set for every possible value of a, and so a strategy must prescribe an action for each of them. Thus, $q_1(a) = a/3$ means: "At the information set following action a by firm 1, set the quantity equal to $a/3$." To report less than this function would be failing to completely specify the subgame perfect equilibrium.

Incidentally, it should be obvious that the strategy identified is a Nash equilibrium in the game. Because the strategy specifies a Nash equilibrium in the proper subgames, no player wishes to deviate in such a subgame. Furthermore, player 1 has no incentive to deviate at the beginning of the game, either by selecting a different level of advertising or by combining such a deviation with a different quantity choice in the Cournot phase of the game.

A MODEL OF LIMIT CAPACITY

In 1945, the Aluminum Company of America (Alcoa) dominated aluminum production in the United States, controlling 90 percent of the raw ingot market. As a result of this supremacy, one of the seminal antitrust cases of the post–World War II era, *United States* v. *Alcoa*, was initiated and considered by the Supreme Court. In his decision to break up the aluminum giant, Judge Learned Hand ruled that Alcoa was indeed guilty of anticompetitive practices. Central to Hand's argument was Alcoa's rapid accumulation of capacity for aluminum production, exceeding the levels that demand for its output seemingly warranted. This excess capacity, it was argued, was intended to thwart the entry efforts of Alcoa's potential competitors. In essence, Alcoa was sacrificing some profitability by overdeveloping its production facilities to maintain industrial dominance.[1]

A game-theoretic model demonstrates how excess capacity can limit entry in an industry.[2] Suppose two firms are considering whether and how to enter a new industry in which a specialized electronic component will be produced. Industry demand is given by the inverse demand function $p = 900 - q_1 - q_2$, where p is the market price, q_1 is the quantity produced by firm 1, and q_2 is the quantity produced by firm 2. To enter the industry, a firm must build a production facility. Two types of facility can be built: small and large. A small facility requires an investment of \$50,000, and it allows the firm to produce as many as 100 units of the good at zero marginal cost. Alternatively, the firm can pay \$175,000 to construct a large facility that will allow the firm to produce *any* number of units at zero marginal cost. A firm with a small production facility is called *capacity constrained*; a firm with a large facility is called *unconstrained*.

The firms make their entry decisions sequentially. First, firm 1 must choose among staying out of the industry, building a small facility, and building a large facility. Then, after observing firm 1's action, firm 2 must choose from the same alternatives. If only one of the firms is in the industry, then it selects a quantity and sells its product at the price dictated by market demand. If both firms are in the industry, then they compete by selecting quantities (as in the Cournot model). All output decisions are subject to capacity constraints in that a firm with a small production facility cannot produce more than 100 units.

[1]See *United States* v. *Aluminum Co. of America*, 148 F.2d 416 (2d Cir. 1945) or read any good industrial organization textbook.

[2]The model analyzed here is derived from Heinrich von Stackelberg's dynamic model, which appeared in *Marktform un Gleichgewicht* (Vienna: Springer, 1934). A translation is *The Theory of the Market Economy*, trans. A. T. Peacock (London: William Hodge, 1952). Exercise 4 at the end of this chapter asks you to study such a model.

To find the subgame perfect equilibrium of this game, let us begin by analyzing the subgames following the firms' entry decisions. First suppose that only firm i is in the industry. Holding aside the firm's sunk entry cost, this firm obtains revenue of $(900 - q_i)q_i$ by producing q_i. This is maximized by selecting $q_i = 450$, which yields revenue of \$202,500. (You should check this.) Of course, firm i can produce 450 units *only* if it had earlier invested in a large production facility. Otherwise, it can produce only 100 units, which generates revenue of $\$(900 - 100)(100) = \$80,000$. Factoring in the entry cost, if firm i builds a large production facility and is alone in the industry, then it obtains a profit of $\$202,500 - 175,000 = \$27,500$. If firm i builds a small facility and is alone in the industry, then it gets $\$80,000 - 50,000 = \$30,000$.

Next consider quantity decisions when both firms are in the industry. Net of its entry cost, firm i's revenue is $(900 - q_i - q_j)q_i$. Here q_i is the quantity selected by firm i, and q_j is the quantity selected by firm j. Firm i's best-response function is found by taking the derivative with respect to q_i while holding q_j fixed. This yields

$$BR_i(q_j) = 450 - \frac{q_j}{2}.$$

If neither firm has a capacity constraint, then the Nash equilibrium of this Cournot game is $q_1 = q_2 = 300$ (we get this result by solving $q_1 = 450 - q_2/2$ and $q_2 = 450 - q_1/2$), which yields each firm revenue of $\$(900 - 300 - 300)(300) = \$90,000$.

Yet, one or both firms may be capacity constrained. If both firms have small production facilities, then each will produce only 100 units (neither has an incentive to produce less when the other produces 100), yielding each firm $\$(900 - 100 - 100)(100) = \$70,000$. Finally, if one firm is capacity constrained, then it will produce 100 while the other will produce the unconstrained best response to 100, which is $450 - 100/2 = 400$. In this case, the price is $900 - 100 - 400 = 400$, the constrained firm gets a revenue of \$40,000, and the unconstrained firm earns a revenue of \$160,000.

In summary, with the entry cost factored in, if both firms have large production facilities, then each firm obtains a profit of $\$90,000 - 175,000 = -\$85,000$. If both firms have small production facilities, then each firm gets $\$70,000 - 50,000 = \$20,000$. If firm i has a small facility and firm j has a large facility, then firm i earns $\$40,000 - 50,000 = -\$10,000$ and firm j gets $\$160,000 - 175,000 = -\$15,000$.

The foregoing analysis fully characterizes the firms' equilibrium behavior in the subgames that follow their entry decisions. We next turn to the entry decisions, which can be analyzed with the help of Figure 16.1. The figure represents the first part of the extensive-form game, in which the firms decide whether to enter the industry. In Figure 16.1, N stands for "not enter," S means build a small production facility, and L means build a large production facility. Note that, because we have solved for the equilibrium quantities, we know what payoffs we will obtain following the entry decisions. These payoffs are written directly in Figure 16.1, in units representing thousands. For example, as already computed, if both firms enter with large production facilities, then equilibrium in the Cournot phase implies that each firm obtains −$85,000.

To finish the analysis, we can solve the extensive form in Figure 16.1 by using backward induction. This method is justified because in Figure 16.1 every node initiates a subgame. Note that S is firm 2's optimal action at its top decision node, S′ is its best alternative at the middle decision node, and N″ is its preference at the bottom decision node. In words, it is optimal for firm 2 to enter and build a small production facility if and only if firm 1 either does not enter the industry or enters with a small facility; otherwise, firm 2 would not enter. Thus, at the initial node, firm 1 knows that it will obtain zero if it selects N, 20 if it chooses S, and 27.5 if it selects L. Firm 1's optimal action is thus L. In summary, the subgame perfect equilibrium of this entry game has firm 1 playing L, firm 2 choosing SS′N″, and the firms selecting quantities as heretofore described.

In the end, firm 1 invests in a large production facility and firm 2 decides not to enter the industry. This result is interesting because, without competition in the industry, firm 1 would have been better off investing in the small production facility (yielding $30,000 instead of $27,500). That is, in the absence of firm 2's entry threat, firm 1 would prefer to hold a monopoly in the industry

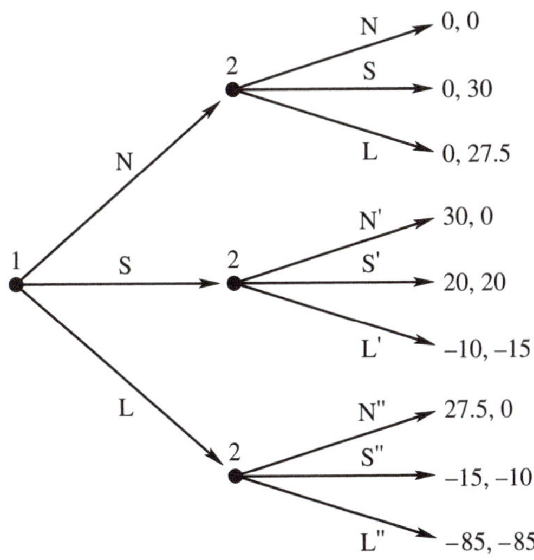

FIGURE 16.1
Abbreviated version of the entry game.

with a small production facility. But firm 1 *must* consider the possibility that firm 2 will enter. A rational firm 1 calculates that overinvesting in capacity—by building a large facility—commits it to being a tough competitor in the event that firm 2 decides to enter the industry. This makes entry unprofitable for firm 2.

DYNAMIC MONOPOLY

Retail firms frequently adjust prices. Sometimes price adjustments are due to changes in wholesale prices, but more often the adjustments have to do with customer demand. Consider the market for flat-panel LCD computer monitors. Not all consumers have the same interest in buying such a product; in other words, demand is differentiated. The "high-value" customers are people either who have a real need for the LCD technology (such as business people in environments where space is a major constraint) or who just love to buy the newest gadget on the market. "Low-value" customers would be interested in purchasing LCD monitors, but only if the monitors are priced to compete with standard CRT displays. Facing this heterogeneous consumer population—as is the case in most markets—retailers wish to extract a high price from the high-value customers and a lower price from the low-value customers. Although legal constraints and arbitrage force retailers to charge the same price to all people, a retailer can use time to its advantage. A common scheme is to initially set a high price that is intended to attract only the high-value customers. Then, after selling to the high-value folks, the retailer will drop the price in the hopes of capturing the demand of low-value customers.

Superficially, this scheme appears to give the retailer a very high profit by allowing it to extract the surplus of trade from both the high- and the low-value customers. In other words, the retailer gets both types of customers to pay the most that they are willing to pay for the product. But the retailer does face some constraints. First, competition with other retailers may exert downward pressure on prices. Second, the retailer must realize that customers will behave strategically—in particular, high-value customers may delay purchasing LCD monitors if they anticipate that the price will fall over time. In this section, I sketch a simple model that helps us explore the second strategic aspect.[3] The model focuses on a monopoly setting (where a single firm is in the industry), and therefore it ignores the matter of competition between firms.

[3]Game-theoretic analysis of this issue first appeared in J. I. Bulow's "Durable-Goods Monopolists," *Journal of Political Economy* 90(1982):314–332, and N. L. Stokey's "Rational Expectations and Durable Goods Pricing," *Bell Journal of Economics* 12(1982):112–128.

FIGURE 16.2
The customers' values of owning an LCD monitor.

	Period 1	Period 2
Benefit to Hal and Hilbert	1200	500
Benefit to Laurie and Lauren	500	200

Suppose a single firm sells LCD monitors. Call the manager of this firm Tony. To keep things simple, suppose there are only four potential customers: Hal, Hilbert, Laurie, and Lauren. These five players interact over two periods of time, which I call period 1 and period 2. Think of period 1 as a quarter of the year (say, January through March) and period 2 as the following quarter. Each customer's benefit from owning an LCD monitor depends on which periods he or she owns the product. Hal could really use such a monitor at work—he is a *high type*. Hal would get a benefit of $1,200 by using an LCD monitor in period 1, and he would get a benefit of $500 from its use in period 2. Laurie, on the other hand, doesn't really need an LCD monitor—she is a *low type*. In period 1, Laurie would obtain a benefit of $500 from an LCD monitor; in period 2, her benefit would be $200. Hilbert is a high type just like Hal; Lauren is a low type just like Laurie. The customers' values of owning an LCD monitor are summarized in Figure 16.2.

To be clear about the benefits, note that, if Hilbert owns a monitor for both periods 1 and 2, then his total ownership value is $1,700. Therefore, if Hilbert pays p for the monitor at the beginning of period 1, then his payoff is $1700 - p$. If Hilbert waits until period 2 to purchase the monitor at price p, then Hilbert's payoff is $500 - p$ (because he does not own the monitor in period 1, he gets no benefit in this period). Hal's payoff is defined the same way. Laurie's and Lauren's payoffs are defined analogously, given their individual ownership benefits of $500 in period 1 and $200 in period 2. Tony's payoff is his total revenue from the sale of LCD monitors, minus some fixed production cost, which I ignore.

The game is played as follows. At the beginning of period 1, Tony selects a price p_1. After observing the price, Hal, Hilbert, Laurie, and Lauren simultaneously decide whether to purchase monitors at this price. After a customer has purchased a monitor, he or she has no need to make another purchase (a monitor purchased in period 1 is used in both periods 1 and 2). Further, Tony has four units available, so the customers are not competing with one another for a scarce product. At the beginning of period 2, Tony selects another price p_2. After observing this price, any customer who did not purchase a monitor in period 1 may then buy one at price p_2.

This is a complicated game, with many (an infinite number of) subgames and a very large extensive-form tree representation. The game has a unique subgame perfect equilibrium, which you can find by working backward through the extensive form. But rather than subject you to the complete analysis, which can be confusing, I shall present a few pricing strategies for Tony and explain why one of them characterizes the subgame perfect equilibrium.

I start with a useful fact:

> Regardless of p_1, if at least one of the high-type customers (Hal, Hilbert) and neither of the low types (Laurie, Lauren) purchases in period 1, then it is optimal for Tony to select $p_2 = 200$ in period 2.

To understand why this is true, first note that Tony would never choose $p_2 < 200$ as long as there is a sale to be made. For example, suppose Laurie does not purchase a monitor in period 1 and then Tony picks $p_2 = 150$. Laurie will definitely buy the monitor at this price, for doing so would give her a payoff of $200 - 150 = 50$, whereas not purchasing it would give her 0 (because the game ends at the end of period 2). In fact, in period 2 both Laurie and Lauren will buy a monitor *at any price less than or equal to* \$200. If $p_2 = 200$, then the low types will be indifferent between buying and not buying; I assume that they will each decide to purchase in this case.[4] Likewise, Hal and Hilbert will purchase monitors in period 2 at any price less than or equal to \$500, assuming that they did not purchase in period 1.

Second, note that $p_2 = 200$ is obviously best if both Hal and Hilbert already purchased in period 1. Further, if one of the high-type customers did not purchase in period 1, then Tony has a simple choice: set $p_2 = 200$ and sell to all three remaining customers or set $p_2 = 500$ and sell only to the single high-type customer. The former option generates a profit of \$600, whereas the latter yields \$500. Thus, $p_2 = 200$ is best.

Pricing Scheme A: Induce all of the customers to purchase in period 1. Let us determine the highest price Tony can charge in period 1 that will guarantee that Hal, Hilbert, Laurie, and Lauren all purchase in this period. Certainly Laurie and Lauren will not purchase monitors in period 1 if the price is greater than $\$500 + 200 = \$700$; if either did so, she would obtain a negative payoff, whereas she can always get 0 by sticking to window shopping. In addition, Laurie and Lauren each knows that she cannot obtain a positive

[4]In reality, Tony may have to set a price below \$200 to make sure that the the low types have the incentive to purchase. Our theory says that the price can be just slightly below \$200 (such as \$199.99), which can be made as close to \$200 as desired. Thus, Tony can effectively charge $p_2 = 200$. In fact, in equilibrium, the customers will purchase when they are indifferent. This is used later as well.

payoff by waiting until period 2 to buy a monitor (since $p_2 \geq 200$). Thus, each low type will definitely buy a monitor in period 1 as long as the price does not exceed her ownership value of 700. The high types also will buy monitors at this price because it will give them each a payoff of $1700 - 700 = 1000$, which exceeds what they could get by waiting until period 2 (which is less than $500 - 200 = 300$ since $p_2 \geq 200$). Thus, if Tony selects $p_1 = 700$, then all four customers will purchase in period 1. This pricing strategy yields a payoff of 2800 for Tony.

Pricing Scheme B: Induce all of the customers *not* to purchase in period 1. Tony can ensure that no customer buys a monitor in period 1 by setting p_1 strictly above 1700, the ownership value of the high-type customers. Then, Hal, Hilbert, Laurie, and Lauren will all be without a monitor until the beginning of period 2. At this point, Tony could sell to all four customers at the price $p_2 = 200$ or sell only to Hal and Hilbert at the price $p_2 = 500$. The latter yields a payoff of 1000, whereas the former yields 800 for Tony; thus, $p_2 = 500$ is best with scheme B.

Pricing Scheme C: Induce the high types (Hal and Hilbert) to purchase in period 1 and the low types (Laurie and Lauren) to purchase in period 2. By setting p_1 above 700 but not too high, Tony can induce only the high types to purchase in period 1. As already noted, Laurie and Lauren will not purchase at such a price; further, as long as at least one high type purchases in period 1, we know that $p_2 = 200$ will be selected by Tony in period 2. Anticipating $p_2 = 200$, each high type knows that he would obtain $500 - 200 = 300$ by waiting until period 2 to buy a monitor. Purchasing in period 1 at price p_1 gives a high type a payoff of $1200 + 500 - p_1$. Thus, Hal and Hilbert have the incentive to purchase in period 1 if and only if $1700 - p_1 \geq 300$, which simplifies to $p_1 \leq 1400$. The best such price for Tony is $p_1 = 1400$. Thus, when Tony chooses $p_1 = 1400$, he obtains a payoff of $2 \cdot 1400$ in period 1 (from the sales to Hal and Hilbert) and $200 \cdot 2$ in period 2 (from the sales to Laurie and Lauren), for a total of 3200.

The foregoing analysis completely identifies Tony's options.[5] Comparing the payoffs from the various pricing schemes reveals that scheme C is Tony's best course of action. Intuitively, Tony wants to find a way of discriminating between the high- and low-type customers. But, because he must charge them the same price in each period, he can only use time to differentiate between them. Time is useful in this regard because there is a sense in which Hal and

[5] Given the customers' ownership values, there is no way for Tony to induce the low types to purchase in period 1 and the high types to purchase in period 2. In any case, it would not be optimal for Tony to select such a pricing scheme.

Hilbert are less patient than are Laurie and Lauren. The high types' value of owning an LCD monitor from period 1—relative to the value of purchasing it in period 2—is much greater than is the case with the low types. To see this, observe that Hal's and Hilbert's ownership value from period 1 is $1200+500 = 1700$, which is 1200 more than the value of 500 that they would each get by waiting until period 2 to buy a monitor. In contrast, Laurie's and Lauren's value from period 1 is only 500 more than their value from period 2. Because Hal and Hilbert gain very little by waiting until period 2, Tony can extract most of their ownership value by setting a very high period 1 price.

PRICE GUARANTEES AS A COMMITMENT TO HIGH PRICES

As described earlier, under some conditions retail firms have an incentive to decrease prices over time to extract surplus from customers with different valuations. You might think that such a pricing policy is good for a firm—it helps the firm obtain profit from each type of customer. In fact, the incentive to decrease prices over time may actually have the *opposite* effect on profit. To make sense of my claim, let us continue with the dynamic monopoly example.

Consider again the optimal pricing scheme, C. Recall that, when selecting the period 1 price p_1, Tony must consider the customers' strategic calculations. In particular, all of the players know that, if the low-type customers do not purchase and at least one of the high types does purchase a monitor in period 1, then *in period 2 Tony will have the incentive to select* $p_2 = 200$. (You will recall that this is the "useful fact" noted on page 188.) Tony's scheme C pricing policy is therefore a balancing act. On one hand, Tony wants to deal only with Hal and Hilbert in period 1, raising the price as high as they are willing to pay. On the other hand, once Hal and Hilbert have purchased monitors, Tony wants to decrease the price and sell two more units to Laurie and Lauren. Because Hal and Hilbert anticipate the second-period price, Tony can get them to pay only $1,400 in period 1.

By comparison, consider how much Tony would make if he could somehow *commit never to deal with Laurie and Lauren.* For example, suppose that in period 1 Tony could make a legally enforceable promise not to sell monitors in period 2. Then suppose Tony chose a period 1 price of $p_1 = 1700$, or just below this amount. In this case, Hal and Hilbert would purchase monitors in period 1. (By purchasing a monitor, a high type's payoff would be $1700 - p_1$; if he did not purchase a monitor, then the high type would get 0, because no monitors would be sold in period 2.) Laurie and Lauren would not purchase at this high price. However, Tony's payoff would be $3,400, which is greater than the payoff he would get without the ability to commit. Thus, commitment helps Tony.

There are many ways in which firms achieve commitment to prices over time. One of the most common ways is through the use of "price guarantees," whereby a retailer offers to refund the difference between its current price and any lower price that it sets on the same item in the near future. That is, if today the store sells you a television for \$300 and then tomorrow it lowers its price on the same television to \$250, you can claim a \$50 refund. You probably have seen advertisements or been reassured by a salesperson to this effect. Consumers usually think of price guarantees as something favorable to them, but in fact this may not be the case.

The dynamic monopoly model can be used to demonstrate how price guarantees can be good for firms and bad for consumers. Suppose that Tony institutes a price-guarantee policy and selects a period 1 price of $p_1 = 1700$. Suppose that at least one of the high types—Hal, for example—purchases a monitor in period 1. Then, if Tony wants to sell monitors to Laurie and Lauren in period 2, he has to reduce the price to 200. This gives Tony at most 600 in period 2 revenue (sales to Hilbert, Laurie, and Lauren). However, Hal will claim a refund of $1700 - 200 = 1500$ from Tony, far offsetting the sales gain. Thus, the price guarantee helps Tony commit not to reduce the price, thus allowing him to achieve the payoff of 3400.

Intuitively, a price guarantee works like any other contractual commitment—it helps the retailer tie its hands in the future. To the extent that the retailer controls the terms of the sales contract, we should expect it to include such contractual provisions when they benefit the retailer. The consumers' remedy lies in seizing more control of the contract specification process or relying on the legal system to limit retailers' use of commitment mechanisms.

GUIDED EXERCISE

Problem: A manufacturer of automobile tires produces at a cost of \$10 per tire. It sells units to a retailer who in turn sells the tires to consumers. Imagine that the retailer faces the inverse demand curve $p = 200 - (q/100)$. That is, if the retailer brings q tires to the market, then these tires will be sold at a price of $p = 200 - (q/100)$. The retailer has no cost of production, other than whatever it must pay to the manufacturer for the tires.

(a) Suppose that the manufacturer and retailer interact as follows. First, the manufacturer sets a price x that the retailer must pay for each tire. Then, the retailer decides how many tires q to purchase from the manufacturer and sell to consumers. The manufacturer's payoff (profit) is $q(x - 10)$, whereas the retailer's profit is

$$\left(200 - \frac{q}{100}\right) q - xq = 200q - \frac{q^2}{100} - xq.$$

Calculate the subgame perfect equilibrium of this game.

(b) Suppose that the manufacturer sells its tires directly to consumers, bypassing the retailer. Thus, the manufacturer can sell q tires at price $p = 200 - (q/100)$. Calculate the manufacturer's profit-maximizing choice of q in this case.

(c) Compare the joint profit of the manufacturer and retailer in part (a) with the manufacturer's profit in part (b). Explain why there is a difference. This is called the *double-marginalization problem.*

Solution:

(a) Start by calculating the retailer's optimal quantity q as a function of the manufacturer's choice of x. Given x, the retailer selects q to maximize

$$200q - \frac{q^2}{100} - xq.$$

The first-order condition for optimization implies that the retailer chooses quantity $q^*(x) = 10{,}000 - 50x$. Thus, the manufacturer can anticipate selling exactly $q^*(x)$ units if it prices at x, which means that the manufacturer's payoff, as a function of x, will be

$$q^*(x)(x - 10) = (10{,}000 - 50x)(x - 10) = 10{,}500x - 50x^2 - 100{,}000.$$

Taking the derivative of this expression and setting it equal to zero yields the first-order condition for the manufacturer's optimization problem at the beginning of the game. Solving for x, we find that the manufacturer sets $x^* = 105$. Thus, the equilibrium quantity is $q = q^*(105) = 4750$. This implies $p = 152.50$.

(b) In this case, the manufacturer's profit is

$$\left(200 - \frac{q}{100}\right) q - 10q.$$

Taking the derivative of this expression and setting it equal to zero yields the manufacturer's optimal quantity, which is $\hat{q} = 9500$. This implies $p = 105$.

(c) Calculating the firms' equilibrium payoffs in part (a), we see that the joint profit is

$$(152.50 - 10) \cdot 4{,}750 = 676{,}875.$$

The manufacturer's profit in part (b) is

$$95 \cdot 9{,}500 = 902{,}500.$$

The difference arises because, in a market in which a monopoly firm must set a single price per unit of output, the monopolist optimally raises the price above marginal cost. This implies an inefficiently low level of trade. In part (a), such a monopoly distortion occurs *twice*—by the manufacturer to the retailer, and again by the retailer to the consumers. But in part (b), distortion occurs just once because the monopolist manufacturer sells directly to consumers.

EXERCISES

1. Consider the model of advertising and Cournot competition analyzed in this chapter. Suppose the two firms could write an externally enforced contract that specifies an advertising level a and a monetary transfer m from firm 2 to firm 1. Would the firms write a contract that specifies $a = 3$? If not, to what level of advertising would they commit?

2. Continuing with the advertising model, suppose the firms compete on price rather than quantity. That is, quantity demanded is given by $Q = a - p$, where p is the price consumers face. After firm 1's selection of the level of advertising, the firms simultaneously and independently select prices p_1 and p_2. The firm with the lowest price obtains all of the market demand at this price. If the firms charge the same price, then the market demand is split equally between them. (To refresh your memory of the price-competition model, review the analysis of Bertrand competition in Chapter 10.) Find the subgame perfect equilibrium of this game and explain why firm 1 advertises at the level that you compute.

3. Consider a variation of the limit-capacity model analyzed in this chapter. Suppose that, instead of the firms' entry decisions occurring sequentially, the firms act simultaneously. After observing each other's entry decisions, market interaction proceeds as in the original model. Find the subgame perfect Nash equilibria of this new model and compare it/them to the subgame perfect equilibrium of the original model.

4. [von Stackelberg duopoly model] Imagine a market in which two firms compete by selecting quantities q_1 and q_2, respectively, with the market price given by $p = 1000 - 3q_1 - 3q_2$. Firm 1 (the incumbent) is already in the market. Firm 2 (the potential entrant) must decide whether or not to enter and, if she enters, how much to produce. First the incumbent commits to its production level, q_1. Then the potential entrant, having seen q_1, decides whether to enter the industry. If firm 2 chooses to enter, then it selects its production level q_2. Both firms have the cost function $c(q_i) = 100q_i + F$, where F is a constant fixed cost. If firm 2 decides not to enter, then it obtains a payoff of 0. Otherwise, it pays the cost of production, including the fixed cost. Note that firm i in the market earns a payoff of $pq_i - c(q_i)$.

   (a) What is firm 2's optimal quantity as a function of q_1, conditional on entry?

   (b) Suppose $F = 0$. Compute the subgame perfect Nash equilibrium of this game. Report equilibrium strategies as well as the outputs, profits, and price realized in equilibrium. This is called the *Stackelberg* or *entry-accommodating outcome*.

   (c) Now suppose $F > 0$. Compute, as a function of F, the level of q_1 that would make entry unprofitable for firm 2. This is called the *limit quantity*.

   (d) Find the incumbent's optimal choice of output and the outcome of the game in the following cases: (i) $F = 18{,}723$, (ii) $F = 8{,}112$, (iii) $F = 1{,}728$, and (iv) $F = 108$. It will be easiest to use your answers from parts (b) and (c) here; in each case, compare firm 1's profit from limiting entry to its profit from accommodating entry.

5. Consider a slight variation of the dynamic monopoly game analyzed in this chapter. Suppose there is only one high-type customer (Hal) and only one low-type customer (Laurie).

   (a) Analyze this game and explain why $p_2 = 200$ is not optimal if Hal does not purchase a monitor in period 1. Find the optimal pricing scheme for Tony. Discuss whether Tony would gain from being able to commit not to sell monitors in period 2.

   (b) Finally, analyze the game with one of each type of customer and ownership benefits given in the following figure. In this case, would Tony gain from being able to commit not to sell monitors in period 2?

	Period 1	Period 2
Benefit to Hal	1200	300
Benefit to Laurie	500	200

6. Imagine a market setting with three firms. Firms 2 and 3 are already operating as monopolists in two different industries (they are not competitors). Firm 1 must decide whether to enter firm 2's industry and thus compete with firm 2, or enter firm 3's industry and thus compete with firm 3. Production in firm 2's industry occurs at zero cost, whereas the cost of production in firm 3's industry is 2 per unit. Demand in firm 2's industry is given by $p = 9 - Q$, whereas demand in firm 3's industry is given by $p' = 14 - Q'$, where p and Q denote the price and total quantity in firm 2's industry and p' and Q' denote the price and total quantity in firm 3's industry.

The game runs as follows: First, firm 1 chooses between E^2 and E^3. (E^2 means "enter firm 2's industry" and E^3 means "enter firm 3's industry.") This choice is observed by firms 2 and 3. Then, if firm 1 chooses E^2, firms 1 and 2 compete as Cournot duopolists, where they select quantities q_1 and q_2 simultaneously. In this case, firm 3 automatically gets the monopoly profit of 36 in its own industry. On the other hand, if firm 1 chooses E^3, then firms 1 and 3 compete as Cournot duopolists, where they select quantities q_1' and q_3' simultaneously; and in this case, firm 2 automatically gets its monopoly profit of 81/4.

(a) Calculate and report the subgame perfect Nash equilibrium of this game. In the equilibrium, does firm 1 enter firm 2's industry or firm 3's industry?

(b) Is there a Nash equilibrium (not necessarily subgame perfect) in which firm 1 selects E^2? If so, describe it. If not, briefly explain why.

7. Consider the location game (in Chapter 8) with nine possible regions at which vendors may locate. Suppose that, rather than the players moving simultaneously and independently, they move sequentially. First, vendor 1 selects a location. Then, after observing the decision of vendor 1, vendor 2 chooses where to locate. Use backward induction to solve this game (and identify the subgame perfect Nash equilibrium). Remember that you need to specify the second vendor's sequentially optimal strategy (his best move conditional on every different action of vendor 1).

8. Consider the following market game: An incumbent firm, called firm 3, is already in an industry. Two potential entrants, called firms 1 and 2, can each enter the industry by paying the entry cost of 10. First, firm 1 decides whether to enter or not. Then, after observing firm 1's choice, firm 2 decides whether to enter or not. Every firm, including firm 3, observes the choices of firms 1 and 2. After this, all of the firms in the industry (including firm 3) compete in a Cournot oligopoly, where they simultaneously and independently select quantities. The price is determined by the inverse demand curve $p = 12 - Q$, where Q is the total quantity produced in the industry. Assume that the firms produce at no cost in this Cournot game. Thus, if firm i is in the industry and produces q_i,

then it earns a gross profit of $(12 - Q)q_i$ in the Cournot phase. (Remember that firms 1 and 2 have to pay the fixed cost 10 to enter.)

(a) Compute the subgame perfect equilibrium of this market game. Do so by first finding the equilibrium quantities and profits in the Cournot subgames. Show your answer by designating optimal actions on the tree and writing the complete subgame perfect equilibrium strategy profile. [Hint: In an n-firm Cournot oligopoly with demand $p = 12 - Q$ and 0 costs, the Nash equilibrium entails each firm producing the quantity $q = 12/(n + 1)$.]

(b) In the subgame perfect equilibrium, which firms (if any) enter the industry?

9. This exercise will help you think about the relation between inflation and output in the macroeconomy. Suppose that the government of Tritonland can fix the inflation level $\dot{p}$ by an appropriate choice of monetary policy. The rate of nominal wage increase, $\dot{W}$, however, is set not by the government but by an employer–union federation known as the ASE. The ASE would like *real* wages to remain constant. That is, if it could, it would set $\dot{W} = \dot{p}$. Specifically, given $\dot{W}$ and $\dot{p}$, the payoff of the ASE is given by $u(\dot{W}, \dot{p}) = -(\dot{W} - \dot{p})^2$. Real output y in Tritonland is given by the equation $y = 30 + (\dot{p} - \dot{W})$. The government, perhaps representing its electorate, likes output more than it dislikes inflation. Given y and $\dot{p}$, the government's payoff is $v(y, \dot{p}) = y - \dot{p}/2 - 30$. The government and the ASE interact as follows. First, the ASE selects the rate of nominal wage increase. Then the government chooses its monetary policy (and hence sets inflation) after observing the nominal wage increases set by the ASE. Assume that $0 \leq \dot{W} \leq 10$ and $0 \leq \dot{p} \leq 10$.

(a) Use backward induction to find the level of inflation $\dot{p}$, nominal wage growth $\dot{W}$, and output y, that will prevail in Tritonland. If you are familiar with macroeconomics, explain the relationship between backward induction and "rational expectations" here.

(b) Suppose that the government could commit to a particular monetary policy (and hence inflation rate) ahead of time. What inflation rate would the government set? How would the utilities of the government and the ASE compare in this case with that in part (a)?

(c) In the "real world," how have governments attempted to commit to particular monetary policies? What are the risks associated with fixing monetary policy before learning about important events, such as the outcomes of wage negotiations?

10. Regarding the dynamic monopoly game, can you find ownership values for Hal and Laurie that make scheme A optimal? Can you find values that make scheme B optimal?

PARLOR GAMES 17

As a kid, you probably played tic-tac-toe with your friends or siblings. In this game, two players take turns claiming cells of a 3×3 matrix (a matrix with three rows and three columns). Player 1 claims a cell by writing an "X" in it; player 2 writes an "O" in each cell that he claims. A player can claim only one cell per turn. The game ends when all of the cells have been claimed or when one of the players has claimed three cells in a line (by claiming an entire row, column, or diagonal). If a player has claimed three cells in a line, then he wins and the other player loses. If all of the cells are claimed and neither player has a line, then the game is declared a tie. Figure 17.1 depicts the playing of a game in which player 1 wins.

Tic-tac-toe is a game of skill—nothing in the game is left to chance (no coins are flipped, no dice rolled, etc.). It is also a game of perfect information, because players move sequentially and observe each others' choices. Further, it is a finite game. These facts imply that the result on page 170 (Chapter 15)

FIGURE 17.1 A game of tic-tac-toe.

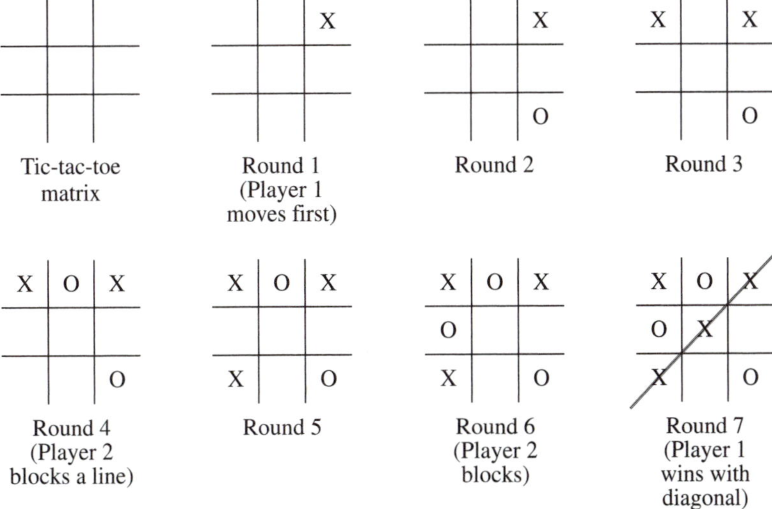

holds for tic-tac-toe: this game has a pure-strategy Nash equilibrium. What else do we know about tic-tac-toe? It is a two-player, strictly competitive game: either player 1 wins and player 2 loses, or player 2 wins and player 1 loses, or the players tie. Thus, the result on page 135 (Chapter 12) also applies, meaning that each player's Nash equilibrium strategy is a security strategy for him.

Let's review. One of the outcomes (1 wins, 2 wins, tie) occurs in a Nash equilibrium. Because the equilibrium strategies are security strategies, each player can guarantee himself this equilibrium outcome. For example, suppose "1 wins" is the equilibrium outcome. Then our mathematical analysis proves that player 1 has a *winning strategy,* which guarantees that he will win the game *regardless of what his opponent does*. On the other side of the table, player 2 has a strategy that guarantees at least a loss, regardless of what player 1 does. In this case, player 2's security strategy is not very helpful to her, but player 1's security strategy is quite helpful to him. The game is "solved" as long as player 1 is smart enough to calculate his security strategy.

In fact, "1 wins" is *not* the equilibrium outcome of tic-tac-toe. The Nash equilibrium actually leads to the tie outcome. Thus, our mathematical analysis implies that each player has a strategy guaranteeing a tie. In other words, when smart, rational players engage in tic-tac-toe, the game always ends in a tie. You probably knew this already, because most people figure it out relatively early in life. As a matter of fact, discovering the solution to tic-tac-toe is one of the major stepping stones of childhood development. I am sure child psychologists put it just before the dreaded Santa Claus realization on the time line of youth, at least for kids destined to become game theorists.

The neat thing about the results in Chapters 12 and 15 is that they apply generally. There are lots of two-player, strictly competitive, perfect-information, finite games. On one end of the complexity spectrum, we have tic-tac-toe; on the other end are games such as chess, which also ends with either 1 wins, 2 wins, or tie (also called "a draw").[1] Putting the results together, we get:

> **Result:** Take any two-player, extensive form game that (a) is finite, (b) has perfect information, and (c) is strictly competitive.
>
> - If the possible outcomes of the game are "1 wins," "2 wins," and "tie," then either one of the players has a winning strategy or both players have strategies guaranteeing at least a tie.
> - If the possible outcomes of the game are only "1 wins" and "2 wins," then either player 1 has a winning strategy or player 2 has a winning strategy.

[1] The rules of chess, including the "50 moves rule," make this game finite.

This result proves something about chess. In chess, either one of the players has a winning strategy or both players have a strategy guaranteeing a tie.[2] At this point, no one knows *which* of these statements is true; chess is too complicated to solve by finding a Nash equilibrium. The best human chess players rely on experience and the ability to look ahead a few turns to beat their opponents. Computer programs such as IBM's Deep Blue can now play at the grandmaster level and will soon be able to beat any unassisted human. But, as Garry Kasparov observes, chess will remain fascinating as the focus shifts to competition between computer programs.[3]

You might find it useful to make a list of games that you have played to which the preceding result applies. Then see what you can deduce about the games by using the theory you have learned. The following exercises will push you in this direction.

GUIDED EXERCISE

Problem: Suppose there are two players and two baskets of colored balls. The red basket contains m red balls. The blue basket contains n blue balls. The numbers n and m are common knowledge between the players. Consider a perfect-information game in which, starting with player 1, the players take turns individually removing one ball from one of the baskets. The player who removes the last red ball loses the game. Demonstrate that player 1 has a winning strategy if $m + n$ is an even number and player 2 has a winning strategy if $m + n$ is an odd number.

Solution: To win, a player must leave her opponent with one red ball and no blue balls (an odd total). Consider the strategy of always removing a blue ball as long as at least one blue ball remains, and otherwise removing a red ball. If $m + n$ is even, then the first player is guaranteed to win with this strategy. If $m + n$ is odd, then player 2 wins with this strategy.

EXERCISES

1. Consider the following game between two players. The players take turns moving a rock among cells of an $m \times n$ matrix. At the beginning of the game (be-

[2]Mathematician Ernst Zermelo is credited with doing some of the early game-theoretic work on chess. The result presented here uses the two building-block results on pages 135 and 170, which are due to John von Neumann and Harold Kuhn, respectively.

[3]The subject is discussed in M. Newborn, *Kasparov versus Deep Blue,* (Montreal: McGill University Press, 1996).

fore the first move), the rock is placed in the bottom-right cell of the matrix [cell (m, n)]. Player 1 goes first. At each turn, the player with the move must push the rock into one of the three cells above or to the left (or both) of the cell where the rock currently sits. That is, the player may move the rock to the cell above the current position, to the cell to the left of the current position, or to the cell diagonal to the current position in the up-left direction, as pictured below.

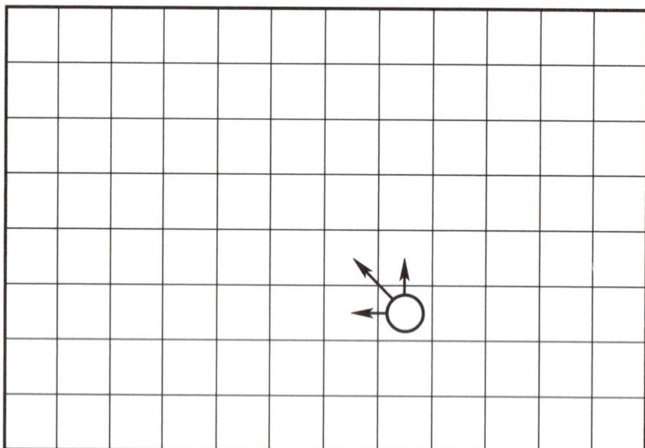

A player may not move the rock outside of the matrix. The player who is forced to move the rock into the top-left cell of the matrix [cell (1, 1)] loses the game.

(a) Suppose the dimensions of the matrix are 5×7. Does either player have a strategy that guarantees a victory? If so, which of the two players has such a strategy?

(b) In general, under what conditions on m and n does player 1 have a winning strategy and under what conditions does player 2 have a winning strategy?

2. The game Cliff runs as follows. There are two players, each of whom has a pocketful of pennies, and there is an empty jar. The players take turns tossing pennies into the jar, with player 1 moving first. There are two rules: (a) When a player is on the move, he must put between one and four pennies in the jar (that is, he must toss at least one penny in the jar, but he cannot toss more than four pennies in the jar), and (b) the game ends as soon as there are sixteen or more pennies in the jar. The player who moved last (the one who caused the number of pennies to exceed fifteen) wins the game. Determine which of the players has a strategy that guarantees victory, and describe the winning strategy.

3. Consider Marienbad, a variation of the game described in the Guided Exercise of this chapter. As before, players take turns removing balls from the baskets, except that players are allowed to remove as many balls as they wish, provided that each player removes balls from only *one* of the baskets in each round. For example, player 1 might remove three red balls in one round and four blue balls in another round. Each player must remove at least one ball when it is his turn to move. The player who is forced to remove the last ball (whichever basket it is in) is declared the loser. As the result of this chapter establishes, one of the players has a winning strategy in this game. Explain which player has the winning strategy and how the identity of the winning player depends on m and n. If you can, also describe the winning strategy.

4. Consider a variant of the game in Exercise 1, in which the player who moves the rock into cell (1, 1) *wins* the game.

(a) Does one of the players have a strategy that guarantees him a win? If so, which player has a winning strategy?

(b) Now consider a *three*-player version of the game, whereby play rotates between the players, starting with player 1. That is, player 1 moves first, followed by player 2, then player 3, then back to player 1, and so on. Players are allowed to move the rock just as described above. The player who moves the rock into cell (1, 1) wins and gets a payoff of 1; the other two players lose, each obtaining 0. Is there a subgame perfect equilibrium in which player 1 wins? Is there a subgame perfect equilibrium in which player 2 wins? Is there a subgame perfect equilibrium in which player 3 wins?

5. Chomp is a game in which two players take turns choosing cells of an $m \times n$ matrix, with the rule that, if a cell has been selected, then it and all cells below and/or to the right of it are removed from consideration (graphically, filled in) and cannot be selected in the remainder of the game. That is, if cell (j, k) is selected, then one fills in all cells of the form (j', k') with $j' \geq j$ and $k' \geq k$. The player who is forced to pick the top-left corner cell [cell (1, 1)] loses; the other player wins. Player 1 moves first. Analyze this game and determine which player has a strategy guaranteeing victory. Explain how the identity of the player with the winning strategy depends on m and n. Can you calculate the winning strategy, for at least some cases of m and n?[4]

6. Consider a three-player version of Chomp. Play rotates between the three players, starting with player 1. That is, player 1 moves first, followed by player 2, then player 3, then back to player 1, and so on. The player who is forced to select cell (1, 1) loses the game and gets a payoff of 0. The player who moved *immediately before* the losing player obtains 1, whereas the other player wins with a payoff of 2.

(a) Does this game have a subgame perfect Nash equilibrium?

---

[4]Chomp was invented by mathematician David Gale. You can read about it and hundreds of other parlor games in D. Gale, *Tracking the Automatic Ant* (New York: Springer Verlag, 1998) and E. R. Berlekamp, J. H. Conway, and R. K. Guy, *Winning Ways for Your Mathematical Plays,* vols. 1 and 2 (New York: Academic Press, 1982).

(b) Do you think any one of the players has a strategy that guarantees him a win (a payoff of 2)?

(c) Can you prove that player 1 can guarantee himself any particular payoff? Sketch your idea.

18 BARGAINING PROBLEMS

Contracting is a fundamental part of everyday economic life. Firms and their employees, consultants and clients, doctors and patients, children and parents, and members of a team are examples of relations governed by contract. Many things may be covered in a contract. A sales agreement, for example, may specify a quantity of items, a price to be paid for them, and terms of delivery. An employment contract specifies the worker's duties, the firm's responsibilities, and a wage or salary. An agreement between team members or family members determines how they will coordinate their behavior over time.

Recall that any contractual relationship can be divided into two phases: a contracting phase, in which players set the terms of their contract, and an implementation phase, in which the contract is carried out and enforced. The nature of contract and forms of enforcement were addressed in Chapter 13 and are further discussed in Chapter 20. In this chapter, I begin the analysis of how the problem of contract selection is resolved. I focus on the two-player case.

Contracts are usually determined through a negotiation process. Businesses negotiate over the prices and specifications of intermediate goods (those traded between firms), people bargain over the price to be paid for articles at garage sales (tag sales if you live in New England), and smart shoppers try to get discounts at stores by offering prices slightly below those marked. People buying or selling a house usually engage in a long bargaining process. First-graders even know how to shrewdly negotiate the trade of lunch items. As you can see, negotiation is a fundamental process in everyday economic life.

BARGAINING: VALUE CREATION AND DIVISION

Many contracts that people negotiate have to do with the trade of goods, services, and money. Economists, who view themselves as undisputed experts on the topic of trade, have always emphasized one important insight: people want to trade because, in a society in which goods are scarce (that is, not available in quantities that would satisfy everyone), trade can create value.[1] Suppose I have a pizza and you have a television. If we each consume what we individually own, then you will be hungry for food and I will be hungry for entertainment. On the other hand, we could agree to a trade whereby you let me watch television in exchange for half of my pizza. The trade may make us both better off, in which case we say that the trade *creates value.*

Creating value is one thing—dividing value is another. What if I suggest giving you one-quarter of the pizza in exchange for viewing the television? This trade may still make us both better off than if no trade took place, but now the *terms of trade* are more favorable to me than before. When you and I negotiate, not only do we have to look for valuable trades, but we also have to jointly decide how to *divide* the value. Our inherent bargaining strengths, the procedure by which we negotiate, and the general contracting environment all contribute to the terms of our final agreement. For example, if you are a tough negotiator, then perhaps you will force me to accept a trade giving you most of the pizza in exchange for the right to watch television.

To divide value, people often utilize a *divisible good.* In the pizza–television example, pizza most easily serves this role: the pizza can be cut to achieve any proportional split between you and me. If we want an outcome in which I get most of the value of trade, then we can specify that I obtain most of the pizza. On the other hand, if we want you to get most of the value, then we should specify that you get most of the pizza. In principle, we can also "divide" the television by, say, allowing me to watch only the first half of a movie that I wish to see. But dividing the movie can destroy its value (Who would want to watch only the first half of a movie, unless it is a dud?), which makes it a poor candidate for dividing value.

There is a special good that facilitates arbitrary division of the gains from trade: money. In trade, money is often exchanged for some other good or service. In a sales context, the *price* (money transfered from the buyer to the seller) determines the division of value.[2] In the labor market, wages may play this role.[3]

Note that negotiation and value creation do not have to entail a direct exchange. For example, a firm and worker might negotiate over whether to

[1] Economists are well known for studying trade in *markets,* where people and firms meet to buy, sell, or barter. An active market clearly indicates value from trade.

[2] The archetypical market setting is one in which there is a single market price for a particular good being sold. This price is taken as given by firms wishing to sell the product and by consumers wishing to purchase it. Such a price is thought to prevail under the ideal conditions of perfect competition—conditions that most students of intermediate microeconomics can recall. In this chapter and the next, we develop an understanding of how the price is determined through negotiation.

[3] Money is not very useful when people have liquidity constraints or when their preferences over money differ substantially.

pursue business tactic A (tailoring the firm to a specific group of consumers, for example) over business tactic B (emphasizing a quick response to competitors' market behavior). Both players care about the business tactic, and implementing it may require the consent and coordination of both the firm and the worker. Yet, it may be a stretch to think of this selection as a direct trade because nothing changes hands.

In summary, the main point here is that bargaining can be usefully viewed in terms of value creation and value division. If you understand these two components, you will have gone a long way toward learning how best to negotiate and how bargaining is resolved in the real world. You will also be ready to think about bargaining problems in the abstract.

AN ABSTRACT REPRESENTATION OF BARGAINING PROBLEMS

A simple way of mathematically representing a bargaining problem is to describe the alternatives available to the parties—that is, the various contracts they can make—and to describe what happens if the parties fail to reach an agreement. With an understanding of how the relationship will proceed after negotiation (and how contracts will be enforced), all of these things can be put in terms of payoff vectors. For example, imagine that two players are trying to decide whether to start a business partnership. Suppose that, if they initiate the partnership, then the business will yield a utility of 4 for player 1 and 6 for player 2; if they do not form a partnership, then they will each get a payoff of 2. Then, instead of thinking about their negotiation in terms of "form a partnership" versus "do not form a partnership," we can put it in terms of the end result: payoff vector (4, 6) versus payoff vector (2, 2). In other words, we can think of the players as negotiating directly over the payoffs.

Let V denote the set of payoff vectors defining the players' alternatives for a given bargaining problem. For example, in the partnership story of the preceding paragraph, $V = \{(4, 6), (2, 2)\}$. V is called the *bargaining set*. Let d denote the payoff vector associated with the *default outcome,* which describes what happens if the players fail to reach an agreement. The default outcome is also called the *disagreement point,* and it is an element of V. In the partnership story, the disagreement point is given by $d = (2, 2)$, because the partnership is not formed unless the parties agree to it. In general, each player can unilaterally induce the default outcome by withholding his consent on any contract proposal.

In many bargaining problems, players can agree to a monetary transfer as a part of the contract. The transfer may signify a wage, salary, or price, or it may simply be an up-front payment made by one party to the other. I generally assume that money enters the players' payoffs in an *additive* way. For example, consider a bargaining problem between players 1 and 2. Suppose the players negotiate over a monetary transfer from player 2 to player 1 as well as other items, such as whether to form a partnership, task assignment, and so forth. Let t denote the contracted monetary transfer and let z represent the other items. A positive value of t indicates a transfer from player 2 to player 1, whereas a negative value indicates a transfer in the other direction. Money is said to enter additively if player 1's payoff can be written $u_1 = v_1(z) + t$ and player 2's payoff can be written $u_2 = v_2(z) - t$, for some functions v_1 and v_2. In other words, $v_i(z)$ is player i's benefit of z in monetary terms, which can then be added to the amount of money that this player receives or gives up. When payoffs are additive in money, we say that the setting is one of *transferable utility,* because utility can be transferred between players on a one-to-one basis with the use of money.

With transferable utility, the bargaining set can be graphed as a collection of diagonal lines, each with a slope of -1. For example, suppose that, in the preceding partnership story, the players' alternatives are expanded to include monetary transfers. Observe how this bargaining problem can be specified by using the v, z, t notation. First, let $z = 1$ represent forming a partnership and let $z = 0$ represent no partnership. Note that $v_1(0) = v_2(0) = 2$, $v_1(1) = 4$, and $v_2(1) = 6$. Thus, if bargaining is resolved with $z = 0$, then player 1 gets $2+t$ and player 2 gets $2-t$. If the players agree to $z = 1$, then player 1 obtains $4+t$ and player 2 receives $6-t$. Payoff vector (4, 6) is in the bargaining set, as before, because $t = 0$ is always feasible. Vector (6, 4) also is in the bargaining set because it can be achieved by selecting $z = 1$ and $t = 2$. In fact, all vectors of the form $(4 + t, 6 - t)$ and $(2 + t, 2 - t)$ are in the bargaining set. Varying t traces out the lines shown in Figure 18.1; these lines constitute the bargaining set. The disagreement point is (2, 2) because each player can unilaterally impose that no partnership be formed and that no transfer be made.

As the picture makes clear, whenever there is transferable utility, the set of efficient outcomes is precisely those that maximize the players' *joint value*.[4] The joint value is defined as the sum of the players' payoffs—in other words, the total payoff. To develop your intuition, note that the bargaining set in Figure 18.1 contains no points upward and to the right of the line through (4, 6). Thus, starting at any point on this line, there is no way to increase one player's payoff without decreasing the other's payoff. On the other hand, all of the points on the line through (2, 2) are inefficient, because, for each of them, one can find a point on the outer line that is more efficient. For example, payoff

[4]Recall that an outcome is called *efficient* if there is no other outcome that makes one player better off without making another player worse off.

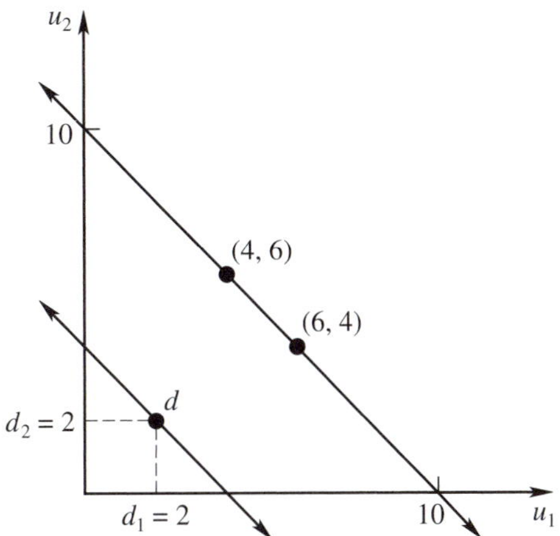

FIGURE 18.1
The bargaining set in the partnership example.

vector $(5, -1)$ can be attained by setting $z = 0$ and $t = 3$. However, the players would both get higher payoffs—in particular $(6, 4)$—if they chose $z = 1$ and $t = 2$.

Because t is merely a transfer between the players, it does not figure into the joint-value calculation. Mathematically, for any z and t, the joint value is given by

$$[v_1(z) + t] + [v_2(z) - t] = v_1(z) + v_2(z).$$

The *surplus* of an agreement is defined as the difference between the joint value of the contract and the joint value that would have been obtained had the players not reached an agreement. The latter value is just the total payoff of the default outcome, $d_1 + d_2$. Thus, the surplus is equal to

$$v_1(z) + v_2(z) - d_1 - d_2.$$

AN EXAMPLE

Rosemary chairs the English department at a prominent high school; Jerry, a former computer specialist and now a professional actor, is interested in working at the school. These two people have to make a joint decision. First, they have to decide whether to initiate an employment relationship (that is, whether Jerry will work for the high school). Further, two aspects of employment are on the bargaining table: Jerry's responsibilities on the job (represented by x) and Jerry's salary t. I use x instead of z here because, in this specialized application, the variable has to do with items corresponding only to agreement between the players (not the default outcome).

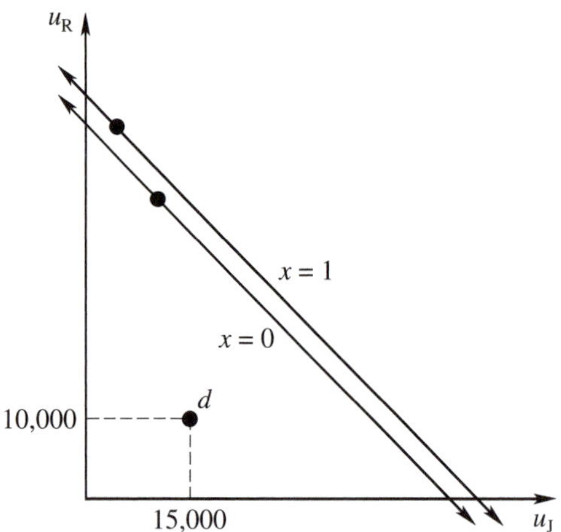

FIGURE 18.2
The bargaining problem of Jerry and Rosemary.

Suppose there are two possibilities for job duties: Jerry could be made responsible for only the drama courses ($x = 0$), or Jerry could also be put in charge of the softball team ($x = 1$). Suppose Jerry gets a personal value of \$10,000 when employed at the high school, owing to the happiness he obtains from service work as a drama teacher. But if he coaches the softball team, this value decreases by \$3,000 because of the effort he must expend in the evenings. Rosemary and the school value Jerry's work as a drama teacher in the amount \$40,000; she assesses the additional value of Jerry's work as softball coach to be \$5,000. Thus, if the two parties agree to employment, Jerry's payoff will be $10{,}000 - 3{,}000x + t$ and Rosemary's payoff will be $40{,}000 + 5{,}000x - t$. Note that, using the notation defined in the preceding section, we can write Jerry's payoff as $u_J = v_J(x) + t$ and Rosemary's payoff as $u_R = v_R(x) - t$, where

$$v_J(x) = 10{,}000 - 3{,}000x \quad \text{and} \quad v_R(x) = 40{,}000 + 5{,}000x.$$

If the parties do not initiate an employment relationship, then Jerry gets \$15,000 from working on his own and Rosemary gets \$10,000, which is her value of hiring a less-qualified applicant. (Assume that no transfer can be made in this case.) The disagreement point is therefore $d = (d_J, d_R)$, where $d_J = 15{,}000$ and $d_R = 10{,}000$. The bargaining problem is pictured in Figure 18.2, in which the two points noted on the $x = 0$ and $x = 1$ lines denote the payoff vectors associated with $t = 0$.

In this bargaining problem, an agreement to employ Jerry at the high school yields the parties a joint value of

$$u_J(x) + u_R(x) = [v_J(x) + t] + [v_R(x) - t].$$

Using the exact form of the payoff functions, we have

$$[10{,}000 - 3{,}000x] + [40{,}000 + 5{,}000x] = 50{,}000 + 2{,}000x.$$

Jerry and Rosemary have a joint interest in selecting x to maximize this value. In Figure 18.2, this corresponds to selecting the diagonal line that is farthest out—that is, choosing $x = 1$ to get the joint value \$52,000. Note that increasing x from 0 to 1 makes Jerry worse off, but it makes Rosemary so much better off that the joint value of the relationship rises. Jerry can be compensated by way of a salary increase, making both parties happy. In this example, the players' joint value from the default outcome is

$$d_J + d_R = 15{,}000 + 10{,}000 = 25{,}000.$$

Thus, the agreement to set $x = 1$ generates a surplus of $52{,}000 - 25{,}000 = 27{,}000$.

THE STANDARD BARGAINING SOLUTION

Sometimes it is useful to discuss how bargaining problems are resolved, with the use of terms such as "efficiency" and "bargaining power." For example, we may say that a particular player is expected to obtain most of the surplus because she has a lot of bargaining power. Or we may say that players are generally expected to bargain efficiently. We can explore these notions by studying their noncooperative foundations—that is, by examining how they are related to the specific bargaining game that is played. You will be happy to know, in fact, that noncooperative analysis of bargaining is the topic of Chapter 19. At this point, however, I want you to see how bargaining power and efficiency are defined by using the abstract representation of bargaining problems just presented.

Efficiency is an important criterion with which to judge the outcome of a negotiation process. As already noted, in settings with transferable utility, an outcome is efficient if and only if it maximizes the players' joint value. Thus, it is easy to determine which outcomes are efficient—simply find the value(s) of x that yields the largest sum $v_1(x) + v_2(x)$. Let us denote by v^* the maximized joint value for any given bargaining game.[5] As an example, recall that, in the bargaining problem faced by Jerry and Rosemary, the parties' joint value is maximized by employing Jerry and having him coach softball in addition to teaching drama ($x = 1$); thus, $v^* = 52{,}000$. In general, efficiency

[5]There are cases in which default yields the greatest joint value, so v^* is the default payoff.

is an attractive property and, on normative grounds, you would advise players to achieve it. On the positive side, however, you might doubt whether people actually realize efficient outcomes in some cases.

Bargaining power is associated with how players divide the value of their contract. To assess the scope of bargaining power, first recall that each player can unilaterally induce the default outcome by refusing to reach an agreement. Thus, no rational player would accept an agreement that gives her less than her default payoff. As a result, the players really do not negotiate over v^*; they negotiate over the *surplus* $v^* - d_1 - d_2$. In the job-negotiation example, Jerry will not accept an agreement that gives him less than 15,000 and Rosemary will not accept an agreement yielding her less than 10,000. Therefore, they actually negotiate over the difference between the maximized joint value and the default joint value, which is $52{,}000 - 25{,}000 = 27{,}000$.

It is useful to summarize bargaining powers by using *bargaining weights* π_1 and π_2, where $\pi_1, \pi_2 \geq 0$ and $\pi_1 + \pi_2 = 1$. We interpret π_i as the proportion of the surplus obtained by player i. For example, in the bargaining problem of Jerry and Rosemary, Jerry obtains proportion π_J of the \$27,000 surplus, whereas Rosemary obtains proportion π_R.

The **standard bargaining solution** is a mathematical representation of efficiency and proportional division.[6] Each player is assumed to obtain his default payoff, plus his share of the surplus. For example, suppose Jerry's bargaining weight is $\pi_J = 1/3$ and Rosemary's bargaining weight is $\pi_R = 2/3$. Then we expect the players to reach an agreement in which Jerry obtains the payoff

$$u_J^* = d_J + \pi_J(v^* - d_J - d_R) = 15{,}000 + (1/3) \cdot 27{,}000 = 24{,}000$$

and Rosemary obtains the payoff

$$u_R^* = d_R + \pi_R(v^* - d_J - d_R) = 10{,}000 + (2/3) \cdot 27{,}000 = 28{,}000.$$

Note that, because $\pi_J + \pi_R = 1$, the sum of these payoffs is the joint value $v^* = 52{,}000$. We know that $x = 1$ is selected. Further, it must be that t is chosen to yield the values u_J^* and u_R^*. Looking at Jerry's payoff, we see that t solves

$$u_J^* = v_J(1) + t = 24{,}000;$$

[6]The standard bargaining solution is an offshoot of John Nash's seminal treatment of bargaining problems, which is contained in his article "The Bargaining Problem," *Econometrica* 18(1950):155–162. For an overview of the literature on solutions to bargaining problems, see W. L. Thomson, *Bargaining Theory: The Axiomatic Approach* (San Diego: Academic Press, forthcoming).

that is,

$$10{,}000 - 3{,}000 + t = 24{,}000,$$

which simplifies to $t = 17{,}000$. (We get the same result by looking at Rosemary's payoff.) In the end, the standard bargaining solution predicts that Jerry and Rosemary will agree to put him in charge of both the drama courses and the softball team ($x = 1$) and pay him a salary of $t = 17{,}000$. This leads to the payoff $u_J^* = 24{,}000$ for Jerry and $u_R^* = 28{,}000$ for Rosemary.

In a general bargaining problem, you can compute the standard bargaining solution as follows.

Procedure for calculating the standard bargaining solution:

1. Calculate the maximized joint value v^* by determining the value x^* that maximizes $v_1(x) + v_2(x)$.
2. Note that, with the standard bargaining solution, player i obtains the payoff $d_i + \pi_i(v^* - d_1 - d_2)$, where d_i is player i's default payoff. Thus, the transfer t will satisfy

$$d_1 + \pi_1(v^* - d_1 - d_2) = v_1(x^*) + t$$

   for player 1 and

$$d_2 + \pi_2(v^* - d_1 - d_2) = v_2(x^*) - t$$

   for player 2. These two equations are equivalent.
3. Solve one of these equations for t to find the transfer that achieves the required split of the surplus.

GUIDED EXERCISE

Problem: John is a computer expert who is negotiating an employment contract with a prospective employer (the firm). The contract specifies two things: (1) John's job description (programmer or manager), and (2) John's salary t. If John works as a programmer for the firm, then John's payoff is $t - 10{,}000$ (in dollars) and the firm's payoff is $100{,}000 - t$. If John works as a manager, then his payoff is $t - 40{,}000$ and the firm's payoff is $x - t$, where x is a constant parameter. Assume $x > 150{,}000$. If John and the firm fail to reach an agreement, then the firm gets zero and John obtains w. In other words, the default outcome of this negotiation problem leads to the payoff vector $(w, 0)$, where

John's payoff is listed first. The value w is due to John's outside opportunity, which is to work on his own as a computer consultant.

Solve this bargaining problem by using the standard bargaining solution, under the assumption that John's bargaining power is π_J and the firm's bargaining power is π_F. Assume that $w < 90{,}000$.

Solution: Note that the surplus with John working as a programmer is

$$(t - 10{,}000) + (100{,}000 - t) - w = 90{,}000 - w.$$

The surplus with him working as a manager is

$$(t - 40{,}000) + (x - t) - w = x - 40{,}000 - w,$$

which exceeds $110{,}000 - w$ by the assumption on x. Thus, the maximal joint value is attained by having John work as a manager. The standard bargaining solution makes John's overall payoff equal to

$$w + \pi_J(x - 40{,}000 - w) = (1 - \pi_J)w + \pi_J(x - 40{,}000).$$

Setting this equal to $t - 40{,}000$ (the direct definition of John's payoff when he works as a manager) and solving for t yields

$$t = (1 - \pi_J)(w + 40{,}000) + \pi_J x.$$

EXERCISES

1. Calculate the standard bargaining solution for the following variations of the Jerry–Rosemary example in this chapter. In each case, graph the bargaining set, find the maximized joint value, determine the players' individual values, and compute the transfer t that the players select.

   (a) $d_J = 0$, $d_R = 0$, $v_J(x) = 10{,}000 - 6{,}000x$, $v_R(x) = 40{,}000 + 4{,}000x$, $x \in \{0, 1\}$, $\pi_J = 1/2$, and $\pi_R = 1/2$.

   (b) $d_J = 0$, $d_R = 0$, $v_J(x) = 60{,}000 - x^2$, $v_R(x) = 800x$, x is any positive number, $\pi_J = 1/2$, and $\pi_R = 1/2$.

   (c) $d_J = 40{,}000$, $d_R = 20{,}000$, $v_J(x) = 60{,}000 - x^2$, $v_R(x) = 800x$, x is any positive number, $\pi_J = 1/4$, and $\pi_R = 3/4$.

2. Consider the setting of this chapter's Guided Exercise. Suppose John can invest some of his free time either enhancing his productivity with the firm (increasing x) or raising his productivity as an individual computer consultant (increasing

w). How would you recommend that John spend his free time? (Hint: The answer depends on π_J.)

3. Use the standard bargaining solution to find the outcomes of the following bargaining problems. Player 1's payoff is $u_1 = v_1(x) + t$ and player 2's payoff is $u_2 = v_2(x) - t$. In each case, graph the maximized joint value and the default outcome; report the chosen x, t as well as the players' individual payoffs.

   (a) $x \in \{5, 10, 15\}$, $v_1(x) = x$, $v_2(x) = x$, $d_1 = 0$, $d_2 = 0$, $\pi_1 = 1/2$, and $\pi_2 = 1/2$.

   (b) $x \in \{5, 10, 15\}$, $v_1(x) = x$, $v_2(x) = x$, $d_1 = 2$, $d_2 = 4$, $\pi_1 = 1/2$, and $\pi_2 = 1/2$.

   (c) $x \in \{5, 10, 15\}$, $v_1(x) = x$, $v_2(x) = x$, $d_1 = 2$, $d_2 = 4$, $\pi_1 = 1/4$, and $\pi_2 = 3/4$.

   (d) $x \in \{5, 10, 15\}$, $v_1(x) = 20x$, $v_2(x) = -x^2$, $d_1 = 0$, $d_2 = 0$, $\pi_1 = 1/4$, and $\pi_2 = 3/4$.

   (e) $x \in (-\infty, \infty)$, $v_1(x) = 16x + x^2$, $v_2(x) = 8x - 2x^2$, $d_1 = 0$, $d_2 = 0$, π_1 and π_2 arbitrary.

4. Suppose that you must bargain with another party over how to realize a large joint value v^*. Explain why you care about the *other* party's disagreement payoff.

5. Suppose that you must negotiate with another party and that you have an opportunity to either raise your disagreement payoff by 10 units or raise the maximal joint value v^* by 10 units. Which should you choose? Is your choice efficient?

6. Discuss a real-world example of negotiation in terms of maximized joint value, bargaining weights, and a disagreement point.

19 ANALYSIS OF SIMPLE BARGAINING GAMES

The preceding chapter developed some of the basic concepts and language for studying bargaining problems. In this chapter, I explain how bargaining is studied by using noncooperative game theory. The program analyzes specific bargaining procedures as represented by extensive-form games. I evaluate the games by using the concept of subgame perfect Nash equilibrium. The modeling exercises allow us to generate intuition on whether efficiency is achieved and on the determinants of bargaining power. With reference to the standard bargaining solution developed in Chapter 18, the goal here is to develop insights on what determines a player's bargaining weight π_i. As you shall see, at the least we can trace bargaining strength to a player's patience and his ability to make offers at strategically important points in time.

ULTIMATUM GAMES: POWER TO THE PROPOSER

The ultimatum bargaining game is perhaps the simplest of bargaining models. In the example presented in Chapter 14 (see Figure 14.5), a buyer and a seller negotiate the price of a painting. The seller offers a price and then the buyer accepts or rejects it, ending the game. Because the painting is worth \$100 to the buyer and nothing to the seller, trade will generate a surplus of \$100. The price determines how this surplus is divided between the buyer and the seller; that is, it describes the terms of trade. From the Guided Exercise of Chapter 15, you have already seen that there is a subgame perfect equilibrium of this game in which trade takes place at a price of 100, meaning that the seller gets the entire surplus. In this section, I expand the analysis to show that this equilibrium is unique.

It will be helpful to study the ultimatum bargaining game in the abstract setting in which the surplus is normalized to 1. This normalization helps us to concentrate on the players' shares of the monetary surplus; for example, if player 1 obtains 1/4, then it means that player 1 gets 25 percent of the surplus. As discussed in Chapter 18, we assume transferable utility so that the price divides the surplus in a linear fashion; thus, if one player gets m, then the

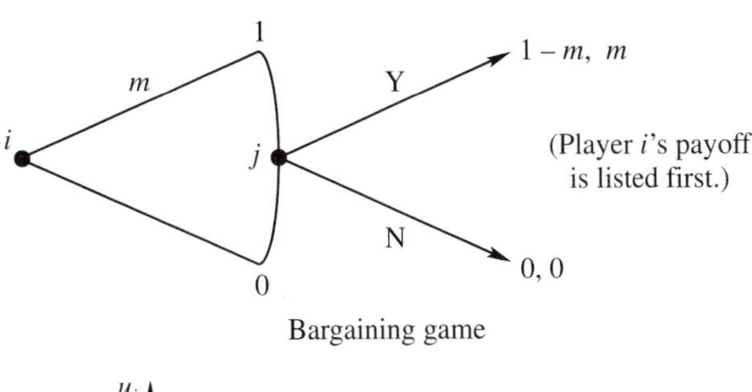

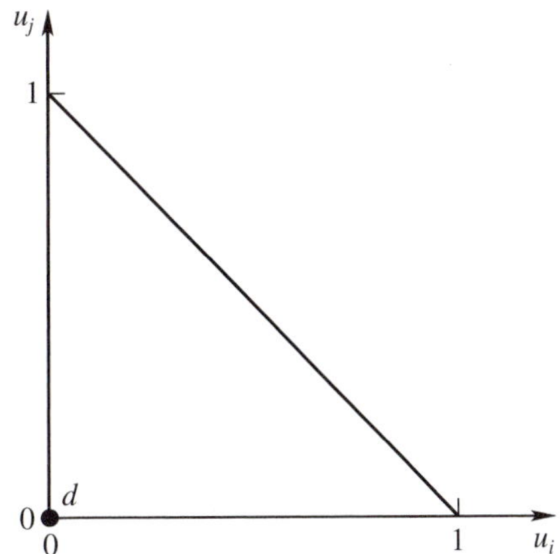

FIGURE 19.1
Ultimatum bargaining.

other gets $1 - m$. It will also be helpful to name the players i and j, allowing us to put either player 1 or player 2 into the role of making the offer. This standard ultimatum game is pictured in Figure 19.1.[1] Also pictured are the bargaining set and the disagreement point corresponding to this game. The disagreement point is (0, 0) because this payoff occurs when the responder rejects the proposer's offer.

To find the subgame perfect equilibrium of the ultimatum game, begin by observing that the game has an infinite number of subgames. In particular, every decision node for player j initiates a subgame (that ends following his decision). This should be obvious because player j observes the offer of player i; all information sets consist of single nodes. Consider the subgame following any particular offer of m by player i, where $m > 0$. If player j accepts the offer, he obtains m. If he rejects the offer, then he gets 0. Therefore, player j's best action is to accept. Only when $m = 0$ can rejection be an opti-

[1] Note that, according to the extensive form pictured, player i cannot make an offer of $m > 1$ or $m < 0$. In fact, the results do not change if one leaves m unrestricted. Intuitively, it is not rational for player i to offer $m > 1$ and it is not rational for player j to accept an offer of $m < 0$. For simplicity, I just impose $m \in [0, 1]$.

mal response for player j, and in this case acceptance also is optimal (because player j is indifferent between the two actions). The analysis thus indicates that player j has only two sequentially rational strategies: (s_j^*) accept all offers, and ($\hat{s}_j$) accept all offers of $m > 0$ and reject the offer of $m = 0$. These are the only strategies that specify a Nash equilibrium for each of the proper subgames.

Having identified player j's sequentially rational strategies, one can easily determine the subgame perfect equilibrium of the ultimatum game. The equilibrium must include either strategy s_j^* or strategy $\hat{s}_j$ on the part of player j. Let us evaluate which of these strategies might be part of a subgame perfect equilibrium. First, note that the strategy profile in which player i picks $m = 0$ and player j plays s_j^* is a Nash equilibrium of the game. Player i obviously has no incentive to deviate from $m = 0$, given that he obtains the full surplus under the prescribed strategy profile. Second, observe that there is no Nash equilibrium of the game in which player j adopts $\hat{s}_j$. To see this, note that player i has *no* well-defined best response to $\hat{s}_j$. If player i chooses $m > 0$, then he obtains $1 - m$, given player j's strategy. Therefore, player i would like to select the smallest possible m. But $m = 0$ yields a payoff of 0 against $\hat{s}_j$.

These facts imply that there is a single subgame perfect equilibrium of the ultimatum game: player i selects $m = 0$, and player j selects s_j^*. The equilibrium payoff is 1 for player i and 0 for player j. The equilibrium outcome is efficient because the parties realize the gains from trade; their joint value is maximized.

The ultimatum game demonstrates how a great deal of bargaining power is wielded by a person in the position of making a take-it-or-leave-it offer; in terms of the standard bargaining solution, $\pi_i = 1$. If you can put yourself in this position (say, by somehow committing to terminate negotiation after consideration of your offer), you would be wise to do so.[2]

TWO-PERIOD, ALTERNATING-OFFER GAMES: POWER TO THE PATIENT

The ultimatum game is instructive and applicable, but it is too simplistic a model of most real-world negotiation. Bargaining generally follows a more in-

[2]To some, it seems that offering 0 to player j is a bit extreme and maybe even risky. Perhaps player i should at least make it worthwhile for player j to accept by offering a small positive amount. Intuitively, this might be the case. However, whatever is required to get player j to accept is supposed to already be embedded in the payoffs. That is, zero represents the amount that player j requires to accept the offer. If this still seems a bit unintuitive, you may be comforted to know that, in games with a smallest monetary unit, there are subgame perfect equilibria in which the proposer offers a small, positive amount to the responder. For example, if player i cannot offer fractions of pennies, then the offer and acceptance of one cent is a subgame perfect equilibrium outcome.

tricate process in reality. The theory can be expanded in many ways, perhaps the most obvious of which is to explicitly model multiple offers and counteroffers by the parties over time. In fact, in many real settings, the sides alternate making offers until one is accepted. For example, real estate agents routinely force home sales into the following procedure: the seller posts a price, the prospective buyer makes an offer that differs from the seller's asking price, the seller makes a counteroffer, and so forth.

Offers and counteroffers take time. In home sales, one party may wait a week or more for an offer to be considered and a counteroffer returned. "Time is money," an agent may say between utterances of the "location, location, location" mantra. In fact, the agent is correct to the extent that people are impatient or forego productive opportunities during each day spent in negotiation. Most people are impatient to some degree. Most people prefer not to endure protracted negotiation procedures. Most people *discount* the future relative to the present.

How a person discounts the future may affect his or her bargaining position. It seems reasonable to expect that a very patient bargainer—or someone who has nothing else to do and nowhere else to go—should be able to win a greater share of the surplus than should an impatient one. To incorporate discounting into game-theoretic models, we use the notion of a *discount factor*. A discount factor δ_i for player i is a number used to deflate a payoff received tomorrow so that it can be compared with a payoff received today.

To make sense of this idea, fix a length of time that we call the *period length*. For simplicity, suppose that a period corresponds to 1 week. If given the choice, most people would prefer receiving x dollars today (in the current period) rather than being assured of receiving x in one week (in the next period). People exhibit this preference because people are generally impatient, perhaps because they would like the opportunity to spend money earlier rather than later. In the least, a person could take the x dollars received today and deposit it in a bank account to be withdrawn next week with interest.

We could conduct a little experiment by asking a person what amount of money received in the current period would be worth about the same to her as receiving x in the next period. Call this amount y. Because of discounting, it will be the case that $y < x$. Then we can define δ_i so that $y = \delta_i x$. Thus, δ_i is the multiplicative factor by which any amount received in the following period is discounted to make it comparable to payoffs in the current period. The discount factor δ_i is generally a number between 0 and 1, where larger values correspond to greater patience.

Consider a two-period, alternating-offer bargaining game. This game begins in the first period, at which time player 1 makes an offer m^1. After observing the offer, player 2 decides whether to accept or reject it. If player 1's offer is

FIGURE 19.2 Bargaining set and disagreement point for two-period game.

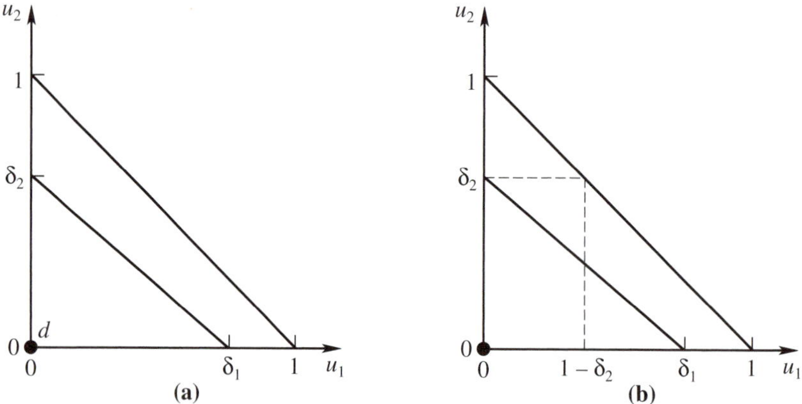

accepted, the game ends with player 1 receiving $1 - m^1$ and player 2 obtaining m^1. If player 2 rejects the offer, then interaction continues in the second period with the roles reversed. Player 2 makes a counteroffer m^2, which player 1 either accepts or rejects.[3] Note that, if the initial offer is rejected, time elapses before player 2 can make her counteroffer. To compare payoffs received in the first period with those received in the second period, we multiply the latter by the players' discount factors. Specifically, if player 1 accepts player 2's offer in the second period, then player 1 gets $\delta_1 m^2$ and player 2 gets $\delta_2(1 - m^2)$. If player 1 rejects player 2's offer, then the game ends and both players obtain 0.

Observe that the two-period, alternating-offer game consists basically of repetition of the ultimatum-game structure. In the first period, the players interact as in the ultimatum game, with player 1 playing the role of i and player 2 the role of j. If the offer is rejected, then time passes into the second period, where players interact in an actual ultimatum game with the roles reversed. The game ends after two periods. If an agreement is reached in the second period, then the payoffs are discounted relative to the first period.

The bargaining set and disagreement point associated with this bargaining game are pictured in Figure 19.2(a). The extensive-form game itself is not pictured; as an exercise, you might draw the extensive form yourself. In Figure 19.2(a), the outer line represents payoff vectors possible if the players reach an agreement in the first period. The inner line, from δ_1 on the x-axis to δ_2 on the y-axis, represents payoff vectors that can be achieved if the players reach an agreement in the second period. The disagreement point $(0, 0)$ is the payoff that would result if the players never reach an agreement.

---

[3]The analysis of finite-period games such as this one were reported in I. Stahl, *Bargaining Theory* (Stockholm: Economics Research Institute at the Stockholm School of Economics, 1972).

Computing the subgame perfect equilibrium of this game is straightforward once one realizes that the subgames starting in the second period have already been solved by our ultimatum game analysis. If player 2 rejects m^1, then she effectively induces play of the ultimatum game in the following period. To convince yourself of this fact, simply observe that, starting in period 2, the game looks exactly like the ultimatum game, with player 2 playing the role of i. Then note that subgame perfection implies that player 1 accepts all counteroffers and player 2 chooses $m^2 = 0$. The nondiscounted payoffs in these subgames are 1 for player 2 and 0 for player 1.

We have completely characterized equilibrium behavior in the second period. With this characterization in mind, we can easily determine player 2's optimal decision in response to player 1's initial offer. Both players know that, if player 2 rejects player 1's first-period offer, then player 2 will obtain the full surplus available in the next period (given the equilibrium continuation). From the point of view of the first period, player 2 thus would receive $\delta_2 \cdot 1 = \delta_2$ by rejecting player 1's offer. In other words, player 2 can always reject and guarantee himself the full surplus in the following period, which is worth δ_2 in comparison with a payoff received in the first period. Therefore, player 2 must accept player 1's offer of m^1 if $m^1 > \delta_2$. On the other hand, rejection is the only rational move if $m^1 < \delta_2$. Player 2 is indifferent between accepting and rejecting the offer $m^1 = \delta_2$.

The preceding analysis implies that only one thing has yet to be determined regarding behavior following player 1's initial offer: whether player 2 accepts or rejects the offer $m^1 = \delta_2$. As with the ultimatum game, the only Nash equilibrium has player 2 accepting the offer that makes him indifferent; furthermore, player 1 makes this offer. Thus, there is a unique subgame perfect equilibrium of the two-period game. Player 1 offers $m^1 = \delta_2$, player 2 accepts if and only if $m^1 \geq \delta_2$, player 2 always offers $m^2 = 0$ at the start of the second period, and player 1 accepts every offer in the second period. The equilibrium yields a payoff of $1 - \delta_2$ to player 1 and δ_2 to player 2. Figure 19.2(b) gives a graphical account of the analysis.

A few comments are in order. First, observe that patience is positively related to bargaining power. In particular, if player 2 is impatient (represented by a small value of δ_2), then his equilibrium payoff is close to 0. In terms of the standard bargaining solution, π_2 is close to 0, so player 1 obtains most of the surplus to be divided. In general, with the equilibrium in the two-period, alternating-offer game, we interpret player 1's bargaining weight to be $\pi_1 = 1 - \delta_2$ and player 2's to be $\pi_2 = \delta_2$.

The relation between patience and bargaining power is supported more generally. For example, one can study a T-period version of the game where the players alternate making offers until either an offer is accepted or T periods have elapsed. Note that, if T is even, then player 2 has the last offer in such a game; if T is odd, then player 1 has the final offer. In the T-period game, if agreement is reached in some period t, then payoffs are discounted by the factors δ_1^{t-1} and δ_2^{t-1}. That is, the discount factor is applied for every consecutive period that one moves away from the first period. The T-period game can be analyzed in the same way that the two-period game is studied. One simply inserts the equilibrium payoffs from the $(T-1)$-period contest, appropriately discounted, as the value of rejection in the first period of the T-period game. For example, rejection by player 2 in the first period of the three-period game results in the discounted payoffs of the two-period game, which are δ_1^2 for player 1 and $\delta_2(1-\delta_1)$ for player 2. I elaborate on this with the Guided Exercise of this chapter. Exercise 3 at the end of this chapter asks you to characterize the subgame perfect equilibria of the T-period game.

The equilibrium of the alternating-offer game entails agreement made in the first period, so the players realize a joint value of 1. If they had reached an agreement only in the second period, then the joint value would have been $\delta_1 m^2 + \delta_2(1-m^2)$, which is strictly less than 1. If they had completely failed to reach an agreement in either period, then the joint value would have been 0. Thus, the outcome of the subgame perfect equilibrium is efficient. In fact, the equilibrium of the general T-period game also features agreement without delay. Thus, although the bargaining games studied thus far provide a theory of bargaining power, they do not yield a theory of bargaining delay or inefficiency.

This conclusion may be a bit unsatisfying, because delay and inefficiency are regular occurrences in real bargaining settings. For example, almost every year there is a highly publicized case in which a union and a firm fail to reach an agreement in contract negotiations, leading to a strike that lasts a substantial number of days. Perhaps the main difference between the real world and the models just examined is that people do not always have complete information about each other in the real world, whereas both parties are assumed here to know exactly what game is being played (including knowing each other's discount factors). Part IV of this text introduces games with incomplete information and demonstrates how informational asymmetry can lead to inefficiency.

INFINITE-PERIOD, ALTERNATING-OFFER GAME

Next consider an alternating-offer game that runs for a potentially infinite number of periods. That is, there is no final period in which bargaining takes place. This game has a special property: the subgames starting from any period

t are exactly the same as the subgames starting from period $t + 2$. In other words, every subgame starting in an odd period looks like every other such subgame—these are infinite-period, alternating-offer games in which player 1 makes the first offer. Likewise, the subgames starting from even periods are identical. Although the infinite game seems much more complex than the ones studied so far in this chapter, these facts about subgames serve to simplify the search for a subgame perfect equilibrium.[4]

Recall that a subgame perfect equilibrium specifies an equilibrium on every subgame. Let us look for an equilibrium of the infinite-period game that is "stationary" in the sense that a player makes the same offer whenever she is on the move. Also suppose that these offers are accepted. To find such an equilibrium, if one exists, let m_j denote the offer made by player i whenever on the move. If such an offer is accepted, then player i expects to get $1 - m_j$ in any subgame in which i makes the first offer. Intuitively, it should be the case that player i's equilibrium offer puts player j in a position of being indifferent between accepting it and rejecting it in favor of going on to the next period. Accepting the offer would give player j the payoff m_j, discounted to the current period. Rejecting the offer would yield a payoff of $\delta_j(1 - m_i)$, because player j expects to obtain $1 - m_i$ in the next period when the current offer is rejected. Indifference means $\delta_j(1 - m_i) = m_j$. Because this equation holds for both players, we have the following system of equations: $\delta_1(1 - m_2) = m_1$ and $\delta_2(1 - m_1) = m_2$. Solving these equations yields

$$m_1 = \frac{\delta_1(1 - \delta_2)}{1 - \delta_1\delta_2} \quad \text{and} \quad m_2 = \frac{\delta_2(1 - \delta_1)}{1 - \delta_2\delta_1}.$$

One can prove that the outcome sketched in the last paragraph is the unique subgame perfect equilibrium outcome in the infinite-period game. The equilibrium features agreement in the first period and yields a payoff of

$$\frac{1 - \delta_2}{1 - \delta_1\delta_2}$$

to player 1 and

$$\frac{\delta_2(1 - \delta_1)}{1 - \delta_1\delta_2}$$

[4]The first analysis of this game was reported in A. Rubinstein, "Perfect Equilibrium in a Bargaining Model" *Econometrica* 50(1982):97–110. My calculations of the equilibrium follow the method of A. Shaked and J. Sutton, "Involuntary Unemployment as a Perfect Equilibrium in a Bargaining Model," *Econometrica* 52(1984):1351–1364. For an overview of noncooperative models of bargaining, see A. Muthoo, *Bargaining Theory with Applications* (New York: Cambridge University Press, 1999).

to player 2. You should verify that player i's equilibrium payoff increases as δ_i rises or δ_j falls or both. Furthermore, using L'Hospital's derivative rule, one can verify that, for $\delta_1 = \delta_2 = \delta$, these payoffs both converge to $1/2$ as δ approaches 1. In other words, if the players are equally very patient, then they split the surplus evenly. This outcome is characterized by equal bargaining weights ($\pi_1 = \pi_2 = 1/2$) in the language of the standard bargaining solution.

MULTILATERAL BARGAINING

Sometimes negotiation involves more than two parties. With many players at the bargaining table, there is an endless variety of bargaining protocols and rules for agreement. For example, the members of a legislature must bargain over the specifications and passage of new laws. The legislature's procedural rules describe (1) how individual members can be recognized to make proposals, (2) whether other members are allowed to offer amendments, and (3) a voting rule that determines how members can respond to proposals and that gives the criteria for passage into law.

Let us analyze a simple model of legislative bargaining, where the legislature has three members (players 1–3) and there is an infinite number of periods. Imagine that the three players represent three districts in a small metropolitan region. There is an opportunity to fund a new project (an emergency-response system, for instance) that will benefit all of the regions equally. To keep things simple, suppose that the benefit of the project to each region is 1, for a total benefit of 3. The cost of the project is 2, which can be divided in any way between the regions. Note that the project creates value, and thus efficiency dictates that it be funded. Each player wants to maximize the gain to her own region and therefore wants to lower the portion of the project's cost borne by her constituents. This is essentially a setting of three-person bargaining, whereby the players negotiate over how to split the project's surplus of 1.

Suppose that legislative interaction takes place as follows. In a given period, one player has the right to make a proposal $x = (x_1, x_2, x_3)$, where x_i denotes the amount offered to player i. For example, if $x_1 = 1/4$ then it means that player 1's district is asked to pay $3/4$ to help fund the project, so its gain is $1 - (3/4)$. Player 1 makes the proposal in period 1, player 2 makes the proposal in period 2, player 3 makes the proposal in period 3, and the order cycles accordingly in future periods. After a proposal is made in a given period, the other two players simultaneously vote for or against it. Assume that the voting rule is unanimity. If both players vote in favor, then the proposal passes, and the game ends. If one or both players vote against the proposal, then the game continues with another proposal in the next period. Amendments are not allowed, so we have a so-called *closed rule legislature*. Assume that players discount the future using the shared discount factor δ.

Let us look for a stationary subgame perfect equilibrium in the way that we analyzed the infinite-period, alternating-offer game in the previous section. Suppose that, conditional on reaching any particular period, the proposer will offer x^{n} to the player who would be next to make a proposal, and will offer x^{l} to the other player who would make the proposal two periods hence. Let x^{p} be the amount that the proposer offers for herself (that is, her own district). Because the players have the same discount factor, it makes sense to conjecture that these equilibrium amounts will not depend on the identity of the proposer. Suppose also that the equilibrium proposal is accepted by the other players. We next examine how these values must be related in equilibrium.

First note that, under our working hypothesis, equilibrium offers are accepted, and we have

$$x^{\mathrm{p}} + x^{\mathrm{n}} + x^{\mathrm{l}} = 1$$

because the players are bargaining over a total value of 1. Next, notice that the proposer must offer at least δx^{p} to the player who would make the proposal in the next period; otherwise, this player would vote against the proposal and get x^{p} in the following period when she can make the proposal and expect that it be passed. Thus, we have $x^{\mathrm{n}} \geq \delta x^{\mathrm{p}}$. Likewise, the proposer must offer at least $\delta^2 x^{\mathrm{p}}$ to the player who would make the proposal two periods hence, because this player has the option of rejecting proposals until it is her turn to make the offer. Thus, we have $x^{\mathrm{l}} \geq \delta^2 x^{\mathrm{p}}$.

The proposing player would like to obtain the most possible for her district, which means lowering x^{n} and x^{l} until these two inequalities bind (which means that they became equalities—that is, $x^{\mathrm{n}} \geq \delta x^{\mathrm{p}}$ becomes $x^{\mathrm{n}} = \delta x^{\mathrm{p}}$). We conclude that $x^{\mathrm{n}} = \delta x^{\mathrm{p}}$ and $x^{\mathrm{l}} = \delta^2 x^{\mathrm{p}}$. Using this substitution in the first equation of the previous paragraph, we get

$$x^{\mathrm{p}} + x^{\mathrm{n}} + x^{\mathrm{l}} = x^{\mathrm{p}} + \delta x^{\mathrm{p}} + \delta^2 x^{\mathrm{p}} = 1,$$

which simplifies to

$$x^{\mathrm{p}} = \frac{1}{1+\delta+\delta^2}, \quad x^{\mathrm{n}} = \frac{\delta}{1+\delta+\delta^2}, \quad \text{and} \quad x^{\mathrm{l}} = \frac{\delta^2}{1+\delta+\delta^2}.$$

As a numerical example, take $\delta = 1/2$, in which case $x^{\mathrm{p}} = 4/7$, $x^{\mathrm{n}} = 2/7$, and $x^{\mathrm{l}} = 1/7$. Then, in the subgame perfect equilibrium, player 1's offer in the first period is $x^1 = (4/7, 2/7, 1/7)$, which is accepted by the other players.

In the out-of-equilibrium contingency in which period 2 is reached, player 2's equilibrium offer would be $x^2 = (1/7, 4/7, 2/7)$.

This simple model of legislative bargaining shows that the intuition from the two-player, infinite-period model carries over to the general setting of multilateral bargaining. The player who gets to make the first proposal fares better than do the others, but equity arises as the discount factor converges to 1. You can quickly verify that, as δ becomes close to 1, x^{p}, x^{n}, and x^{l} all converge to 1/3. Important issues still on the table are: (1) whether there are other equilibria in addition to the stationary one just characterized, and (2) how the equilibrium outcome might change under different procedural rules. These issues have been probed in the scientific literature.[5]

GUIDED EXERCISE

Problem: Consider the three-period, alternating-offer bargaining game. Suppose that the discount factor for both players is δ, where $0 < \delta < 1$. In this game, player 1 makes the first offer, in period 1. If player 2 rejects this offer, then the game continues in period 2, where player 2 makes the offer and player 1 decides whether to accept or reject it. If player 1 rejects this offer, then the game continues in period 3, where player 1 makes the final offer and player 2 accepts or rejects. Compute the subgame perfect equilibrium of this game.

Solution: Consider the subgames from the beginning of period 3, where player 1 makes an offer m^3 and player 2 accepts or rejects it. Since this is an ultimatum-offer subgame, we know its equilibrium: player 1 chooses $m^3 = 0$ and player 2 accepts all offers. To see this, note that it would be irrational for player 2 to reject a positive offer. As a consequence, it is not rational for player 1 to select some $\hat{m} > 0$ because, given that player 2 is sure to accept it, player 1 gains by selecting a lower offer (such as $m = \hat{m}/2$). Thus, if the game reaches period 3, then sequential rationality dictates that player 1 will obtain the entire surplus from that point.

Next, examine the end of period 2, where player 1 is considering whether to accept an offer of m^2 from player 2. If player 1 accepts this offer, then she will obtain m^2 in period 2. It will be useful *not* to write this payoff in period 1 terms, so let us not multiply it by the discount factor. Importantly, player 1 must compare m^2 to the payoff she would receive by rejecting player 2's offer and waiting until period 3. By waiting, player 1 will obtain the discounted

---

[5] If you are interested in the analysis of legislative processes, a good place to start is with D. P. Baron and J. A. Ferejohn, "Bargaining in Legislatures," *The American Political Science Review* 83(1989):1181–1206.

value of the outcome from period 3, which is $\delta \cdot 1 = \delta$. Note that this is discounted because player 1 gets all of the surplus in the *next* period. Also note that discounting occurs just once since the payoffs are put in terms of period 2 (which we are relating to period 3). We conclude that, in period 2, player 1 will accept m^2 if and only if $m^2 \geq \delta$. There is the issue of player 1 being indifferent between accepting and rejecting if $m^2 = \delta$. But it will turn out that she will have to accept in this case, for otherwise, player 2 would have the incentive to offer "the smallest amount that is strictly greater than δ," which does not exist.

Moving to the beginning of period 2, note that player 2 expects to get nothing if he offers less than δ to player 1, for in this case player 1 would reject the offer. If player 2 offers at least δ, then player 1 will accept. Thus, player 2's optimal offer is $m^2 = \delta$. In summary, if the game reaches period 2, then in the subgame perfect equilibrium player 1 gets δ and player 2 gets $1-\delta$.

The foregoing basically repeated the calculations for the two-period game discussed earlier in this chapter. Analysis of behavior in period 1 works the same way. If player 2 rejects player 1's offer of m^1, then player 2 will get $1 - \delta$ from the next period. Discounting this amount to compare it to a payoff received in period 1, we see that player 2 will accept player 1's offer if and only if $m^1 \geq \delta(1 - \delta)$. Player 1's optimal offer is thus $m^1 = \delta(1 - \delta)$.

In summary, the unique subgame perfect equilibrium of the three-period, alternating-offer bargaining game features agreement in the first period. Player 1 obtains $1 - \delta(1 - \delta)$ and player 2 obtains $\delta(1 - \delta)$. The outcome is efficient. Player 1's payoff is large when the players are very patient (because she can always wait until the third period, where she enjoys having the final offer), as well as when the players are very impatient (because then she can take advantage of player 2's incentive to accept almost anything).

EXERCISES

1. Suppose the president of the local teachers' union bargains with the superintendent of schools over teachers' salaries. Assume the salary is a number between 0 and 1, 1 being the teachers' preferred amount and 0 being the superintendent's preferred amount.

(a) Model this bargaining problem by using a simple ultimatum game. The superintendent picks a number x, between 0 and 1, which we interpret as his offer. After observing this offer, the president of the union says "yes" or "no." If she says "yes," then an agreement is reached; in this case, the superintendent (and the administration that she represents) receives $1 - x$ and the president (and the union) receives x. If the president says "no," then both parties receive

0. Using the concept of backward induction, what would you predict in this game?

(b) Let us enrich the model. Suppose that, before the negotiation takes place, the president of the union meets with the teachers and promises to hold out for an agreement of at least a salary of z. Suppose also that both the superintendent and the president of the union understand that the president will be fired as union leader if she accepts an offer $x < z$. They also understand that the president of the union values both the salary and her job as the leader of the union and that she will suffer a great personal cost if she is dismissed as president. To be precise, suppose that the president suffers a cost of y utility units if she is fired (this is subtracted from whatever salary amount is reached through negotiation, 0 if the president rejects the salary offer). That is, if the president accepts an offer of x, then she receives $x - y$ in the event that $x < z$, and x in the event that $x \geq z$. If the president rejects the offer, then she obtains a payoff of 0. In the bargaining game, what offers of the superintendent will the president accept? (For what values of x will the president say "yes"? Your answer should be a function of y and z.)

(c) Given your answer to part (b), what is the outcome of the game? (What is the superintendent's offer and what is the president's response?) Comment on how the union's final salary depends on y.

(d) Given your answer to part (b), what kind of promise should the president make?

2. Suppose that you are attempting to buy a house, and you are bargaining with the current owner over the sale price. The house is of value \$200,000 to you and \$100,000 to the current owner; so, if the price is between \$100,000 and \$200,000, then you would both be better off with the sale. Assume that bargaining takes place with alternating offers and that each stage of bargaining (an offer and a response) takes a full day to complete. If agreement is not reached after ten days of bargaining, then the opportunity for the sale disappears (you will have no house and the current owner has to keep the house forever). Suppose that you and the current owner discount the future according to the discount factor δ per day. The real estate agent has allowed you to decide whether you will make the first offer.

(a) Suppose that δ is small; in particular $\delta < 1/2$. Should you make the first offer or let the current owner make the first offer? Why?

(b) Suppose that δ is close to 1; in particular $\delta > \sqrt[9]{1/2}$ (which means that $\delta^9 > 1/2$). Should you make the first offer or let the current owner make the first offer? Why?

3. Consider the alternating-offer bargaining game with T stages, in which the players discount payoffs received in successive stages according to the discount factor δ. That is, the two players take turns making offers. In each stage, one of the players makes an offer, followed by acceptance or rejection of the offer by the other player. Acceptance ends the game, and the surplus is divided as agreed. Offers and counteroffers continue as long as the players reject each other's offers. For example, if the players agree in the tth stage that player 1 shall have m share of the surplus, then player 1 gets a payoff of $\delta^{t-1}m$. If the offer in stage T is rejected, then the game ends with both players receiving 0. For simplicity, assume the surplus to be divided is one unit of utility.

Analyze this game by computing the subgame perfect Nash equilibrium. Start by restating the equilibrium derived in the text for $T = 1$ and $T = 2$. Then analyze the three-, four-, and five-period games in order. Write the payoff vectors corresponding to the equilibria of these games to reveal a pattern. Can you tell to what the payoff vector of the T-stage game converges as T becomes large?

4. [Nash's demand game[6]] Compute and describe the Nash equilibria of the following static bargaining game. Simultaneously and independently, players 1 and 2 make demands m_1 and m_2. These numbers are required to be between 0 and 1. If $m_1 + m_2 \leq 1$ (compatible demands, given that the surplus to be divided equals 1), then player 1 obtains the payoff m_1 and player 2 obtains m_2. On the other hand, if $m_1 + m_2 > 1$ (incompatible demands), then both players get 0. In addition to describing the set of equilibria, offer an interpretation in terms of bargaining weights as in the standard bargaining solution.

5. See if you can draw a graph like the one in Figure 19.2 to represent the subgame perfect equilibrium calculation for the infinite-period alternating-offer bargaining game.

6. Consider the following discounted, three-period bargaining game. The discount factor is δ, where $0 < \delta < 1$. In this game, player 1 makes the first offer. If player 2 rejects this offer, then player 1 makes *another* offer. If player 2 rejects the second offer, then player 2 makes the final offer. In other words, player 1 makes the offers in periods 1 and 2, whereas player 2 makes the offer only in period 3. Compute the subgame perfect equilibrium of this game.

7. Consider a three-player bargaining game, where the players are negotiating over a surplus of one unit of utility. The game begins with player 1 proposing a three-way split of the surplus. Then player 2 must decide whether to accept the proposal or to substitute for player 1's proposal his own alternative proposal. Finally, player 3 must decide whether to accept or reject the current proposal (whether it is player 1's or player 2's). If he accepts, then the players obtain the specified shares of the surplus. If player 3 rejects, then the players each get 0.

[6] J. F. Nash, "Two Person Cooperative Games," *Econometrica* 21(1953):128–140.

Draw the extensive form of this perfect-information game and determine the subgame perfect equilibria.

8. In experimental tests of the ultimatum bargaining game, subjects who propose the split rarely offer a tiny share of the surplus to the other party. Furthermore, sometimes subjects reject positive offers. These findings seem to contradict our standard analysis of the ultimatum game. Many scholars conclude that the payoffs specified in the basic model do not represent the *actual* preferences of the people who participate in the experiments. In reality, people care about more than their own monetary rewards. For example, people also act on feelings of spite and the ideal of fairness. Suppose that, in the ultimatum game, the responder's payoff is given by $y+a(y-z)$, where y is the responder's monetary reward, z is the offerer's monetary take, and a is a positive constant. That is, the responder cares about how much money he gets *and* he cares about relative monetary amounts (the difference between the money he gets and the money the other player gets). Assume that the offerer's payoff is as in the basic model.

(a) Represent this game in the extensive form, writing the payoffs in terms of m, the monetary offer of the proposer, and the parameter a.

(b) Find and report the subgame perfect equilibrium. Note how equilibrium behavior depends on a.

(c) What is the equilibrium monetary split as a becomes large? Explain why this is the case.

20 GAMES WITH JOINT DECISIONS; NEGOTIATION EQUILIBRIUM

I titled the preceding chapter "Analysis of Simple Bargaining Games" because the models examined there are very crude representations of the complex processes through which negotiation often takes place in the real world. In reality, there is usually more to bargaining than alternating offers and counteroffers. For example, negotiation may be segmented into times in which different aspects of the bargaining problem are discussed separately. The negotiation process may include a brainstorming session. A person may make all sorts of physical gestures intended to strengthen his bargaining position. Parties may make silly threats, such as "I will scream if you don't give in to my demands!" A person may scream or talk softly or insult the party at the other side of the table. None of these options are included in the models of Chapter 19; more realistic models would include them.

Remember, however, that the point of game-theoretic modeling is *not* to completely describe all of the nuances of strategic interaction. Rather, an artful and useful application of game theory isolates just a few strategic elements by using a model that is simple enough to analyze. There is no point in trying to capture all of the aspects of negotiation in one model. In fact, for many purposes, the games in Chapter 19 are *themselves* too complicated. For example, suppose we want to model a situation in which two business partners first bargain over a profit-sharing rule and then engage in productive interaction. On its own, the bargaining component may be modeled by using an alternating-offer game. Likewise, the productive interaction may be modeled by a game in which, say, the parties exert effort on the job. If we combine these two games, we would get a larger, more complicated game that may be difficult to analyze. Furthermore, perhaps a characterization of the players' bargaining powers is all we really want from the bargaining game. That is, we just want to capture the idea that the outcome of the negotiation process is consistent with certain bargaining weights. Then it makes sense to just employ the standard bargaining solution in place of the alternating-offer game.

Let me make this point again and more generally. As I noted already, contractual relationships normally include both (1) phases of interaction in which the parties negotiate over something, and (2) phases in which the parties work independently. As theorists, we often want to focus on aspects of the second type of interaction while summarizing the outcome of the first type in terms of the players' bargaining weights and disagreement points. It is thus helpful to address different components of strategic interaction at different levels of modeling detail. That is, we can study some components of a strategic setting by using a full noncooperative approach and other components by using an abbreviated approach. In this chapter, I explain how to use the standard bargaining solution as an abbreviated model of negotiation.

JOINT DECISIONS

A simple way of inserting a "summary" negotiation component into a noncooperative game is to include *joint-decision nodes* in the game tree.[1] A joint-decision node is an abbreviated description of negotiation between players over some tangible objects, such as profit-sharing rules, monetary transfers, or whether to form a partnership. Thus, a joint-decision node represents a place in the game where players negotiate and establish a contract. We specify a joint decision when we do not want to create a full noncooperative model of the negotiation process and when we have a simple theory of how negotiation is resolved (by using, for example, the standard bargaining solution).

To represent joint decisions in a tree, we can employ the same devices currently used to specify individual decisions. We simply allow some decision nodes to be designated as joint-decision nodes. The joint-decision nodes are graphically represented by double circles, to differentiate them from individual decision nodes. Furthermore, we label a joint-decision node with the set of players who are called on to make the joint decision. Branches represent the alternatives available to the players, as is the case with individual decision nodes. In addition, wherever there is a joint-decision node, we must designate one of the branches as the *default decision,* which is assumed to go into effect in the event that the players do not reach an agreement.[2]

A game with joint decisions is illustrated in Figure 20.1, which is a simple model of contracting between a supplier firm and a buyer firm. First, the firms jointly determine whether to contract and, if so, what damages c to specify if

[1]The game-theoretic framework with joint decisions is covered more deeply in J. Watson, "Contract and Game Theory: Basic Concepts for Settings with Finite Horizons," University of California, San Diego, Working Paper, 2005.

[2]Theories of joint decision making, such as the standard bargaining solution, may generally conclude that the agents avoid negotiation impasse. Nonetheless, it is important to specify a default decision because it influences the relative bargaining strengths of the parties and thus the outcome of negotiation. We implicitly assume that each party to a joint decision can unilaterally induce the default decision (by withholding agreement).

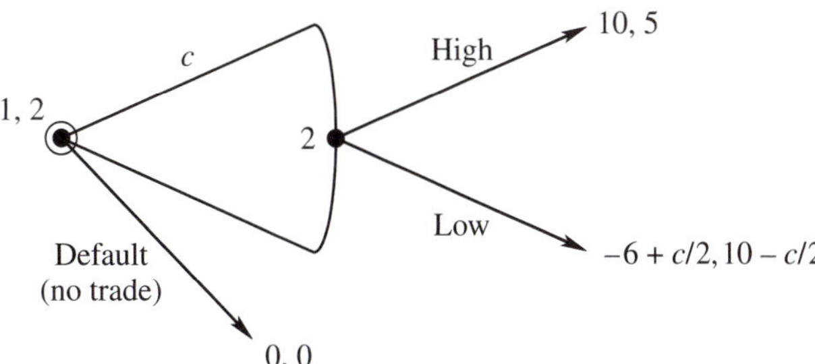

FIGURE 20.1
Representing joint decisions in a simple contract game.

the supplier (player 2) is caught providing a low-quality intermediate good. If they choose not to contract (which is the default decision), then the game ends and each receives nothing. If they contract and firm 2 then provides a high-quality good, payoffs are 10 for the buyer and 5 for the supplier. By providing a low-quality good, firm 2 saves money. However, the low-quality good is useless to the buyer. (These ideas are captured by the payoff numbers −6 and 10.) But with probability 1/2, the supplier is caught and damages are awarded by a court. (This is a payment of c from the supplier to the buyer.)

For another example, take the bargaining problem of Jerry and Rosemary that was discussed in Chapter 18. This problem is depicted in Figure 20.2 as a game with joint decisions. Note that, in this example, the tree has just one decision node—the joint-decision node representing the negotiation between the players. The default decision is no employment.

It is possible for a joint decision to be made at an information set containing more than one node. One can simply connect with dashed lines (enclose in the same information set) several nodes specifying joint decisions. This implies that all of the players participating in a joint decision have the same

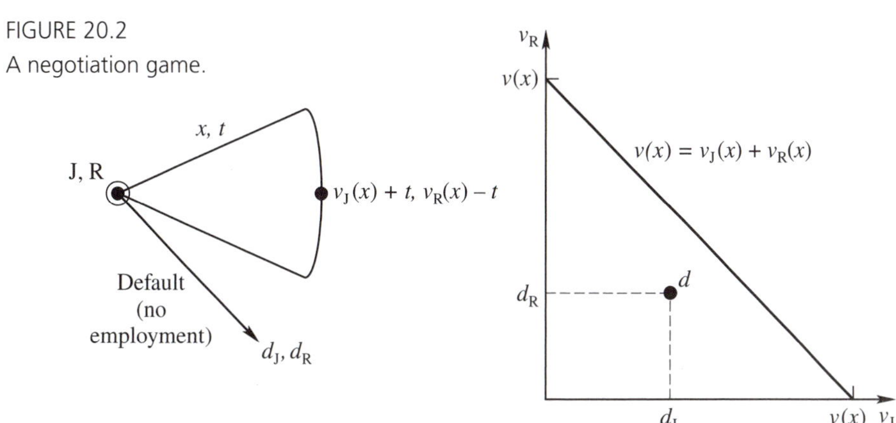

FIGURE 20.2
A negotiation game.

information when the decision is to be made.[3] While on this subject of information, let us revisit the tree rules stated in Chapter 14. For games with joint decisions, Tree Rules 1, 2, 3, and 5 are prescribed as before. Tree Rule 4 is replaced by:

> **Tree Rule 6** *For each information set, all included nodes must be decision nodes for the same subset of players—that is, belonging to either a single player or, in a joint decision, the same group of players.*

Before I move on to the analysis of games with joint decisions, let me make a few more comments on the interpretation of joint-decision nodes. At a joint-decision node, the branches correspond to tangible items over which the parties *spot contract*. By "spot contract," I mean that these items are automatically enforced as a part of the agreement. In regard to things such as profit-sharing rules, wage rates, or salaries, the spot contract amounts to a document that both parties sign. Of course, the document (specifying the profit-sharing scheme, wage, or salary) certainly may affect future behavior because it may directly affect the players' payoffs or future options. Usually the future effect is related to external enforcement; for example, the document may be submitted to a court. This is the case for the game in Figure 20.1, in which the damage award c is imposed by the court.

Overall, there are two reasons why joint decisions make a good modeling tool. First, by using an abbreviated model of negotiation, the theorist can emphasize other aspects of a strategic situation while still capturing the intuitive notion of bargaining power in the resolution of bargaining problems. This has the further benefit of helping us differentiate between the process of negotiation and what the players are negotiating over. Second, joint decisions mark the places in a game where contracting takes place.[4]

NEGOTIATION EQUILIBRIUM

To analyze general games with joint decisions, we combine backward induction (more specifically, subgame perfection) with the standard bargaining solution; the former pins down behavior at individual decision nodes, whereas the latter identifies behavior at joint-decision nodes.

[3]The simple framework described here does not consider contracting under asymmetric information.

[4]In fact, in addition to representing spot contracting on tangible items, joint-decision nodes may designate where players engage in "meta-level" contracting, where they actively coordinate their future behavior by selecting among self-enforceable alternatives. At this point, I shall not address meta-level contracting in the context of games with joint decisions. The subject is important but beyond the scope of this book.

Given an extensive-form game with joint decisions, a specification of behavior at every information set is called a **regime**. This is simply a generalization of the "strategy" concept to include joint decisions. I use the following equilibrium definition:

> A regime is called a **negotiation equilibrium** if its description of behavior at individual decision nodes is consistent with sequential rationality and its specification of joint decisions is consistent with the standard bargaining solution, for given bargaining weights.

This definition is not precise enough to be clear-cut in every game with joint decisions. In particular, we can run into two problems when trying to construct a negotiation equilibrium. First, we have to decide what is meant by "sequential rationality." For example, we could use backward induction or subgame perfection. Second, how to apply the standard bargaining solution in some contexts may not be obvious, in particular where there is not transferable utility. I avoid these problems by focusing on games in which backward induction or subgame perfection can be easily employed (this is a wide class, by the way) and by assuming that players can transfer money whenever they negotiate. You can leave to interested hot-shot theorists the task of navigating the labyrinthine esoterica of more general application.

EXAMPLE: CONTRACTING FOR HIGH-POWERED INCENTIVES

To illustrate the negotiation equilibrium concept, consider a contract example. Carina and Wendy, two bright and motivated people, wish to initiate a business partnership. They have decided to open and run a bookstore specializing in sports, games, literature, and dance. Wendy is an investment wizard, so she will handle the financial side of the company. She also plans to promote the store by carrying a banner during her weekly running races and triathlons. Carina, disgruntled after years of underappreciated professional work in the educational system, is looking forward to handling the day-to-day operations of the store.

Witty and erudite, Carina will have no problem directing customers to challenging new works of literature. On the other hand, as a liberal benefactor of all humanity, Carina finds it difficult to ask people to actually *pay* for books. In particular, when a customer asks for a price reduction on a book, holding to the regular price makes Carina suffer some disutility. However, Wendy notes that the disutility is more than offset by the gain in store profit from maintaining a "fixed-price" policy. Wendy therefore suggests that Carina consider a compensation package rewarding her for administering the fixed-price policy.

FIGURE 20.3 Bookstore example.

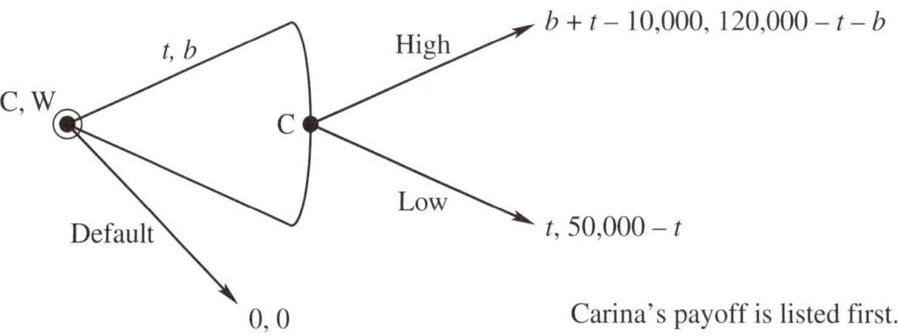

The interaction between Wendy and Carina is modeled by the game in Figure 20.3. First, the partners jointly decide on a compensation package consisting of Carina's salary t and a bonus b, the latter of which is paid only if Carina administers the fixed-price policy. Wendy obtains the remainder of the firm's revenue. Note that we are implicitly assuming that whether or not Carina administers the pricing policy can be verified to the court; thus, transfers on the basis of Carina's pricing decisions are externally enforced. The default decision is that Carina and Wendy do not open the store, leading to a payoff of zero for both of them. If the players reach an agreement, then Carina chooses between high effort and low effort on the job. High effort means implementing the fixed-price policy at a personal cost of \$10,000, whereas low effort means giving in to her instinct to be more generous with customers. High effort implies revenues of \$120,000 for the firm; low effort implies revenues of \$50,000.

To solve this game, we start by analyzing Carina's effort decision. Note that there is an infinite number of decision nodes for Carina, one corresponding to each of the infinite number of salary–bonus combinations. Given t and b, Carina has the incentive to exert high effort if and only if

$$b + t - 10{,}000 \geq t.$$

Thus, Carina selects high effort if and only if $b \geq 10{,}000$. In words, Carina will not exert high effort unless her bonus for doing so sufficiently covers her disutility (a high salary does not work). This is a simple but important idea in the area of contract and incentives: so-called *high-powered incentives*—which compensate people conditional on their direct contribution to output—work well if a person's contribution is verifiable.

Next, we move to the joint-decision node, which we solve by using the standard bargaining solution. Note that, if Carina and Wendy select $b \geq 10{,}000$, in which case Carina will subsequently choose high effort, then their joint value is

$$(b + t - 10{,}000) + (120{,}000 - t - b) = 110{,}000.$$

If they select a bonus that is less than 10,000, then their joint value is 50,000. Because they jointly prefer the higher value, they will choose a bonus of at least 10,000. Further, the salary and bonus serve to divide the surplus of 110,000 according to the players' bargaining weights π_C and π_W so that Carina obtains $110{,}000\pi_C$ and Wendy gets $110{,}000\pi_W$. For example, suppose Carina and Wendy have equal bargaining powers. Then they might choose $b = 10{,}000$ and $t = 55{,}000$. In general, with $b = 10{,}000$, the salary is $t = 110{,}000\pi_C$. Any higher bonus is also fine, but it would be combined with an offsetting lower salary.

GUIDED EXERCISE

Problem: Suppose Frank, an aspiring chef, has a brilliant plan to open a musical-themed, neo-Italian restaurant in the hip Clairemont district of San Diego. For the restaurant to be successful, Frank must employ two people: Gwen, a free-spirited singer, and Cathy, a savvy market analyst and legal consultant. With the help of Gwen and Cathy, Frank's new business will generate a gross profit of $300,000. But unless both Gwen and Cathy assist Frank, his business will fail, yielding zero profit. The three parties negotiate over monetary transfers and over whether Gwen and Cathy will participate in Frank's project.

(a) Consider a situation in which the three players negotiate all at once. Let x denote a transfer from Frank to Gwen, and let y denote a transfer from Frank to Cathy. Thus, if the parties agree to start the business and to make transfers x and y, then Frank obtains $300{,}000 - x - y$, Gwen gets x, and Cathy receives y. Under the assumption that the players have equal bargaining weights and the disagreement point is $(0, 0, 0)$, what is the outcome of negotiation? Use the standard bargaining solution here.

(b) Next suppose that Frank negotiates with Gwen and Cathy *sequentially*. First, Frank and Gwen make a joint decision determining whether Gwen commits to participate in the business. They also agree to an immediate transfer t from Frank to Gwen, as well as a payment x contingent on the event that Cathy also agrees to participate in the business. Then Frank and Cathy negotiate, with complete knowledge of the agreement between Frank and Gwen. Frank and Cathy jointly determine a transfer y

and whether Cathy will join the project. Disagreement in the negotiation between Frank and Gwen yields 0 to each player. Disagreement between Frank and Cathy yields an outcome in which Frank loses t (it was a sunk payment), Gwen gains t, and Cathy gets 0. This game can be represented as shown here.

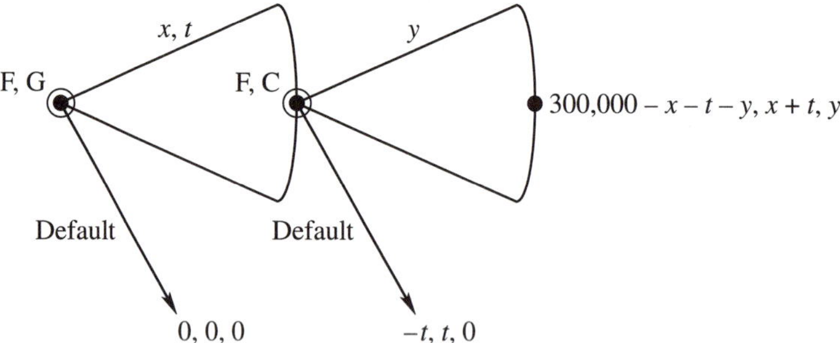

Under the assumption that the players have equal bargaining weights, compute the negotiation equilibrium of this game.

(c) What are the players' payoffs in the outcome of part (b)? Explain how the contract between Frank and Gwen affects the negotiation between Frank and Cathy. Is the outcome efficient?

Solution:

(a) The surplus of negotiation is clearly 300,000. The default outcome yields a payoff of 0 for everyone. Because the players have equal bargaining power and negotiate all at once, the standard bargaining solution implies that each player gets a third of the surplus, so each player gets a payoff of 100,000. To achieve this, it must be that $x = y = 100{,}000$.

(b) Consider first the negotiation between Frank and Cathy, under the assumption that Frank and Gwen reached an agreement (specified x and t) earlier. Examining the payoffs of Frank and Cathy, observe that the surplus of their negotiation is

$$(300{,}000 - x - t - y) + y - [-t + 0] = 300{,}000 - x.$$

Frank's disagreement payoff is $-t$, whereas Cathy's disagreement payoff is 0. Because Frank and Cathy have equal bargaining weights, the standard bargaining solution implies that their negotiation will be resolved in a way that makes Frank's payoff $150{,}000 - (x/2) - t$ and Cathy's payoff $150{,}000 - (x/2)$. Anticipating this, at the beginning of the game, Frank and Gwen realize that Cathy will extract $150{,}000 - (x/2)$. Subtracting this amount from 300,000, we find that the negotiation surplus for Frank and Gwen at the start of the game is $150{,}000 + (x/2)$. Note that Frank and Gwen have to pick $x \leq 300{,}000$ in order to induce Cathy to sign on as well. Thus, Frank and Gwen optimally select $x = 300{,}000$, which yields a surplus of 300,000 for them. They split the surplus evenly, which implies $t = -150{,}000$. Note that Frank and Cathy then agree to a contract specifying $y = 0$.

(c) In the negotiation equilibrium in part (b), Frank and Gwen each receive 150,000, while Cathy obtains 0. By contracting before Cathy is on the scene, Frank and Gwen are able to manipulate the way that Frank contracts with Cathy. Basically, Frank and Gwen agree that, if Frank and Cathy later reach an agreement, then Frank will have to transfer the value of the agreement to Gwen; this lowers to 0 the value of any agreement between Frank and Cathy. Gwen's lump-sum payment of 150,000 to Frank divides the surplus.

EXERCISES

1. A manager (M) and a worker (W) interact as follows: First, the players make a joint decision, in which they select a production technology x, a bonus payment b, and a salary t. The variable x can be any number that the players jointly choose. The bonus is an amount paid only if the worker exerts high effort on the job, whereas the salary is paid regardless of the worker's effort. The default decision is "no employment," which yields a payoff of 0 to both players.

If the players make an agreement on x, b, and t, then the worker chooses between low effort (L) and high effort (H). If the worker selects L, then the manager gets $-4 - t$ and the worker gets t. On the other hand, if the worker selects H, then the manager gets $8x - b - t$ and the worker gets $b + t - x^2$. The interpretation of these payoffs is that $8x$ is the firm's revenue and x^2 is the worker's cost of high effort, given production technology x.

(a) Represent this game as an extensive form with *joint decisions* (draw the game tree). Your tree should show and properly label the joint-decision node, as well as the worker's individual decision node. Clearly represent the default outcome and payoff for the joint-decision node.

(b) Given x, b, and t, under what conditions does the worker have the incentive to choose H?

(c) Determine the negotiation equilibrium of this game, under the assumption that the players have equal bargaining weights. Start by calculating the maximized joint value of the relationship (call it v^*), the surplus, and the players' equilibrium payoffs. What are the equilibrium values of x, b, and t?

(d) In this setting, is it appropriate to say that the worker's effort is *verifiable* or *unverifiable*? Why?

2. Consider another version of the game between Carina and Wendy, where Carina selects any effort level e on the job. Assume $e \geq 0$. The revenue of the firm is equal to \$800e. Carina's disutility of effort is e^2 (measured in dollars). Carina and Wendy interact as before; first, they jointly determine Carina's compensation package and, then (if they agree), Carina selects her level of effort.

(a) Suppose Carina's effort is not verifiable, so Carina and Wendy can write a contract specifying only a salary t for Carina. Assume Carina and Wendy have equal bargaining weights. Draw the extensive-form game and compute the negotiation equilibrium. Does Carina expend effort?

(b) Next, suppose that Carina and Wendy are constrained to linear "revenue-sharing" contracts. Such a compensation package states that Carina gets a fraction x of the revenue of the firm [leaving the fraction $(1 - x)$ to Wendy]. Calculate Carina's optimal effort level as a function of x. Then, under the assumption that Wendy has all of the bargaining power, calculate the value of x that maximizes Wendy's payoff. (Do not graph the bargaining set; this setting does not fit very well into our bargaining theory.)

(c) Now suppose that the contract is of the form $w(p) = xp + t$, where w is the amount paid to Carina and p is the revenue of the firm. That is, the contract specifies that Carina receive some base salary t and, in addition, a fraction x of the firm's revenue. Assume the players have equal bargaining weights. Calculate the negotiation equilibrium of this game. (Start by finding Carina's optimal effort decision, given t and x. Then, holding t fixed, determine the number x that maximizes the players' joint value. Finally, determine the players' negotiation values and find the salary t that achieves this split of the surplus.)

3. A manager (M) and a worker (W) interact as follows: First, the players make a joint decision, in which they select a bonus parameter p and a salary t. The salary can be any number (positive or negative). The bonus parameter p must be between 0 and 1; it is the proportion of the firm's revenue that the worker gets. The default decision is "no employment," which yields a payoff of 0 to both players.

If the players make an agreement on p and t, then, simultaneously and independently, the worker chooses an effort level x and the manager chooses an effort level y. Assume that $x \geq 0$ and $y \geq 0$. The revenue of the firm is then $r = 20x + 10y$. The worker's effort cost is x^2, whereas the manager's effort cost is y^2. Each player gets his share of the revenue and his transfer, minus his cost of effort. The players have equal bargaining weights (1/2 and 1/2). The game is depicted at the top of the next page.

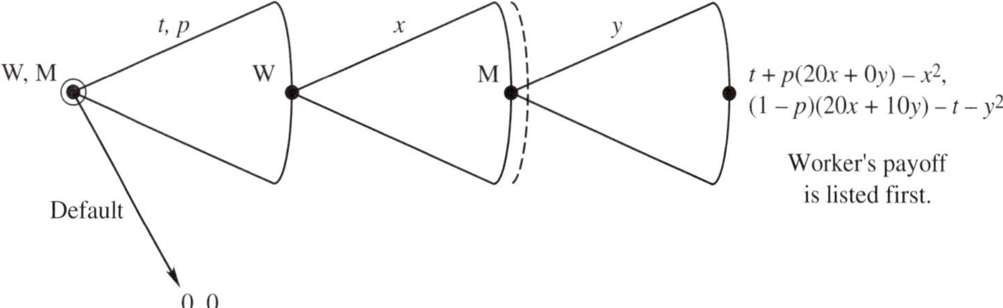

Compute the negotiation equilibrium of this game by answering the following questions:

(a) Given p and t, calculate the players' best-response functions and the Nash equilibrium of the effort-selection subgame.

(b) Finish the calculation of the negotiation equilibrium by calculating the maximized joint value of the relationship (call it v^*), the surplus, and the players' equilibrium payoffs. What are the equilibrium values of p, t, x, and y?

4. Consider a contractual setting with two players, where the underlying strategic interaction is given by the following "partnership game":

1 \ 2	H	L
H	10, 6	2, 10
L	8, 4	3, 5

This matrix describes the technology of the relationship. Suppose that, before noncooperative interaction, the players sign an externally enforced contract specifying a transfer t from player 2 to player 1 in the event that either (H, L) or (L, H) is played. The interpretation here is that the court can observe the total output of the relationship (which is either 8, 12, or 16) and is willing to enforce a transfer if output is 12. However, the court enforces the transfer only if one or both players request enforcement. Each player who requests enforcement must pay a legal fee of c. The court only allows transfers less than or equal to 10 (from one player to the other), because this is the greatest payoff a single player can earn in the partnership.

Consider a "grand game" in which the partnership game is one component. Play in the grand game runs as follows. First, the players interact by playing the illustrated partnership game. Then, if either (H, H) or (L, L) is the outcome, the grand game ends. On the other hand, if (H, L) or (L, H) is the outcome, then the players interact further by playing the following *enforcement* game:

1 \ 2	Enforce	Not
Enforce	$t-c, -t-c$	$t-c, -t$
Not	$t, -t-c$	$0, 0$

The players' payoffs in the enforcement game are added to their payoffs from the partnership game.

(a) Suppose you are a policy maker who determines the fee c and the contract parameter t. Under what conditions on c and t will (H, H) be played in a subgame perfect equilibrium of this game? That is, for what values of c is there a number t that induces cooperation? Remember that t is constrained to be between -10 and 10.

(b) Explain your reasoning for your answer to part (a). In particular, what happens when $c = 0$?

(c) Suppose that between the partnership game and the enforcement game the players make a joint decision about whether to continue with the enforcement game or "settle out of court"; they can also make a monetary transfer at this time. When the players settle out of court, they avoid the court fee. The default decision is to continue with the enforcement game. Do you think (H, H) can be supported in the negotiation equilibrium of this game? Explain your reasoning without explicitly solving this game.

5. Construct a game with joint decisions by augmenting the Cournot duopoly game in the following way: Suppose the two firms have a profit-sharing pact that is enforced by the government (legal collusion). First, the firms simultaneously select quantities, q_1 and q_2, at a marginal cost of 10. The price is determined from the inverse demand curve, $p = 100 - q_1 - q_2$, and the total revenue $p(q_1 + q_2)$ is deposited into a joint bank account. Then the firms must negotiate over how to share the revenue. Model the negotiation as a joint decision over firm 1's share of the revenue, m, so that agreement yields a payoff of $m - 10q_1$ to firm 1 and $p(q_1 + q_2) - m - 10q_2$ to firm 2. If they disagree then neither firm obtains any revenue, so firm 1 gets $-10q_1$ and firm 2 gets $-10q_2$.

6. This question leads you through the computation of the negotiation equilibrium of the game described in Exercise 5.

(a) Start by analyzing the joint decisions made at the end of the game. For given production levels q_1 and q_2, calculate the players' payoffs by using the standard bargaining solution. Write the payoffs in terms of arbitrary bargaining weights π_1 and π_2.

(b) What are the firms' payoffs as functions of q_1 and q_2 for $\pi_1 = 1/2$ and $\pi_2 = 1/2$?

(c) While continuing to assume $\pi_1 = \pi_2 = 1/2$, solve the Cournot component of the model by using Nash equilibrium. Explain why there are multiple equilibria. How does the outcome compare with that of the basic Cournot model? Is the outcome efficient?

(d) Next suppose that $\pi_1 \neq \pi_2$. Find the players' Nash equilibrium output levels. Remember that q_1 and q_2 are required to be greater than or equal to 0. [Hint: The equilibrium is a corner solution, where at least one of the inequalities binds ($q_1 = 0$ and/or $q_2 = 0$), so calculus cannot be used.]

(e) Discuss how the firms' quantity choices depend on their bargaining weights and explain the difference between the results of parts (c) and (d).

21 UNVERIFIABLE INVESTMENT, HOLD UP, OPTIONS, AND OWNERSHIP

Many economic examples are interesting precisely because of the tensions inherent in strategic situations.[1] In particular, a tension between individual and joint interests exists when a player's private costs and benefits are not equal to joint costs and benefits. This tension can have serious effects if there are limits on external enforcement, such as if the players' actions cannot be verified to an enforcement authority.[2]

Dynamic contract settings often exhibit a particular tension between individual and joint interests that is due to the timing of investments and negotiation. The basic idea is that the success of a particular project relies on an investment by person A as well as on the productive input of person B. The project requires person A to make the investment first. Only later does person B come onto the scene. The key problem is person A's incentives, for she can gain the returns of her investment only by contracting with person B later. If person B will utilize bargaining power to extract a share of the returns, so that person A does not obtain the full value of her investment, then person A may not have the incentive to invest optimally in the first place. This is called the "hold-up" problem, because person B extracts part of the returns under the threat of holding up production.

More generally, the hold-up problem can arise even if parties contract before investments are made, in particular if the investments are unverifiable. This is because the parties may want to write a contract that punishes them jointly if one or both of them fail to invest; however, if the undesired contingency actually arises, then the players would have the incentive to renegotiate their contract to avoid the joint punishment. In this sense, the punishment becomes incredible and, therefore, players may lack incentives to invest efficiently.

[1] Recall the tensions that I have presented: (1) the clash between individual and group incentives, (2) strategic uncertainty, (3) inefficient coordination, and variants of these tensions on the level of contract.

[2] Classical economic theory tends to view this tension as arising because of informational asymmetries between contracting parties or because of "missing markets." The latter refers to interaction between people concerning a good (such as air quality) that is not traded on a market. The term *externality* is often used in relation to such a setting. The theory of strategy may allow a deeper understanding by removing the competitive market from the center of consideration.

In this chapter, I review some of the basic insights on contracting with unverifiable investments and hold up. Of particular interest are contractual methods of alleviating the hold-up problem and encouraging efficient investment (specifically, the judicious use of option contracts and asset ownership).[3]

HOLD-UP EXAMPLE

For an illustration of the hold-up problem, consider a setting in which two people—JD and Brynn—interact in the development of a new product. JD is a scientist who can, with effort, design a new device for treating a particular medical condition. He is the only person with a deep knowledge of both the medical condition and physics sufficient to develop the innovative design. But JD does not have the engineering expertise or resources that are needed to construct the device. Brynn is the CEO of an engineering company; she is capable of implementing the design and creating a marketable product. Thus, success relies on JD's and Brynn's combined contributions.

Suppose JD and Brynn interact over three dates as follows. At Date 1, JD decides how much to invest in the design of the medical device. His investment—in fact, the complete design specifications—is observed by Brynn. Then, at Date 2, JD and Brynn meet to negotiate a contract that sets conditions (a price) under which Brynn can produce and market the device at Date 3. Commercial production of the device will generate revenue for Brynn, and this revenue will be a function of JD's initial investment level. The key issue is whether JD has the incentive to invest efficiently, given that he has to later negotiate with Brynn to obtain the fruits of his investment.

Here, a bit more formally, is a description of the sequence of events: At Date 1, JD selects between "high investment" (abbreviated H), "low investment" (L), and "no investment." If he chooses not to invest, then the game ends, and both parties get a payoff of 0. On the other hand, if JD chooses L or H, then JD pays a personal investment cost, and the game continues at Date 2. JD's cost of low investment is 1, whereas his cost of high investment is 10. Assume that JD's investment choice is observed by Brynn but is not verifiable to the court, so that the investment cannot directly influence a legal action.

[3]Early analysis of the hold-up problem appears in O. Williamson, *Markets and Hierarchies: Analysis and Antitrust Implications* (New York: Free Press, 1975); P. Grout, "Investment and Wages in the Absence of Binding Contracts: A Nash Bargaining Approach," *Econometrica* 52(1984):449–460; and S. Grossman and O. D. Hart, "The Costs and Benefits of Ownership: A Theory of Vertical and Lateral Integration," *The Journal of Political Economy* 94(1986):691–719. The examples that I present here are in the formulation of "Contract, Mechanism Design, and Technological Detail," by J. Watson, *Econometrica*, 75(2007):55–81.

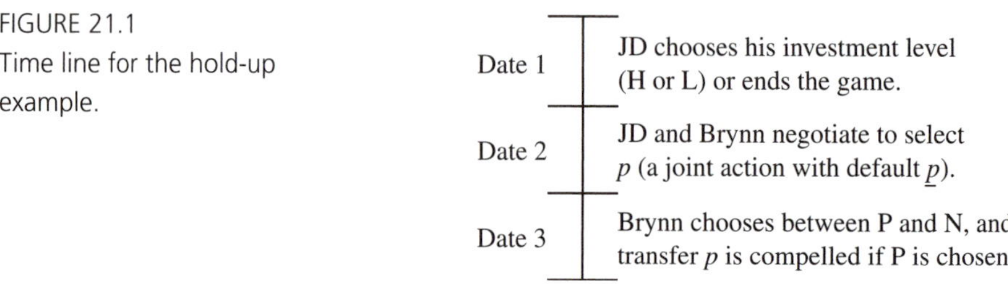

FIGURE 21.1
Time line for the hold-up example.

At Date 2, JD and Brynn negotiate over contracted monetary transfer p, which is a transfer from Brynn to JD to be compelled by the external enforcer (the court) if and only if Brynn elects to produce at Date 3. The default price is $\underline{p}$, which represents the legal default rule in case JD and Brynn do not establish an agreement. Assume that the court always compels a transfer of 0 if Brynn selects N.[4] Also assume that the players have equal bargaining weights, so $\pi_{JD} = \pi_B = 1/2$. At Date 3, Brynn chooses whether to "produce" (P) or not (N). If Brynn chooses to produce, then p is the amount transfered from Brynn to JD; if Brynn chooses not to produce, then the transfer is 0. Thus, Brynn's choice of whether to produce is verifiable, and the contract simply prescribes the transfer as a function of this selection. The time line of the game is pictured in Figure 21.1; note that this is not the extensive-form diagram, which you can draw as an exercise.

In the end, JD's payoff is whatever transfer he gets from Brynn minus his investment cost. Brynn's payoff is her revenue from production (if she chooses P) minus the amount she transfers to JD. Suppose that Brynn's revenue from production is 4 in the event that JD chose investment L, whereas her revenue is 18 in the event that JD chose investment H. If Brynn selects N, then her revenue is 0, regardless of JD's investment choice.

Note that it is more costly for JD to select high investment H rather than low investment L, but high investment leads to a better product and enhanced revenues for Brynn on the market for the medical device. In fact, efficiency requires JD to select H and Brynn to choose P, because these actions yield a joint value of $18 - 10 = 8$. By comparison, if JD were to select L and Brynn were to choose P, then the joint value would be $4 - 1 = 3$.

The key question is whether JD has the incentive to select investment H; that is, does JD take the efficient investment action in the negotiation equilibrium of the game? To answer this question, we use backward induction and the standard bargaining solution to calculate the negotiation equilibrium of the game. The first thing to note is that Brynn has the incentive to select P if and

[4]In general, a contract may specify a transfer other than 0 on the condition that Brynn selects N, but there is no benefit of such a transfer in the example here.

only if her revenue from production exceeds p. Thus, if $p > 18$, then Brynn will choose N, regardless of whether JD invested at level L (when the revenue would be 4) or at level H (when the revenue would be 18). If $p \leq 4$, then Brynn has the incentive to produce, regardless of JD's investment choice. Finally, if $p \in [4, 18]$, then it is rational for Brynn to select P in the event that JD chose H and for Brynn to select N in the event that JD chose L.

Let us suppose that, in the absence of a contract between JD and Brynn, the law gives JD the sole right to produce any medical device that is based on his design. Thus, if Brynn selects P without JD's authorization, then the court will punish Brynn in the form of a monetary transfer from Brynn to JD. To represent this legal default rule, assume that $\underline{p} = 20$. Then the only way production can occur in equilibrium is if Brynn contracts with JD.

To analyze the negotiation between JD and Brynn at Date 2, let us consider separately the case in which JD chose L at Date 1 and the case in which JD chose H. If JD had selected L, then the players would be negotiating over a surplus of 4. To see this, notice that with the default outcome of negotiation, where $\underline{p}$ is in force, Brynn would choose not to produce and the game would end with the payoffs -1 for JD (because he paid the cost of low investment) and 0 for Brynn. Thus, the joint value of the default outcome is -1. On the other hand, by agreeing to set p between 0 and 4, the players know that Brynn will choose to produce and that they will obtain the joint value $4 - 1 = 3$. The surplus is the difference between the joint value of contracting and the joint value of the default outcome:

$$3 - (-1) = 4.$$

Equal bargaining weights imply that the surplus is evenly split between JD and Brynn, which is achieved by setting the price $p = 2$. Thus, if JD invests at level L, he will eventually get the payoff $-1 + 2 = 1$, which is his disagreement value plus his share of the surplus. Brynn will get the payoff $4 - 2 = 2$.

Next take the case in which JD chose H at Date 1. In the default outcome of negotiation (where $\underline{p}$ is in force), Brynn would choose not to produce and the game would end with the payoffs -10 for JD (because he paid the cost of high investment) and 0 for Brynn. The joint value of the default outcome is therefore -10. On the other hand, by agreeing to set p between 0 and 18, the players know that Brynn will choose to produce and that they will obtain the joint value $18 - 10 = 8$. Thus, in this case, the players are negotiating over a surplus of

$$8 - (-10) = 18.$$

Equal bargaining weights imply that the surplus is evenly divided between JD and Brynn, which is achieved by setting the price $p = 9$. Thus, if JD invests at level H, he will eventually get the payoff $-10 + 9 = -1$ and Brynn will get the payoff $18 - 9 = 9$.

Moving back to Date 1, we can now understand JD's investment incentives. Interestingly, *his only rational action is to invest at the low level L*, because this is the only action that leads to a positive payoff for him. Thus, the example illustrates the general phenomenon known as the **hold-up problem**. In particular, Brynn exercises her bargaining power to extract a significant share of the benefit of JD's investment, under the threat of "holding up" production. Anticipating not being able to extract the full benefit of his investment, JD has the incentive to invest at less than the efficient level. In other words, to make his investment decision, JD weighs his own cost and benefit. The joint cost is fully borne by him, whereas he retrieves only a fraction of the joint benefit, so his incentives are distorted in relation to what is efficient.

UP-FRONT CONTRACTING AND OPTION CONTRACTS

Key aspects of the hold-up story are that (1) investments are unverifiable, so the court cannot condition transfers directly on these actions, and (2) there is some barrier to the parties writing a comprehensive contract prior to choosing investments. In the example that I just presented, item (2) is represented by the assumption that JD and Brynn meet only *after* JD makes his investment decision. This assumption may be a stretch, for in many real settings the contracting parties can negotiate and form a contract before they are required to make significant investments and take other productive actions.

Let us consider, therefore, a version of the model in which JD and Brynn meet and form a contract at Date 0, with interaction continuing in Dates 1–3 just as described in the previous section. Think of the contract at Date 0 as the "initial contract," and think of any contracting at Date 2 as "renegotiation." At Date 0, JD and Brynn jointly select the value of $\underline{p}$ (the amount Brynn will have to pay JD if she produces at Date 3), and they also may specify an up-front transfer. Then $\underline{p}$ becomes the default value for renegotiation at Date 2. If the players keep $\underline{p}$ in place at Date 2, then we say that renegotiation of the contract did not occur. Otherwise, the parties will have renegotiated their initial contract to alter the production-contingent transfer. Assume that the default decision for negotiation at Date 0 is $\underline{p} = 20$, the legal default rule assumed in the previous section. The time line of the game is pictured in Figure 21.2.

FIGURE 21.2
Time line for the setting with renegotiation.

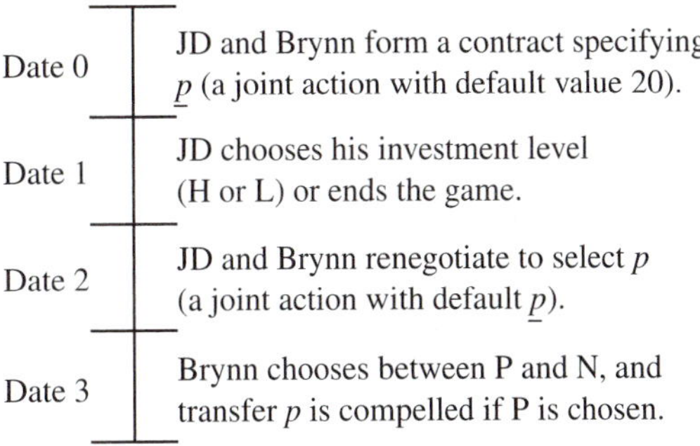

Hold up is an issue even though contracting occurs at Date 0. Here's why. Because JD's investment decision is unverifiable, there is no way for the contract to give JD a high-powered incentive to invest (as was achieved for Carina in the example at the end of Chapter 20; see pp. 233–235). Instead, the contract can only be used to motivate Brynn in her choice between P and N at Date 3 (an action that is verifiable). One hopes that Brynn's action can be made contingent on JD's investment in such a way as to motivate JD to invest. Unfortunately, renegotiation at Date 2 may interfere with the whole plan because it may undo something to which the players wanted to commit at Date 0.

The entire enterprise may seem confusing at this point, because the initial contract has an indirect effect on JD's incentives. But an intuitive and realistic theme will emerge from the analysis: An optimal contract will give Brynn the *option* of producing the device that JD designs, but only at a specific price that Brynn must pay to JD. If the price is set high enough—but not too high—then Brynn will exercise the option precisely in the circumstance in which JD invested at the high level. Thus, JD knows that he will obtain the high price if and only if he chooses H, which motivates him to select H rather than L in the first place.

To analyze the game, we apply the standard bargaining solution to examine the joint decision at Date 0; analysis of the rest of the game proceeds as reported in the previous section. First note that if, at Date 0, the players specify $\underline{p}$ to be a number exceeding 18, then the game would proceed just as derived in the basic hold-up example. Such a high price would basically force Brynn to refrain from producing (regardless of JD's investment choice), and so the players would achieve a surplus only by renegotiating their initial contract. The players anticipate that, at Date 2 renegotiation, Brynn will extract half of the value of production, which will induce JD to select L at Date 1.

Second, note that a joint selection of $\underline{p} < 4$ leads to grim investment incentives as well. With such a low price, Brynn has the incentive to pick P, regardless of JD's investment (L or H), so there is no gain from renegotiation at Date 2. Anticipating a transfer of $\underline{p}$, regardless of his investment choice, JD has no incentive to select H at Date 1. More precisely, in the case of $\underline{p} < 1$, JD rationally chooses no investment and ends the game at Date 1; in the case of $\underline{p} \in [1, 4)$, JD rationally chooses to invest at level L. Either way, the outcome is inefficient.

Contracts that specify $\underline{p} > 18$ or $\underline{p} < 4$ in this example are called **forcing contracts**, because they effectively force Brynn to take the same productive action at Date 3, both in the event that JD invested high and in the event that JD invested low. In contrast, an **option contract** creates different incentives for Brynn, depending on JD's investment. In the example, if we have $\underline{p} \in [4, 18]$, then Brynn has the incentive to choose P in the event that JD invested H, whereas Brynn prefers to choose N in the event that JD invested L.

Consider the particular option contract that specifies $\underline{p} = 14$, and let us see what will occur in Dates 1–3 under this contract. First note what would happen at Date 3 if this contract were not renegotiated at Date 2 (that is, with $p = \underline{p} = 14$). In the event that JD had selected H at Date 1, Brynn will choose P, which gives her a payoff of $18 - 14 = 4$ and gives JD a payoff of $14 - 10 = 4$. Anticipating this efficient outcome, the players have nothing to achieve by renegotiating at Date 2, so indeed renegotiation does not occur. Next consider the case in which JD had selected L at Date 1. In this case, Brynn will choose N at Date 3 because her benefit of production (the amount 4) does not exceed the price she must pay to produce (which is 14). Anticipating this inefficient outcome, the players would renegotiate at Date 2 to pick a price p that is below 4 so that production occurs. As in the basic model, the surplus of negotiation at Date 2 is 4, so equal bargaining weights imply that negotiation is resolved with $p = 2$.

To summarize what we have thus far, if the players initially select $\underline{p} = 14$, then they expect the following in relation to JD's investment decision at Date 1: If JD selects H, then the players will not renegotiate at Date 2, Brynn will choose P at Date 3, JD will get the payoff $14 - 10 = 4$, and Brynn will get the payoff $18 - 14 = 4$. If JD selects L, then the players will renegotiate to the price $p = 2$, Brynn will choose P, JD will obtain the payoff $2 - 1 = 1$, and Brynn will get the payoff $4 - 2 = 2$. Clearly, JD has the incentive to invest H.

Thus, the option contract that specifies $\underline{p} = 14$ induces the players to invest and produce efficiently. Furthermore, this contract achieves an even split of the value of the relationship (4 for each player).

On the calculation of the negotiation equilibrium, one question remains: What contract will the players actually select at Date 0? Although the option contract specifying $\underline{p} = 14$ induces an efficient outcome of the game, it is not quite consistent with the standard bargaining solution applied at Date 0 because the disagreement point is the payoff vector that would result under the default $\underline{p} = 20$—that is, 1 for JD and 2 for Brynn. Subtracting the joint value of default from the joint value of contracting yields a surplus of $8 - 3 = 5$. An even division of this surplus, added to the players' disagreement values, would give JD a payoff of $1 + 2.5$ and Brynn a payoff of $2 + 2.5$ from Date 0. You can easily verify that the contract specifying $\underline{p} = 13.5$ does the trick.

As the example shows, an option contract is quite helpful in situations with unverifiable investments. By giving the "buyer" (Brynn in the example here) the incentive to purchase only a good or service of sufficiently high quality, it gives the "seller" (JD here) the incentive to make investments that ensure high quality. With just a little digging, you will notice option contracts utilized in many real settings, such as professional sports, procurement, entertainment (films and music, in particular), and innovative industries. You will also notice that renegotiation occurs periodically in the real world. Renegotiation can interfere with the initial objectives of contracting, but it can sometimes be dealt with effectively if the initial contract is shrewdly designed. With the optimal option contract in the example here, renegotiation would occur if JD were to select L, but JD is still given the incentive to choose H.

ASSET OWNERSHIP

As you have just seen, a well-designed option contract can sometimes induce efficient investments and production. But option contracts may also be complicated and require many details. Some economists believe that there are barriers to writing detailed contracts, either because composing them is costly (due to lawyer's fees, for instance) or because it is difficult to describe or imagine all of the contingencies that may arise later. A subset of these economists argue that contracting parties can often deal effectively with unverifiable investments and hold up by initially forming simple contracts that judiciously allocate *asset ownership*. They claim that asset ownership can have an important impact on investment incentives and that one can understand ownership patterns in real industries on the basis of optimal responses to hold-up problems.

To see how asset ownership may play a role in investment incentives, consider a simple extension of the JD-Brynn story in which production requires the use of a particular asset, such as a computer-controlled manufacturing device. Suppose that JD's investment amounts to enhancing and configuring this asset. Naturally, the asset is owned by one of the contracting parties. Ownership means that the owner has the exclusive right to use the asset and to obtain the revenue that it generates.

Once we think about the definition of ownership, it is natural to ask whether the identity of the owner influences investment incentives. That is, would JD have the same incentive to invest as the asset owner than he would have if Brynn owned the asset? Suppose that following JD's investment the best use of the asset is by Brynn because her firm has the most relevant expertise among engineering and manufacturing companies. But suppose as well that the asset has an alternative use that also benefits from JD's investment. Think of the alternative use as production by a different engineering company.

In particular, if JD owns the asset and severs his relationship with Brynn, then he can contract with an alternative producer. (If Brynn owns the asset, then JD has no such recourse.) Suppose there are plenty of alternative producers available and that JD can extract the entire surplus of dealing with one of them because they will compete for his business. But an alternative producer generates lower revenue than would Brynn. Assume that revenue with an alternative producer would be 0 if JD invested at level L, whereas the revenue would be y if JD selected H. The value y satisfies $0 < y < 18$. If y is close to 0, then we say that JD's investment is *specific* to Brynn, in that it enhances the value of JD's relationship with Brynn much more than it would enhance a relationship with an alternative producer. On the other hand, if y is larger, then we say that JD's investment is *general.*

Assume that JD and Brynn can contract at Date 0, but that they can then contract only on ownership of the asset. That is, they can agree on who owns the asset, but they cannot specify transfers as a function of Brynn's choice about whether to produce. The rest of the game proceeds as in the basic model, except that if JD and Brynn fail to reach a deal at Date 2 and JD happens to own the asset, then he can contract with an alternative producer. In the event that JD invested H, he will then obtain $y - 10$ (that is, y with the alternative supplier and -10 representing the cost of investment). If JD invested L and owns the asset, his disagreement value from Date 2 is -1 as in the basic model.

The key question is whether giving JD ownership of the asset at Date 0 will encourage him to invest H. The mechanism by which this might occur is that JD's investment affects his disagreement value for negotiation at Date 2. Specifically, if JD selected L at Date 1, then there is no prospect for trading with an alternative producer; his disagreement value is the same as specified in the basic model analyzed at the beginning of this chapter, so JD eventually gets 1. But if JD selected H at Date 1, then his disagreement value is $y - 10$, and the surplus of negotiating with Brynn—the difference between their joint value of forcing production and their joint value of disagreement—is

$$18 - 10 - (y - 10) = 18 - y.$$

Applying the standard bargaining solution with equal bargaining weights, we conclude that JD and Brynn will settle on the price $p = 9 + (y/2)$, which gives JD his disagreement value plus half of the surplus.

Consider JD's incentives at Date 1 when he owns the asset. If he selects L, then he will get 1. On the other hand, if he selects H, then he will obtain

$$p - 10 = 9 + \frac{y}{2} - 10 = \frac{y}{2} - 1.$$

He has the incentive to select H if this amount exceeds 1, which simplifies to $y \geq 4$. Therefore, if JD's investment is sufficiently general, giving him control of the asset will also give him the incentive to make the high investment.

The analysis of asset ownership suggests several lessons for everyday decision making. First, recognize when an investment today affects the value of a future relationship. Be wary of an investment that generates value subject to hold up. Second, understand the difference between specific and general investments. Try to structure your relations so that *you* make general investments and your partners or opponents make specific investments. Distinctions between general and specific investments often emerge in employment relations, where human capital (workers' skills and expertise) is at stake. Third, try to engage in negotiation at times in which you have good outside opportunities.

GUIDED EXERCISE

Problem: Suppose that a worker interacts with a firm as follows: The worker first decides how much to invest in developing his skills. Let x denote the worker's investment, and suppose that the investment entails a personal cost to the worker of x^2; assume $x \geq 0$. If the worker works in the firm, then his investment generates return ax for the firm. If the worker decides to work on his own (separate from the firm), then his investment generates a return of bx that he keeps. The numbers a and b are positive constants satisfying $a > b$ (which means that the investment is more productive in the firm). After the worker chooses his investment, the firm observes it. Then the firm offers the worker a wage w, and the worker accepts or rejects it (ultimatum bargaining). If the worker accepts the wage, then he works in the firm. If he rejects the wage, then he works on his own. Here is the game's extensive form:

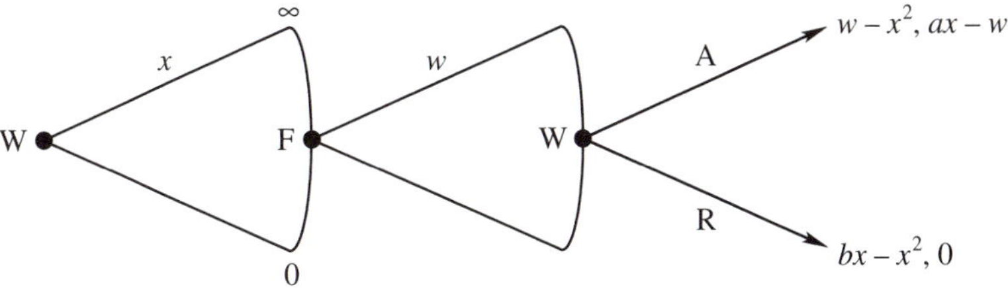

The worker's payoff is listed first.

(a) Find and report the subgame perfect equilibrium of this game. What investment level is selected?

(b) Consider a variant of the game in which the worker gets to make the wage offer. What *investment level* will prevail in equilibrium?

(c) Consider the variant of the game pictured here, where the selection of the wage is modeled as a joint decision of the firm and worker. Solve this game under the assumption that the joint decision is resolved according to the standard bargaining solution. Let π_W be the worker's bargaining weight and let π_F be the firm's bargaining weight. What *investment level* will prevail in equilibrium?

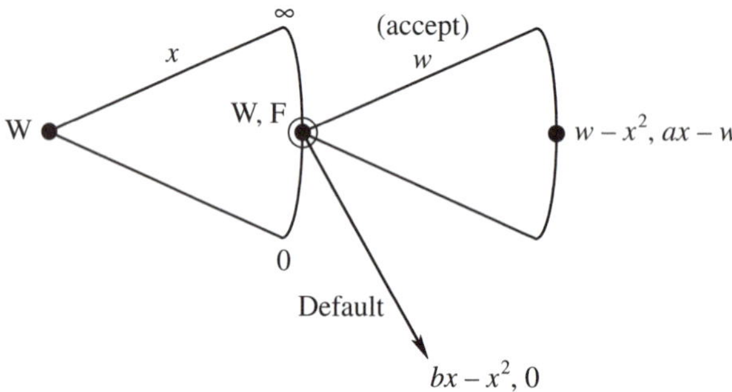

(d) Explain the difference between the outcomes of parts (a), (b), and (c). What determines the level of investment in equilibrium? How do changes in b affect the joint value (total payoff) that is realized? Which of parts (a), (b), and (c) yield the maximum value of the relationship? Why?

Solution:

(a) First, examine the worker's choice of whether to accept the firm's wage offer. Clearly, the worker strictly prefers to select A (accept) if $w > bx$, strictly prefers to select R (reject) if $w < bx$, and is indifferent if $w = bx$. Thus, in equilibrium, the firm offers $w = bx$, and the worker accepts. (There is no equilibrium in which the worker would reject this wage offer, because then there would be no optimal action for the firm in response.) At the initial node, the worker therefore anticipates obtaining a payoff of $bx - x^2$, which he maximizes by investing at level $\hat{x} = b/2$.

(b) In the setting in which the worker makes the ultimatum offer, the firm will accept if $ax \geq w$, so the worker offers $w = ax$ in equilibrium. At the initial node, the worker therefore anticipates obtaining a payoff of $ax - x^2$, which he maximizes by investing at level $x^* = a/2$.

(c) Note that the surplus of negotiation is $(a-b)x$, and the worker's disagreement value is $bx - x^2$. The standard bargaining solution implies that the worker obtains

$$bx - x^2 + \pi_W(a-b)x = (\pi_W a + \pi_F b)x - x^2,$$

which translates into a wage of $w = \pi_W ax + \pi_F bx$. At the initial node, the worker maximizes his payoff by investing at level $\tilde{x} = (\pi_W a + \pi_F b)/2$.

(d) Increasing the worker's bargaining power, which is represented by moving from part (a) to part (b) or increasing π_W in part (c), implies a larger investment by the worker. This is because the worker faces a hold-up problem in that the firm may extract some of the value of the worker's investment. An increase in b raises the equilibrium investment, because it increases the worker's marginal disagreement value as a function of x and thus lessens the hold-up problem. The hold-up problem disappears as either the worker's bargaining weight converges to 1 or b converges to a. The maximum value of the relationship occurs at the efficient investment level x^*, which is the case in part (b).

EXERCISES

1. A bicycle manufacturer (the "buyer," abbreviated B) wishes to procure a new robotic system for the production of mountain-bike frames. The firm contracts with a supplier (S), who will design and construct the robot. The contractual relationship is modeled by the following game:

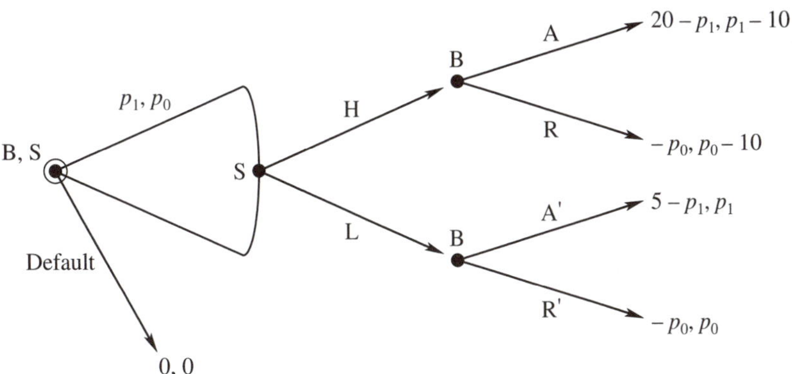

The parties first negotiate a contract specifying an externally enforced price that the buyer must pay. The price is contingent on whether the buyer later accepts delivery of the robot (A) or rejects delivery (R), which is the only event that is verifiable to the court. Specifically, if the buyer accepts delivery, then he must pay p_1; if he rejects delivery, then he pays p_0. After the contract is made, the seller decides whether to invest at a high level (H) or at a low level (L). High investment indicates that the seller has worked diligently to create a high-quality robot—one that meets the buyer's specifications. High investment costs the seller 10. The buyer observes the seller's investment and then decides whether to accept delivery. If the seller selected H and the buyer accepts delivery, then the robot is worth 20 units of revenue to the buyer. If the seller selected L and the buyer accepts delivery, then the robot is only worth 5 to the buyer. If the buyer rejects delivery, then the robot gives him no value.

(a) What is the efficient outcome of this game?

(b) Suppose the parties wish to write a "specific-performance" contract, which mandates that the buyer accept delivery at price p_1. How can p_0 be set so that the buyer has the incentive to accept delivery regardless of the seller's investment? Would the seller choose H in this case?

(c) Under what conditions of p_0 and p_1 would the buyer have the incentive to accept delivery if and only if the seller selects H? Show that the efficient outcome can be obtained through the use of such an "option contract."

(d) Fully describe the negotiation equilibrium of the game, under the assumption that the parties have equal bargaining weights.

2. Estelle has an antique desk that she does not need, whereas Joel and his wife have a new house with no furniture. Estelle and Joel would like to arrange a trade, whereby Joel would get the desk at a price. In addition, the desk could use restoration work, which would enhance its value to Joel. Specifically, the desk is worth 0 to Estelle (its current owner), regardless of whether it is restored. An unrestored desk is worth \$100 to Joel, whereas a restored desk is worth \$900. Neither Joel nor Estelle has the skills to perform the restoration. Jerry, a professional actor and woodworker, can perform the restoration at a personal cost of \$500. Jerry does not need a desk, so his value of owning the restored or unrestored desk is 0.

(a) Suppose Estelle, Jerry, and Joel can meet to negotiate a spot contract specifying transfer of the desk, restoration, and transfer of money. Model this as a three-player, joint-decision problem, and draw the appropriate extensive form. Calculate the outcome by using the standard bargaining solution, under the assumption that the players have equal bargaining weights ($\pi_{\text{E}} = \pi_{\text{Jerry}} = \pi_{\text{Joel}} = 1/3$). Does the desk get traded? Is the desk restored? Is this the efficient outcome?

(b) Suppose spot contracting as in part (a) is not possible. Instead, the players interact in the following way. On Monday, Estelle and Jerry jointly decide whether to have Jerry restore the desk (and at what price to Estelle). If they choose to restore the desk, Jerry performs the work immediately. Then on

Wednesday, regardless of what happened on Monday, Estelle and Joel jointly decide whether to trade the desk for money. Model this game by drawing the extensive form. (Hint: The extensive form only has joint-decision nodes.) Assume the parties have equal bargaining weights at all joint-decision nodes. Determine the negotiation equilibrium. Compare the outcome with that of part (a).

(c) Now suppose the players interact in a different order. On Monday, Estelle and Joel jointly decide whether to trade the desk for money. Trade takes place immediately. On Wednesday, if Joel owns the desk, then he and Jerry jointly decide whether to have Jerry restore the desk (and at what price to Joel). If they choose to restore the desk, Jerry performs the work immediately. Model this game by drawing the extensive form. (Hint: Again, the extensive form only has joint-decision nodes.) Assume the parties have equal bargaining weights at all joint-decision nodes. Determine the negotiation equilibrium. Compare the outcome with that of parts (a) and (b).

(d) Explain the nature of the hold-up problem in this example.

3. Recall that "human capital" refers to skills and expertise that workers develop. *General human capital* is that which makes a worker highly productive in potential jobs with many different employers. *Specific human capital* is that which makes a worker highly productive with only a single employer. What kind of investment in human capital should you make to increase your bargaining power with an employer, general or specific? Why? Do valuable outside options enhance or diminish your bargaining power?

4. This exercise asks you to combine the investment and hold-up issue from this chapter with the "demand" bargaining game explained in Exercise 4 of Chapter 19. Consider an investment and trade game whereby player 1 first must choose an investment level $x \geq 0$ at a cost of x^2. After player 1's investment choice, which player 2 observes, the two players negotiate over how to divide the surplus x. Negotiation is modeled by a demand game, in which the players simultaneously and independently make demands m_1 and m_2. These numbers are required to be between 0 and x. If $m_1 + m_2 \leq x$ (compatible demands, given that the surplus to be divided equals x), then player 1 obtains the payoff $m_1 - x^2$ and player 2 obtains m_2. On the other hand, if $m_1 + m_2 > x$ (incompatible demands), then player 1 gets $-x^2$ and player 2 gets 0. Note that player 1 must pay his investment cost even if the surplus is wasted owing to disagreement.

(a) Compute the efficient level of investment x^*.

(b) Show that there is an equilibrium in which player 1 chooses the efficient level of investment. Completely describe the equilibrium strategies.

(c) Discuss the nature of the hold-up problem in this example. Offer an interpretation of the equilibrium of part (b) in terms of the parties' bargaining weights.

5. Suppose that an entrepreneur is deciding whether or not to build a new high-speed railroad on the West Coast. Building the railroad will require an initial sunk cost F. If operated, the new railroad will generate revenue R. Operating the railroad will cost M in fuel and nw in wages, where n is the number of full-time jobs needed to operate the new railroad and w is the career wage per worker. If a rail worker does not work on the new railroad, then he can get a wage of $\overline{w}$ at some other job. Assume that $R > M + F + n\overline{w}$, so it would be profitable to build and operate the new railroad even if rail workers had to be paid somewhat more than the going rate $\overline{w}$. The entrepreneur, however, must decide whether to invest the initial sunk cost F before knowing the wages she must pay.

(a) Suppose that, if the railroad is built, after F is invested, the local rail workers' union can make a "take it or leave it" wage demand w to the entrepreneur. That is, the entrepreneur can only choose to accept and pay the wage demand w or to shut down. If the railroad shuts down, each worker receives $\overline{w}$. Will the railroad be built? Why?

(b) Next suppose that the wage is jointly selected by the union and the entrepreneur, where the union has bargaining weight π_U and the entrepreneur has bargaining weight $\pi_E = 1 - \pi_U$. Use the concept of negotiation equilibrium to state the conditions under which the railroad will be built.

(c) Explain the nature of the hold-up problem in this example. Discuss why the hold-up problem disappears when the entrepreneur has all of the bargaining power. Finally, describe ways in which people try to avoid the hold-up problem in practice.

6. Describe a real-world setting in which option contracts are used.

7. Suppose that you work for a large corporation and that your job entails many hours of working with a computer. If you treat the computer with care, it will not break down. But if you abuse the computer (a convenience for you), then the computer will frequently need costly service. Describe conditions under which it is best that you, rather than your employer, own the computer. Discuss verifiability and incentives in your answer.

REPEATED GAMES AND REPUTATION 22

People often interact in ongoing relationships. For example, most employment relationships last a long time. Countries competing over tariff levels know that they will be affected by each others' policies far into the future. Firms in an industry recognize that they are not playing a static game but one in which they compete every day over time. In all of these dynamic situations, the way in which a party behaves at any given time is influenced by what this party and others did in the past. In other words, players *condition* their decisions on the history of their relationship. An employee may choose to work diligently only if his employer gave him a good bonus in the preceding month. One country may set a low import tariff only if its trading partners had maintained low tariffs in the past. A firm may elect to match its competitor's price by setting its price each day equal to the competitor's price of the preceding day.

When deciding how to behave in an ongoing relationship, one must consider how one's behavior will influence the actions of others in the future. Suppose I am one of your employees and that our history is one of cooperation. Since you hired me, I have been a loyal and hard-working employee, and you have given me a generous bonus each month (above my salary). Today I am considering whether to work hard or, alternatively, neglect my duties in favor of playing video games on the office computer. For me, shirking has an immediate reward—I get to avoid expending effort on the job. But you will soon learn of my indolence, either through your monitoring activities or through an observed decrease in my productivity. *Your future behavior* (in particular, whether to give me bonuses each month) may very well be influenced by what *I do today*.

For instance, after observing that I have shirked, you might choose to discontinue my monthly bonus. You might say to yourself, "By misbehaving, Watson has lost my trust; I doubt that he will work diligently ever again and therefore I will pay him no more bonuses." Anticipating such a response, I may decide that spending the workday playing chess over the Internet against someone in New Zealand is not such a good idea. Neglecting my duties may yield an immediate gain (relaxation today), but it leads to a greater loss in the future (no more bonuses each month). As this story suggests, people sometimes have an incentive to forego small immediate gains because of the threat of future retaliation by others.

The term "reputation" is often used to describe how a person's past actions affect future beliefs and behavior. If I have always worked diligently on the job, people would say that I have "established a reputation for being a hard worker." If I shirk today, then tomorrow people would say that I have "destroyed my good reputation." Often, those who nurture good reputations are trusted and rewarded; people with bad reputations are punished. As the employment story indicates, the concern for reputation may motivate parties to cooperate with one another, even if such behavior requires foregoing short-term gains. One of the great achievements of game theory is that it provides a framework for understanding how such a reputation mechanism can support cooperation.

The best way to study the interaction between immediate gains and long-term incentives is to examine a *repeated game*. A repeated game is played over discrete periods of time (period 1, period 2, and so on). We let t denote any given period and let T denote the total number of periods in the repeated game. T can be a finite number or it can be infinity, which means the players interact perpetually over time. In each period, the players play a static *stage game,* whereby they simultaneously and independently select actions. These actions lead to a stage-game payoff for the players. The stage game can be denoted by $\{A, u\}$, where

$$A = A_1 \times A_2 \times \cdots \times A_n$$

is the set of action profiles and $u_i(a)$ is player i's stage-game payoff when profile a is played. The same stage game is played in each period. Furthermore, we assume that, in each period t, the players have observed the *history* of play—that is, the sequence of action profiles—from the first period through period $t - 1$. The payoff of the entire game is defined as the sum of the stage-game payoffs in periods 1 through T. We sometimes assume that players discount the future, in which case we include a discount factor in the payoff specification (recall the analysis of discounting in Chapter 19).

A TWO-PERIOD REPEATED GAME

Suppose players 1 and 2 interact over two periods, called periods 1 and 2 (so $T = 2$). In each period, they play the stage game depicted in Figure 22.1. Assume that the payoff for the entire game is the sum of the stage-game

FIGURE 22.1
Stage game, repeated once ($T = 2$).

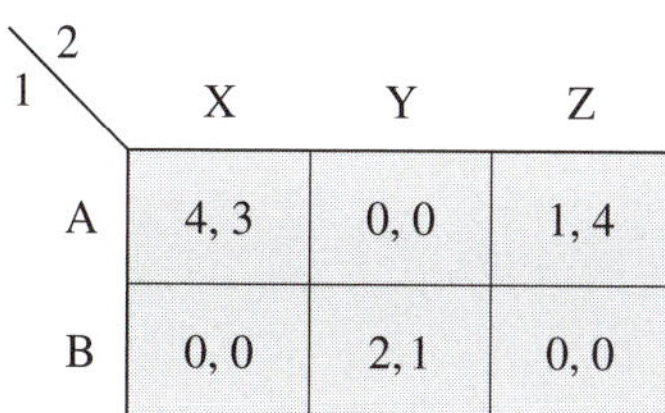

1 \ 2	X	Y	Z
A	4, 3	0, 0	1, 4
B	0, 0	2, 1	0, 0

FIGURE 22.2 Feasible repeated game payoffs.

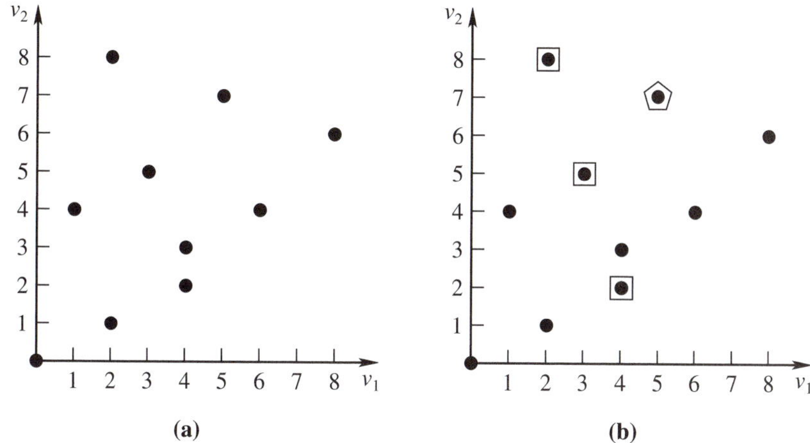

payoffs in the two periods. For instance, if (A, X) is played in the first period and (B, Y) is played in the second period, then player 1's payoff is $4 + 2 = 6$ and player 2's payoff is $3 + 1 = 4$. Figure 22.2(a) graphs the set of possible repeated game payoffs. Every point on the graph corresponds to the sum of two stage-game payoff vectors. For example, the payoff vector (3, 5) can be attained if (A, Z) is played in the first period and (B, Y) is played in the second period; the same payoff results if (B, Y) is played in the first period, followed by (A, Z).

This two-period repeated game has a large extensive-form representation, so I have not drawn it here. The extensive form starts with simultaneous selection of actions in the first period: player 1 chooses between A and B, while player 2 selects X, Y, or Z. Then, having observed each other's first-period actions, the players again select from {A, B} and {X, Y, Z}. Because each player knows what happened in the first period, his choice in the second period can be conditioned on this information. For example, player 1 may decide to pick A in the second period if and only if (A, X) or (B, Y) was played in the first period; otherwise, he picks B. As usual, the players' information is represented by information sets. Because there are six possible outcomes of first-period

FIGURE 22.3
The subgame following (A, Z).

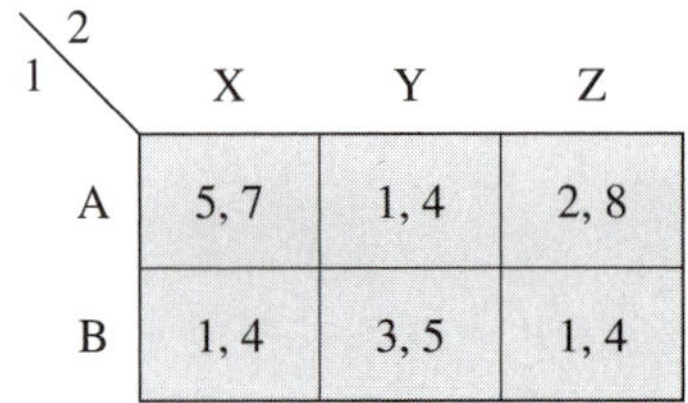

1 \ 2	X	Y	Z
A	5, 7	1, 4	2, 8
B	1, 4	3, 5	1, 4

interaction, each player has six information sets in the second period. In other words, each player has *six* decisions to make in period 2: what to do if the outcome of period 1 was (A, X), what to do if the outcome of period 1 was (A, Y), and so forth.

This game has a large set of strategies, making the analysis of rationality a bit daunting. However, the analysis is quite illuminating, so read on. Let us look for pure-strategy, subgame perfect Nash equilibria. We can simplify the search by recognizing one important thing: every equilibrium must specify that, in the *second* period, the players select an action profile that is a Nash equilibrium of the stage game.

To see why this is so, recall that subgame perfection requires equilibrium in *every subgame*. In the repeated game at hand, a different subgame is initiated following every different action profile in period 1. For example, consider what happens in the event that the players choose (A, Z) in the first period. Then, knowing that (A, Z) was the outcome of first-period interaction, the players proceed to a subgame in which they simultaneously select actions again. Their total payoff will be (1, 4) plus whatever the payoff vector is in the second play of the stage game. Thus, following the play of (A, Z) in the first period, the subgame is described by the matrix in Figure 22.3. I constructed this matrix by adding the payoff vector (1, 4) to each of the cells in the stage game (compare Figures 22.1 and 22.3). You should verify that the subgame has two Nash equilibria, (B, Y) and (A, Z). Therefore, a subgame perfect equilibrium must specify that either (B, Y) or (A, Z) be played in the second period if (A, Z) is played in the first period.

We can say more. Because the subgame matrix is formed by adding the same payoff vector to each cell of the stage-game matrix, the players' preferences over action profiles in the subgame *are exactly the same* as their preferences in an isolated play of the stage game. In other words, the subgame and the stage game are equivalent in this sense, and thus they have exactly the same Nash equilibria.[1] In fact, every subgame starting in period 2 has the same set of Nash equilibria, because these games have matrices like that in Figure 22.3—where the stage-game payoff in the first period is added to every

[1] You should verify that (A, Z) and (B, Y) are the Nash equilibria of the stage game.

cell of the stage game. Another way of thinking about this is that, once the first period is over, the payoffs from the first period are *sunk*. Whatever the players obtain in the second period, it is in addition to what they have already received in the first. Thus a subgame perfect equilibrium must specify that the players select a Nash equilibrium in the stage game in period 2, whatever happens in period 1. To save ink, I use the phrase "stage Nash profile" to refer to a Nash equilibrium in the stage game.

Knowing that a stage Nash profile will be played in the second period, we can turn our attention to two other matters: (1) action choices in the first period, and (2) how behavior in the first period determines *which* of the two stage Nash equilibria will be played in the second period.

First consider subgame perfect equilibria that specify that a stage Nash profile be played in the first period (as well as in the second). Here is one such strategy profile: the players are instructed to choose action profile (A, Z) in the first period and then, regardless of the outcome of the first period, they are to choose (A, Z) in the second period. Because each player has six information sets in the second period (six potential decisions to make), the phrase "regardless of the outcome of the first period" is crucial; it means that, even if one or both of the players deviate from (A, Z) in the first period, they are supposed to play (A, Z) in the second. You can easily verify that this strategy profile is a subgame perfect equilibrium—neither player can gain by deviating in either or both periods, given the other player's strategy. In this equilibrium, player 1 obtains $1 + 1 = 2$ and player 2 gets $4 + 4 = 8$. This payoff vector is one of those boxed in Figure 22.2(b).

Any combination of stage Nash profiles can be supported as a subgame perfect equilibrium outcome. For example, "choose (A, Z) in the first period and then, regardless of the first-period outcome, choose (B, Y) in the second period" is a subgame perfect equilibrium; it yields the payoff vector (3, 5). The payoffs of equilibria that specify stage Nash profiles in both periods are all boxed in Figure 22.2(b). I recommend reviewing the various combinations of stage Nash profiles and verifying that the associated equilibrium payoffs are boxed in Figure 22.2(b). As the example intimates, the following general result holds:

> **Result:** Consider any repeated game. Any sequence of stage Nash profiles can be supported as the outcome of a subgame perfect Nash equilibrium.

It probably does not surprise you that stage Nash profiles can be supported as equilibrium play. A more interesting question is whether there are equilibria stipulating actions that are *not* stage Nash profiles. In fact, the answer is "yes," as the two-period example at hand illustrates. Consider the following strategy profile:

> Select (A, X) in the first period and then, as long as player 2 does not deviate from X, select (A, Z) in the second period; if player 2 deviated by playing Y or Z in the first period, then play (B, Y) in the second period.

This strategy profile prescribes that the players' second-period actions depend on what player 2 did in period 1. By playing X in the first period, player 2 establishes a reputation for cooperating; in this case, he is rewarded in the second period as the players coordinate on the stage Nash profile that is more favorable to him. On the other hand, if player 2 deviates by, say, choosing Z in the first period, then he is branded a "cheater." In this case, his punishment is that the players coordinate on (B, Y) in the second period.

To verify that this strategy profile is a subgame perfect equilibrium, we must check each player's incentives. Suppose player 1 behaves as prescribed and consider the incentives of player 2. If player 2 goes along with the strategy prescription, he obtains 3 in the first period and 4 in the second period. If player 2 deviates in the first period, he can increase his first-period payoff to 4 (by picking Z). But this choice induces player 1 to select B in the second period, where player 2 then best responds with Y. Thus, although a first-period deviation yields an immediate gain of 1, it costs 3 in the second period (4 – 1). This shows that player 2 prefers to behave as prescribed. For his part, player 1 has the incentive to go along with the prescription for play; deviating in either period reduces player 1's payoff. The payoff vector for this subgame perfect equilibrium is enclosed by a pentagon in Figure 22.2(b).

Although the equilibrium construction is a bit complicated, it really is intuitive. Player 2's concern about his reputation and what it implies for his second-period payoff gives him the incentive to forego a short-term gain. If he misbehaves in the first period, his reputation is destroyed and he then suffers in the second period.

Any two-period repeated game can be analyzed as has been done here.[2] Only stage Nash profiles can be played in the second period. However, sometimes reputational equilibria exist whereby the players select non-stage-Nash profiles in the first period. These selections are supported by making the second-period actions contingent on the outcome in the first period (in particular, whether the players cheat or not). The exercises at the end of this chapter will help you better explore the reputation phenomenon.

---

[2] A general analysis of finitely repeated games is reported in J. P. Benoit and V. Krishna, "Finitely Repeated Games," *Econometrica* 53(1985):905–922.

AN INFINITELY REPEATED GAME

Infinitely repeated games are defined by $T = \infty$; that is, the stage game is played each period for an *infinite* number of periods. Although such a game may not seem realistic at first (people do not live forever), infinitely repeated games are useful for modeling some real-world situations. Furthermore, despite the complexity of these games, analysis of their subgame perfect equilibria can actually be quite simple. Consider an infinitely repeated game with *discounting,* whereby the payoffs in the stage game are discounted over time.[3] Let us use δ (a number between 0 and 1) to denote the discount factor for both players. When comparing a payoff received today with a payoff received tomorrow (the next period), we discount tomorrow's payoff by multiplying it by the discount factor. In this way, we say that the stream of payoffs—from today and tomorrow—are "discounted to today." Payoffs obtained two periods from now are discounted by δ^2, payoffs obtained three periods from now are discounted by δ^3, and so on.

For repeated games, we will have to calculate the sum of a stream of discounted payoffs. For example, a player may obtain 1 unit each period for an infinite number of periods. In this case, the sum of his discounted payoff stream is

$$s \equiv 1 + 1\delta + 1\delta^2 + 1\delta^3 + \cdots = 1 + \delta + \delta^2 + \delta^3 + \cdots .$$

We can simplify this expression by noting that

$$\delta + \delta^2 + \delta^3 + \cdots = \delta[1 + \delta + \delta^2 + \delta^3 + \cdots] = \delta s.$$

Therefore, we have

$$s \equiv 1 + \delta s,$$

which means that $s = 1/(1 - \delta)$. In summary,

$$1 + \delta + \delta^2 + \delta^3 + \cdots = \frac{1}{1 - \delta}.$$

This expression will come in handy. Note that, by multiplying both sides by any constant number a, we have

$$a + a\delta + a\delta^2 + a\delta^3 + \cdots = \frac{a}{1 - \delta}.$$

[3]Recall the discussion of discounting in Chapter 19, where the discount factor is used in multistage bargaining games.

FIGURE 22.4
A prisoners' dilemma.

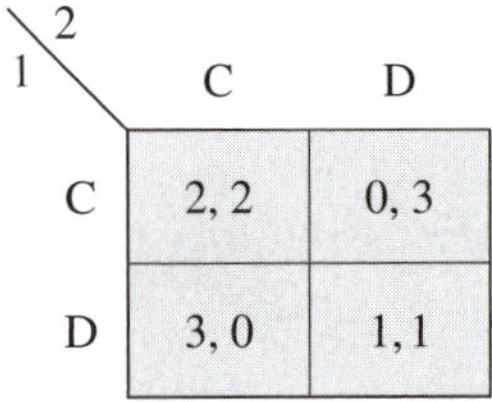

1 \ 2	C	D
C	2, 2	0, 3
D	3, 0	1, 1

The strategies in infinitely repeated games can be exceedingly complex. Recall that, in general, a player's strategy is a full description of what action to take at every information set of the player. In a repeated game, there is a different information set for every period t and every different history of play from the beginning of the game through period $t - 1$. Thus, a strategy prescribes an action for a player to take conditional on everything that took place in the past.

Fortunately, it is often sufficient to consider just a few types of simple strategies in repeated games. The simplest are those that prescribe stage Nash profiles in each period; as noted in the preceding section, we know these constitute subgame perfect equilibria.

To capture the idea of reputation, we can examine another simple type of strategy called a *trigger strategy*. Trigger strategies specifically refer to two action profiles for the stage game: one profile is called the "cooperative profile," and the other is called the "punishment profile." The punishment profile is assumed to be a stage Nash profile. In a trigger-strategy equilibrium, the players are supposed to play the cooperative profile in each period. However, if one or both of them deviate from the cooperative profile, then they play the punishment profile forever after. In other words, deviating from the cooperative profile destroys a player's reputation and triggers the punishment profile for the rest of the game.

To see how this works, consider the infinitely repeated prisoners' dilemma. The stage game is given in Figure 22.4. There is only one stage Nash equilibrium, (D, D), so we use it as the punishment profile. Let (C, C) be the cooperative profile. Our goal is to understand whether the players have the incentive to play (C, C) each period under the threat that they will revert to (D, D) forever if one or both of them cheat. To be precise, the trigger strategy specifies that the players select (C, C) each period as long as this profile was always played in the past; otherwise, they are to play (D, D). This is sometimes called the *grim-trigger* strategy.

Let us evaluate whether the grim-trigger profile is a subgame perfect equilibrium. Consider the incentives of player i ($i = 1, 2$) from the perspective of period 1. Suppose the other player—player j—behaves according to the grim trigger. Player i basically has two options. First, she can herself follow the prescription of the grim trigger, which means cooperating as player j does. In this case, player i obtains a payoff of 2 each period, for a discounted total of

$$2 + 2\delta + 2\delta^2 + 2\delta^3 + \cdots = \frac{2}{1-\delta}.$$

Second, player i could defect in the first period, which yields an immediate payoff of 3 because player j cooperates in the first period. But player i's defection induces player j to defect in each period thereafter, so then the best that i can do is to keep defecting and get 1 each period. Thus, by defecting in period 1, player i obtains the payoff

$$3 + \delta + \delta^2 + \delta^3 + \cdots = 3 + \delta[1 + \delta + \delta^2 + \delta^3 + \cdots] = 3 + \frac{\delta}{1-\delta}.$$

If

$$\frac{2}{1-\delta} \geq 3 + \frac{\delta}{1-\delta},$$

then player i earns a higher payoff by perpetually cooperating against the grim trigger than by defecting in the first period. Simplifying this inequality yields $\delta \geq 1/2$.

So far, we see that the players have no incentive to cheat in the first period as long as $\delta \geq 1/2$. In fact, the same analysis establishes that the players have no incentive to deviate from the grim trigger in *any* period. For example, suppose the players have cooperated through period $t - 1$. Then, because the game is infinitely repeated, the "continuation game" from period t looks just like the game from period 1, so the analysis starting from period t is exactly the same as the analysis at period 1. Discounting the payoffs to period t, we see that cooperating from period t yields each player $2/(1-\delta)$. Defecting against the grim trigger leads to the payoff $3 + \delta/(1-\delta)$. Thus, neither player has an incentive to defect in period t if the discount factor exceeds $1/2$.

In summary, the simple calculation performed in reference to period 1 is enough to establish whether cooperation can be supported in a subgame perfect equilibrium. For the stage game in Figure 22.4, cooperation can be sustained by the reputation mechanism if and only if the discount factor is at least $1/2$.

The grim-trigger analysis applies to any repeated prisoners' dilemma, although the "cutoff discount factor" depends on the payoffs in the stage game. As another example, take the stage game in Figure 22.5 on the next page. To find the conditions under which cooperation can be sustained, as before

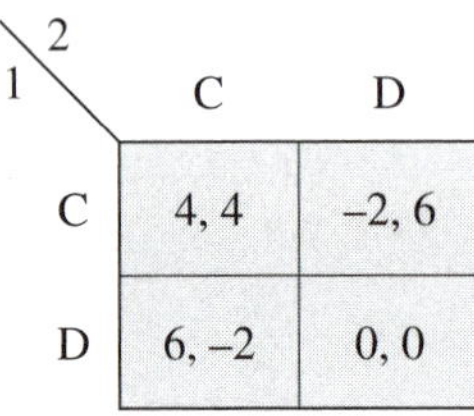

FIGURE 22.5
Another prisoners' dilemma.

1 \ 2	C	D
C	4, 4	–2, 6
D	6, –2	0, 0

we check whether the grim-trigger strategy profile forms a subgame perfect equilibrium. Suppose player j plays the grim-trigger strategy. If player i also adopts the grim trigger (cooperating perpetually), then she obtains 4 each period, for a discounted total of

$$4 + 4\delta + 4\delta^2 + 4\delta^3 + \cdots = \frac{4}{1-\delta}.$$

If player i defects, then she obtains 6 in the period of the defection but only 0 in each succeeding period. Player i has the incentive to cooperate if and only if $4/(1-\delta) \geq 6$. Solving this inequality for δ, we see that the grim-trigger profile is a subgame perfect equilibrium—and cooperation can be sustained—if and only if $\delta \geq 1/3$.

The infinitely repeated game demonstrates that patience—valuing the future—is essential to an effective reputation. When contemplating whether to defect in one period, the players consider the future loss that would result from tarnishing their reputations.[4] Patient players—those with high discount factors—care a lot about payoffs in future periods and therefore they do not want to ruin their reputations for some short-term gain. Thus, there is a sense in which maintaining a reputation is more about the future than the past.[5]

THE EQUILIBRIUM PAYOFF SET WITH LOW DISCOUNTING

The grim-trigger equilibrium discussed in the preceding section is just one of potentially many subgame perfect equilibria in the repeated prisoners' dilemma. We know (from the result on page 261) that the following strategy profile also is an equilibrium regardless of the discount factor: play the stage Nash action profile (D, D) in every period, regardless of the history of play. This equilibrium is not very "cooperative," and it yields a low payoff relative to the grim-trigger profile. In this section, I demonstrate that, depending

[4]The technical term for a player's future payoff is *continuation value*. If a player is patient, then his continuation value is much more important to him than is his payoff in any given period.

[5]One of the early general analyses of discounted, repeated games may be found in D. Abreu, "On the Theory of Infinitely Repeated Games with Discounting," *Econometrica* 56(1988):383–396.

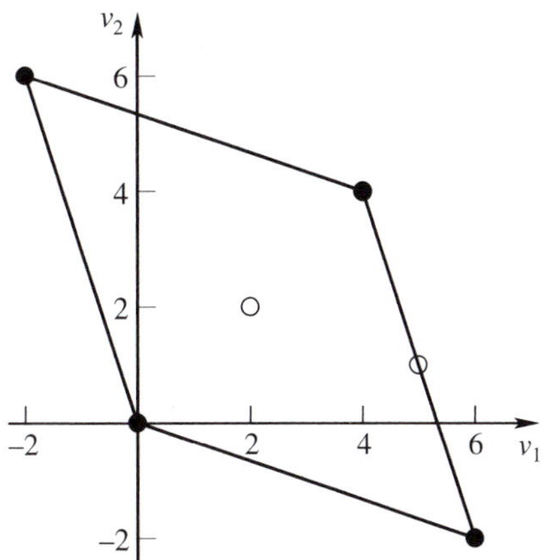

FIGURE 22.6
Possible repeated game payoffs, per period.

on the discount factor, there are many other equilibria exhibiting intermediate amounts of cooperation. The analysis will get a bit technical, but the conclusion at the end is significant.

To develop a picture of the entire set of equilibria in the repeated prisoners' dilemma, consider again the stage game in Figure 22.5. For this stage game, Figure 22.6 on the next page depicts the set of feasible stage-game payoffs. This figure also graphs the possible repeated game payoffs in terms of "average per period," by multiplying the discounted sum payoff by $(1-\delta)$. For example, the point (4, 4) is noted in the picture with a solid circle, which refers to the players obtaining (4, 4) each period in the game [by playing (C, C) each period].[6] The point (6, −2) arises if (D, C) is played each period.

The diamond formed by connecting points (4, 4), (−2, 6), (0, 0), and (6, −2) is important; any payoff vector inside or on the edges of the diamond can be obtained as an average payoff if the players choose the right sequence of actions over time. For instance, consider the point (5, 1) designated by an open circle in Figure 22.6. Suppose the players alternate between (C, C) and (D, C) over time, starting with (C, C) in the first period. For player 1, this sequence of actions yields a discounted payoff of

$$4 + 6\delta + 4\delta^2 + 6\delta^3 + \cdots.$$

Factoring terms, this expression simplifies to

$$4[1 + \delta^2 + \delta^4 + \cdots] + 6\delta[1 + \delta^2 + \delta^4 + \cdots] = \frac{4}{1-\delta^2} + \frac{6\delta}{1-\delta^2}.$$

[6]Technically speaking, each player's payoff is $4/(1-\delta)$ in this case. In Figure 22.6, I have multiplied this payoff by $(1-\delta)$ to put it in terms of per period.

Here I have used the same method as that put to use earlier to calculate the sums.[7] Recognizing that $1 - \delta^2 = (1 + \delta)(1 - \delta)$, we can write player 1's payoff as

$$\frac{4 + 6\delta}{(1 - \delta)(1 + \delta)}.$$

Multiplying by $(1 - \delta)$ puts this in terms of average per period:

$$\frac{4 + 6\delta}{1 + \delta}.$$

Likewise, player 2's per-period average is

$$\frac{4 - 2\delta}{1 + \delta}.$$

Note that, if δ is close to 1, then this average payoff vector is arbitrarily close to (5, 1). Just plug in $\delta = 1$ to see this. You can try other examples to convince yourself that the diamond in Figure 22.6 represents the set of average per-period payoffs that can arise in the repeated game. For instance, determine how an average payoff of (2, 2) can be obtained.

With the set of feasible repeated game payoffs in mind, we can determine whether or not any particular payoff vector can be supported as the result of a subgame perfect equilibrium. For an example, focus on the per-period average vector (5, 1) and consider the following "modified grim-trigger strategy": the players are instructed to alternate between (C, C) and (D, C) over time, starting with (C, C) in the first period; if either or both players has deviated from this prescription in the past, the players are supposed to revert to the stage Nash profile (D, D) forever. To determine whether this strategy profile is an equilibrium, we must compare each player's short-term gain from deviating to the punishment.

Let us begin with the incentives of player 2. First note that, if the players conform to the modified grim trigger, then, starting from any odd-numbered period [in which players select (C, C)], player 2's payoff is

$$4 - 2\delta + 4\delta^2 - 2\delta^3 + \cdots = \frac{4 - 2\delta}{1 - \delta^2}.$$

Starting from any even-numbered period, player 2's payoff is

$$-2 + 4\delta - 2\delta^2 + 4\delta^3 + \cdots = \frac{-2 + 4\delta}{1 - \delta^2}.$$

---

[7] Letting $s = 1 + \delta^2 + \delta^4 + \cdots$, we have $s = 1 + \delta^2 s$. Solving, we get $s = 1/(1 - \delta^2)$.

We need to check whether player 2 has an incentive to cheat in either an odd-numbered or an even-numbered period. Note that, in each odd-numbered period, the players are supposed to select (C, C). If player 2 cheats (by defecting), then he obtains a short-term gain of $6 - 4 = 2$; however, his continuation payoff starting in the next period—an even-numbered period—will be 0 (because the players defect thereafter) rather than $(-2 + 4\delta)/(1 - \delta^2)$. If we discount this future loss to the period in which player 2 cheats, the payoff is $\delta(-2 + 4\delta)/(1 - \delta^2)$. Thus, player 2 prefers to cooperate in odd-numbered periods if and only if the long-term loss of cheating outweighs the short-term gain:

$$\frac{\delta(-2 + 4\delta)}{(1 - \delta^2)} \geq 2. \tag{1}$$

Repeating the calculation for even-numbered periods yields the following inequality:

$$\frac{\delta(4 - 2\delta)}{(1 - \delta^2)} \geq 2. \tag{2}$$

The short-term gain of 2 here is due to player 2 obtaining 0 rather than -2.

In summary, assuming player 1 plays according to the modified grim trigger, player 2 wishes to conform if and only if both inequalities 1 and 2 are satisfied. You can verify with a bit of algebraic manipulation that the first simplifies to $\delta \geq (1 + \sqrt{13})/6$ and the second simplifies to $\delta \geq 1/2$. Because $(1 + \sqrt{13})/6 > 1/2$, the first inequality is more stringent than is the second. Thus, player 2 cooperates as long as $\delta \geq (1 + \sqrt{13})/6$. You can perform the same kind of analysis to find that player 1 will conform to the modified grim trigger as long as $\delta \geq (-3 + \sqrt{21})/6$. Because

$$\frac{1 + \sqrt{13}}{6} > \frac{-3 + \sqrt{21}}{6},$$

we conclude that the modified grim-trigger profile is a subgame perfect equilibrium if and only if $\delta \geq (1 + \sqrt{13})/6$. Furthermore, as already noted, the payoff of this equilibrium is close to (5, 1) if δ is close to 1.

I realize that you might find this analysis complicated. But putting the details of the arguments aside, recognize the general conclusion. I showed that a point on one of the edges of the diamond in Figure 22.6 can be supported as an equilibrium average per-period payoff as long as the players are patient enough. In fact, *any* point on the edge or interior of the diamond can be so supported, as long as two conditions hold: (1) each player obtains more than 0, and (2) the discount factor is close enough to 1. Figure 22.7 on the next page depicts the set of equilibrium payoffs. Here is the result stated for general repeated games:

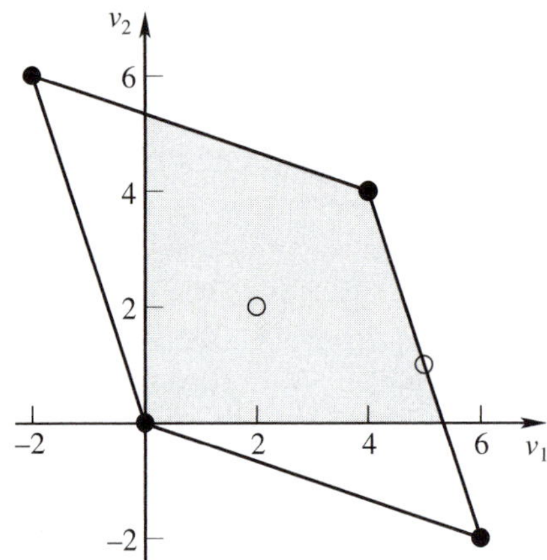

FIGURE 22.7
Equilibrium per-period payoffs for large δ.

> **Result:** Consider any infinitely repeated game. Suppose there is a stage Nash profile that yields payoff vector w (w_i for player i, $i = 1, 2, \ldots, n$). Let v be any feasible average per-period payoff such that $v_i > w_i$ for each player i. The vector v can be supported arbitrarily closely by a subgame perfect Nash equilibrium if δ is close enough to 1.

In other words, with the use of trigger strategies, almost any repeated game payoff can be achieved in equilibrium with patient players.[8]

GUIDED EXERCISE

Problem: Persons 1 and 2 are forming a firm. The value of their relationship depends on the effort that each expends. Suppose that person i's utility from the relationship is $x_j^2 + x_j - x_i x_j$, where x_i is person i's effort and x_j is the effort of the other person ($i = 1, 2$). Assume $x_1, x_2 \geq 0$.

(a) Compute the partners' best-response functions and find the Nash equilibrium of this game. Is the Nash equilibrium efficient?

(b) Now suppose that the partners interact over time, which we model with the infinitely repeated version of the game. Let δ denote the discount factor of

[8]Game theorists call this the *folk theorem* because it was thought to have been a part of the profession's conventional wisdom before versions of the result were formally proved. A general treatment appears in D. Fudenberg and E. Maskin, "The Folk Theorem in Repeated Games with Discounting or with Incomplete Information," *Econometrica* 54(1986):533–554.

the players. Under what conditions can the partners sustain some positive effort level $k = x_1 = x_2$ over time?

(c) Comment on how the maximum sustainable effort depends on the partners' patience.

Solution:

(a) To find the best-response functions, fix player j's effort x_j and consider how player i's effort level x_i affects his payoff $x_j^2 + x_j - x_i x_j$. If $x_j = 0$, then player i's payoff is 0, regardless of his choice, so in this case all of player i's feasible actions are best responses. Observe that, in the case of $x_j > 0$, player i's payoff is strictly decreasing in x_i and so player i's best response is to select the lowest effort $x_i = 0$. The only Nash equilibrium of this game is the profile $(0, 0)$; that is, each player selects the lowest effort level. The equilibrium is not efficient. To see this, note that if both players select effort level $k > 0$, then they each get the payoff $k^2 + k - k^2 = k$, which exceeds the payoff of 0 that they get with profile $(0, 0)$. In other words, profile $(0, 0)$ is less efficient than profile (k, k).

(b) Each player can guarantee himself a payoff of 0 by selecting $x_i = 0$ in every period. Thus, repeated play of the stage Nash profile, which yields a payoff of 0, is the worst punishment that can be levied against a player. With this in mind, consider the grim-trigger strategy profile in which each player selects effort level k in each period, as long as both players had done so in the past. Otherwise, the players revert to the stage Nash profile. When this strategy profile is played, then each player gets k in each period. The discounted sum over an infinite number of periods is $k/(1 - \delta)$. Note that the best way for a player to deviate is to select 0 effort rather than k, which yields a payoff of $k^2 + k$ in the period of the deviation (since the other player selects k). Following the deviation, the players will revert to the stage Nash profile, which gives the deviator a payoff of 0 in all future periods. Thus, for the grim trigger to be an equilibrium, we need $k/(1 - \delta) \geq k^2 + k$, which simplifies to $\delta/(1 - \delta) \geq k$. Rearranging yields $\delta \geq k/(1 + k)$.

(c) The maximal sustainable effort is $\delta/(1 - \delta)$, which is increasing in δ. In other words, more patient players can sustain higher levels of effort.

EXERCISES

1. Consider a repeated game in which the stage game in the following figure is played in each of two periods and there is no discounting.

1 \ 2	L	M	R
U	8, 8	0, 9	0, 0
C	9, 0	0, 0	3, 1
D	0, 0	1, 3	3, 3

Fully describe a subgame perfect equilibrium in which the players select (U, L) in the first period.

2. Find conditions on the discount factor under which cooperation can be supported in the infinitely repeated games with the following stage games.

1 \ 2	C	D
C	2, 2	0, 4
D	4, 0	1, 1

(a)

1 \ 2	C	D
C	3, 4	0, 7
D	5, 0	1, 2

(b)

1 \ 2	C	D
C	3, 2	0, 1
D	7, 0	2, 1

(c)

Use the grim-trigger strategy profile.

3. Consider the following stage game.

1 \ 2	X	Y
A	5, 6	0,0
B	8, 2	2,2

(a) Find and report all of the (pure-strategy) Nash equilibria of this game.

(b) Consider the two-period repeated game in which this stage game is played twice and the repeated game payoffs are simply the sum of the payoffs in each of the two periods. Is there a subgame perfect equilibrium of this repeated game in which (A, X) is played in the first period? If so, fully describe the equilibrium. If not, explain why.

4. If its stage game has exactly one Nash equilibrium, how many subgame perfect equilibria does a two-period, repeated game have? Explain. Would your answer change if there were T periods, where T is any finite integer?

5. Consider the infinitely repeated game where the stage game is the matrix in Exercise 2(c). Under what conditions is there a subgame perfect equilibrium in which the players alternate between (C, C) and (C, D), starting with (C, C) in the first period? Under what conditions is there a subgame perfect equilibrium in which the players alternate between (C, C) and (D, D), starting with (C, C) in the first period? (Use modified trigger strategies.)

6. Which is more important to achieving cooperation through a reputation, a long history together or a long horizon ahead?

7. Consider the following three-player game.

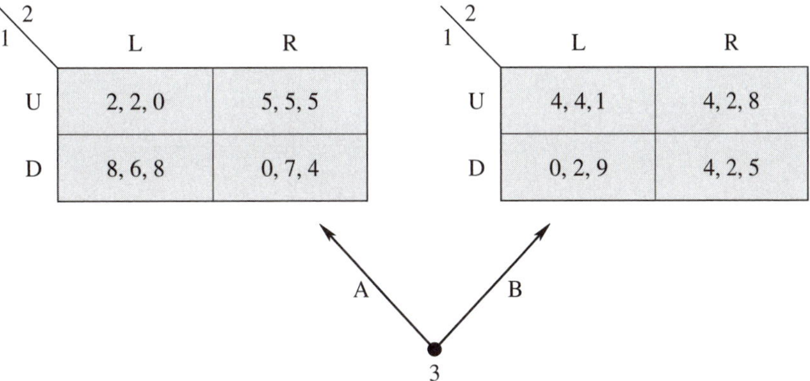

The players make their choices simultaneously and independently. The payoffs are listed in order of the player numbers.

(a) Find the (pure-strategy) Nash equilibria of this game.

(b) Consider the two-period repeated game in which this stage game is played twice and the repeated game payoffs are simply the sum of the payoffs in the two periods. Compute and report all of the subgame perfect equilibria of this repeated game. List the set of subgame perfect equilibrium payoffs.

8. Consider an infinite-period repeated game in which a "long-run player" faces a sequence of "short-run" opponents. Formally, player 1 plays the stage game with a different player 2 in each successive period. Denote by 2^t the player who plays the role of player 2 in the stage game in period t. Assume that all players observe the history of play. Let δ denote the discount factor of player 1. Note that such a game has an infinite number of players.

(a) In any subgame perfect equilibrium, what must be true about the behavior of player 2^t with respect to the action selected by player 1 in period t?

(b) Give an example of a stage-game and subgame perfect equilibrium where the players select an action profile in the stage game that is not a stage Nash equilibrium.

(c) Show by example that a greater range of behavior can be supported when both players are long-run players than when only player 1 is a long-run player.

9. Consider the following "war of attrition" game. Interaction between players 1 and 2 takes place over discrete periods of time, starting in period 1. In each period, players choose between "stop" (S) and "continue" (C) and they receive payoffs given by the following stage-game matrix:

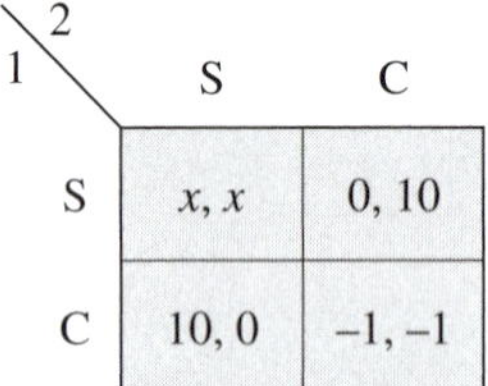

1 \ 2	S	C
S	x, x	0, 10
C	10, 0	–1, –1

The length of the game depends on the players' behavior. Specifically, if one or both players select S in a period, then the game ends at the end of this period. Otherwise, the game continues into the next period. Suppose the players discount payoffs between periods according to discount factor δ. Assume $x < 10$.

(a) Show that this game has a subgame perfect equilibrium in which player 1 chooses S and player 2 chooses C in the first period. Note that, in such an equilibrium, the game ends at the end of period 1.

(b) Assume $x = 0$. Compute the symmetric equilibrium of this game. (Hint: In each period, the players randomize between C and S. Let α denote the probability that each player selects S in a given period.)

(c) Write an expression for the symmetric equilibrium value of α for the case in which x is not equal to 0.

COLLUSION, TRADE AGREEMENTS, AND GOODWILL

23

In this chapter, I sketch three applications of repeated game theory. Two of them elaborate on analysis presented in part II of this book. In particular, to study collusion between firms, I use a repeated version of the Cournot duopoly model; discussion of the enforcement of international trade agreements utilizes a similar repeated game.

DYNAMIC OLIGOPOLY AND COLLUSION

Consider the Cournot duopoly model in Chapter 10, with two firms that each produce at zero cost (which I assume just to make the computations easy), and suppose the market price is given by $p = 1 - q_1 - q_2$. Firm i, which produces q_i, obtains a payoff of $(1 - q_i - q_j)q_i$. Note that the Nash equilibrium of this game is $q_1 = q_2 = 1/3$, yielding a payoff of 1/9 for each firm. As noted in Chapter 10, this outcome is inefficient from the firms' point of view; they would both be better off if they shared the monopoly level of output by each producing 1/4. Sharing the monopoly output yields each firm a payoff of 1/8, which is greater than the Nash equilibrium payoff of 1/9.[1] In the static game, therefore, the firms would like to collude to set $q_1 = q_2 = 1/4$, but this strategy profile cannot be sustained because it is not an equilibrium.

In most industries, firms do not interact in just a single point in time. They interact every day, potentially forever. To model firms' ongoing interaction, we can examine an infinitely repeated version of the Cournot duopoly, where the stage game is defined as the Cournot game described in the preceding paragraph. Analysis of the infinitely repeated game demonstrates that collusion can be sustained in equilibrium, using the reputation mechanism. In particular, let us evaluate the following grim-trigger-strategy profile: Each firm is prescribed to select 1/4 in each period, as long as both firms did so in the past; if one or both players deviates, the firms are supposed to play the stage Nash profile (1/3, 1/3) forever after.

[1] If numbers 1/8 and 1/9 seem insignificant, think of q_i as millions of units, the price in dollars, and the payoff therefore in millions of dollars.

Assume that firm j plays according to this strategy. Then, if firm i cooperates, it gets the discounted sum payoff

$$\frac{1}{8}(1 + \delta + \delta^2 + \cdots) = \frac{1}{8(1 - \delta)}.$$

Firm i can take advantage of firm j in the short run by producing more than $1/4$. However, it will then obtain $1/9$ in all future periods (anticipating that the players revert to the stage Nash profile following the deviation). To check firm i's incentives, note that, to maximize its immediate gain from cheating, firm i chooses q_i to maximize $(1 - 1/4 - q_i)q_i$. You should verify that the maximum is attained by picking $q_i = 3/8$ and that it yields a payoff of $9/64$ in the period of the deviation. Note that $9/64 > 1/8$, meaning that player i has a short-term incentive to deviate. Thus, the most that firm i can get by deviating from the grim trigger is an immediate payoff of $9/64$, plus $1/9$ in all future periods. With appropriate discounting, this payoff stream sums to

$$\frac{9}{64} + \frac{\delta}{9(1 - \delta)}.$$

Collusion can be sustained as a subgame perfect equilibrium if

$$\frac{1}{8(1 - \delta)} \geq \frac{9}{64} + \frac{\delta}{9(1 - \delta)},$$

which simplifies to $\delta \geq 9/17$. In words, if the firms do not discount the future too much, collusion is possible.

This result is a bit disconcerting because economists believe that competition yields many benefits to society, among them efficiency of the economy as a whole.[2] When firms collude, consumers can lose out big time, which is why there are laws against collusion to restrain trade. Government policy is, in this case, best understood in terms of how it restricts contracting. A collusive equilibrium, such as the grim trigger just studied, is a *self-enforced contract.* (Recall the discussion of contract in Chapter 13.) The Sherman Act (passed in the United States in 1890) prohibits such contracts between firms.[3] Thus, it is a no-no for managers of competing firms to meet in smoke-filled rooms and make deals.

Unfortunately, outlawing explicit collusion contracts is not enough, because firms often find ways of colluding without their managers actually having to communicate directly. For example, many firms make a big deal out

---

[2]Do not confuse efficiency of the economy—which takes into consideration all firms, consumers, and markets—with efficiency from the point of view of the two firms in the model presented here.

[3]The Clayton and Federal Trade Commission Acts (1914) extended the law on monopolization practices.

of their commitment to "match competitors' prices," and firms find ways to make this commitment legally binding. Although price-match commitments may seem competitive, in a dynamic setting they can have the opposite effect. By committing to match prices, firms may merely be committing to play the grim-trigger strategy. Thus, although the message to consumers is, "We're competitive," the message between firms may be, "We agree to get into a price war [the stage Nash profile] if any firm deviates from the collusive agreement." Firms with dominant market positions can also facilitate collusion by acting as "market leaders" who set prices expecting other firms to follow suit. When collusion takes place without the firms actively communicating about it, it is called an *implicit contract*. The Sherman Act forbids such tacit collusion, evaluating whether firms engage in parallel conduct that is likely different from what one would expect in a competitive market.

ENFORCING INTERNATIONAL TRADE AGREEMENTS

Whereas self-enforced contracts between colluding firms is undesirable, the opposite is true of contracts between countries. International trade agreements can be very beneficial, but, because there is no strong *external* enforcement institution for interaction between countries, nations must rely on self-enforcement. The reputation mechanism is used to enforce trade agreements.

For example, a significant fraction of the world's nations have agreed to set low tariffs on imports (a reduction from the high tariffs that existed decades ago). Low tariffs are generally efficient in that countries are better off when they all set low tariffs than if they all set high tariffs. However, as you have learned by analyzing the equilibrium of the static tariff game in Chapter 10, low tariffs cannot be sustained as a self-enforced contract when the countries rely on short-term incentives. In other words, low tariffs do not constitute a Nash equilibrium in the static game. Instead, nations utilize the repeated nature of their interaction. They often agree to trigger-strategy equilibria, whereby low tariffs (cooperation) are sustained by the threat of reverting to the high-tariff stage Nash profile. That is, if one country cheats by unilaterally raising a tariff, then it and its trading partners expect low value in the future as play turns to the stage Nash profile.

Self-enforced contracts between countries are quite explicit—they result from active, sometimes intense, negotiation. International institutions facilitate trade agreements by bringing the nations' representatives together, by providing them with a language that fosters mutual understanding, by recording agreements, and by disseminating information. The World Trade Organization (WTO) and its predecessor, the General Agreement on Tariffs and Trade (GATT), have been the focal point for achieving dramatic tariff reductions in the past century. Central to the WTO is the concept of "reciprocity," whereby a country is allowed to retaliate when one of its trading partners raises a tariff level. Reciprocity evokes the notion of trigger strategy.[4]

Owing to uncertainty and information problems, countries periodically get into disputes. For this reason, the WTO encourages a limited trigger-strategy equilibrium in which the punishment phase does not last forever; that is, governments do not exactly play the *grim* trigger, but they play something that delivers moderate punishment for a short time. Countries also renegotiate their contracts over time, to resolve disputes and balance their interests in the rapidly changing world. The "banana trade war" between the United States and the European Union illustrates the manner in which disputes, punishment, and renegotiation take place. In 1998, the United States asked the WTO to force the EU to dismantle favored trading terms given by the EU to banana producers in former European colonies. The WTO sided with the United States. In 1999, the EU responded by relaxing its rules toward imports of U.S. companies such as Chiquita, but not enough to satisfy the United States. With WTO approval, the United States retaliated by raising the tariff rates from 6 to 100 percent on several European luxury goods, such as pecorino cheese, cashmere wool products, and handbags. Negotiations between the United States and the EU are ongoing, including negotiations over issues such as the trade of genetically modified foods.

GOODWILL AND TRADING A REPUTATION

The word "trade" usually makes people think of the exchange of physical goods and services. But some less-tangible assets also are routinely traded. Reputation is one of them. Those who have studied accounting know that "goodwill" is a legitimate and often important item on the asset side of a firm's balance sheet. Goodwill refers to the confidence that consumers have in the firm's integrity, the belief that the firm will provide high-quality goods and services—in other words, the firm's reputation. It is often said that a firm's reputation is its greatest asset. Firms that have well-publicized failures (product recalls, for example) often lose customer confidence and, as a result, profits.

[4]The following recent articles use repeated game theory to study international institutions: K. Bagwell and R. W. Staiger, "An Economic Theory of GATT," *American Economic Review* 89(1999):215–248; G. Maggi, "The Role of Multilateral Institutions in International Trade Cooperation," *American Economic Review* 89(1999):190–214; and M. Klimenko, G. Ramey, and J. Watson, "Recurrent Trade Agreements and the Value of External Enforcement," University of California, San Diego, Discussion Paper 2001–01, 2001.

FIGURE 23.1
Stage game from Chapter 22.

1 \ 2	X	Y	Z
A	4, 3	0, 0	1, 4
B	0, 0	2, 1	0, 0

When a firm is bought or sold, its reputation is part of the deal. The current owners of a firm have an incentive to maintain the firm's good reputation to the extent that it will attract a high price from prospective buyers. This incentive may outweigh short-term desires to take advantage of customers or to do other things that ultimately will injure the firm's good name.

A game-theory model illustrates how reputation is traded.[5] The following game-theoretic example is completely abstract—it is not a model of a firm per se—but it clearly demonstrates how reputation is traded. Consider the two-period repeated game analyzed at the beginning of Chapter 22; the stage game is reproduced in Figure 23.1. Here I add a new twist. Suppose there are *three* players, called player 1, player 2^1, and player 2^2. In the first period, players 1 and 2^1 play the stage game (with player 2^1 playing the role of player 2 in the stage game). Then player 2^1 retires, so he cannot play the stage game with player 1 again in period 2. However, player 2^1 holds the *right* to play in period 2, even though he cannot exercise this right himself. Player 2^1 can sell this right to player 2^2, in which case players 1 and 2^2 play the stage game in the second period.

To be precise, the game begins in the first period, where players 1 and 2^1 play the stage game. Between periods 1 and 2, players 2^1 and 2^2 make a joint decision, determining whether player 2^2 obtains the right to play in period 2 as well as a monetary transfer from player 2^2 to player 2^1. If player 2^2 obtains the right from player 2^1, then players 1 and 2^2 play the stage game in period 2; otherwise, the game ends before the second period. The default outcome at the joint-decision phase is no transfer and no trade of the play right, ending the game. As for payoffs, player 1 obtains the sum of his stage-game payoffs; player 2^1 obtains his stage-game payoff from period 1 plus whatever transfer he negotiates with player 2^2 between periods; and player 2^2 obtains his payoff from the second period stage game (if played) minus the transfer to which he agreed between periods. Note that this is a game with joint decisions.

[5]The model I describe here is inspired by D. M. Kreps, "Corporate Culture and Economic Theory," in *Firms, Organizations and Contracts: A Reader in Industrial Organization,* ed. P. J. Buckley and J. Michie (New York: Oxford University Press, 1996), pp. 221–275. Recent, more rigorous research on this topic is contained in S. Tadelis, "The Market for Reputations as an Incentive Mechanism," *Journal of Political Economy* 92(2002):854–882.

To see how player 2^1's ability to sell the right to player 2^2 affects behavior, let us first solve the version of the game in which there is no joint decision between periods 1 and 2. In this version of the game, players 1 and 2^1 play the stage game in period 1 and then players 1 and 2^2 play the stage game in period 2. Observe that subgame perfection requires either (A, Z) or (B, Y) to be played in each period. To see this, suppose you wanted to sustain (A, X) in the first period. It would be irrational for player 2^1 to follow this prescription, because X is dominated in the stage game and player 2^1's payoff does not depend on anything that happens after the first period.

Now return to the game in which player 2^1 can sell player 2^2 the right to play the stage game in period 2. In this setting, player 2^1 actually *can* be given the incentive to play X. Consider the following regime: Players 1 and 2^1 are prescribed to select (A, X) in the first period. Then, in the event that the second-period stage game is played, the behavior of players 1 and 2^2 depends on the outcome of first-period interaction. If (A, X) was chosen in period 1, then 1 and 2^2 are supposed to choose (A, Z) in period 2; otherwise, they select (B, Y) in period 2. Note that the outcome of period 1 influences the amount that player 2^2 is willing to pay for the right to play. Play in period 2 is worth 4 to player 2^2 if (A, X) was the outcome of period 1; otherwise, the right to play is worth 1.

Assume the joint decision between periods is resolved according to the standard bargaining solution, where players 2^1 and 2^2 divide the surplus in proportion to their relative bargaining powers. Let α be the bargaining weight for player 2^1, so $(1-\alpha)$ is the weight for player 2^2. The disagreement point yields both of these players a payoff of 0, net of whatever player 2^1 received in the first period (which she gets regardless of interaction after the first period). If (A, X) was the outcome in period 1, then players 2^1 and 2^2 negotiate over a surplus of 4 (which is what player 2^2 would obtain by securing the right to play). Thus, conditional on (A, X) occurring in the first period, player 2^1 obtains $\alpha \cdot 4$ and player 2^2 obtains $(1-\alpha) \cdot 4$ from the negotiation phase. These values are achieved by having player 2^2 make a transfer of $\alpha \cdot 4$ to player 2^1 in exchange for the right to play in period 2. By similar reasoning, if (A, Z) was the outcome in the first period, then players 2^1 and 2^2 negotiate over a surplus of 1, yielding $\alpha \cdot 1$ to player 2^1 and $(1-\alpha) \cdot 1$ to player 2^2.

As I have constructed it, the regime under consideration specifies Nash equilibrium behavior in the second period—that is, (A, Z) or (B, Y), depending on the outcome of period 1—and joint decisions consistent with the standard bargaining solution. To complete the analysis, we must check whether players 1 and 2^1 have the unilateral incentive to deviate from playing (A, X) in the first period. If not, the regime is a negotiation equilibrium.

First, observe that player 1 has no incentive to deviate: he could induce (B, Y) to be played in the second period, but only at a first-period cost exceeding his second-period gain. As for player 2^1, if she goes along with (A, X), then her immediate payoff is 3 and she gets 4α through negotiation with player 2^2, for a total of $3 + 4\alpha$. On the other hand, if player 2^1 picks Z, then she would obtain 4 in the first period and α through negotiation, for a total of $4 + \alpha$. Thus, player 2^1 has the incentive to cooperate in the first period if and only if $3 + 4\alpha \geq 4 + \alpha$, which simplifies to $\alpha \geq 1/3$. In conclusion, the regime is a negotiation equilibrium if and only if $\alpha \geq 1/3$.

This analysis demonstrates that a reputation can be established by one party and then transferred to another party who finally exploits it. In the game, player 2^1's incentive to cooperate in period 1 derives entirely from her desire to build a reputation that she can sell to player 2^2. The model also illustrates the hold-up problem first discussed in Chapter 21. If the terms of trade favor player 2^2—represented by $\alpha < 1/3$—then player 2^1 cannot appropriate much of the value of his reputation investment; in this case, (A, X) cannot be supported in the first period.

GUIDED EXERCISE

Problem: Consider the following game:

1 \ 2	x	y	z
x	3, 3	0, 0	0, 0
y	0, 0	5, 5	9, 0
z	0, 0	0, 9	8, 8

(a) What are the Nash equilibria of this game?

(b) If the players could meet and make a self-enforced agreement regarding how to play this game, which of the Nash equilibria would they jointly select?

(c) Suppose the preceding matrix describes the stage game in a two-period repeated game. Show that there is a subgame perfect equilibrium in which (z, z) is played in the first period.

(d) One can interpret the equilibrium from part (c) as a self-enforced, dynamic contract. Suppose that, after they play the stage game in the first period but before they play the stage game in the second period, the players have an opportunity to renegotiate their self-enforced contract. Do you believe the equilibrium from part (c) can be sustained?

Solution:

(a) You can easily verify that the Nash equilibria are (x, x) and (y, y).

(b) Because strategy profile (y, y) is more efficient than profile (x, x), the players would agree to play (y, y).

(c) Consider the following strategy profile for the two-period repeated game: In the first period, the players are supposed to select (z, z). If neither player deviates, then the strategy profile prescribes that stage Nash profile (y, y) be played in the second period. On the other hand, if one or both players deviate in the first period (for example, if one player chooses y for an immediate gain of $9 - 8 = 1$), then the players coordinate on stage Nash profile (x, x) in the second period. To see that this strategy profile constitutes a subgame perfect equilibrium, note that it always prescribes a stage Nash profile in the second-period subgames. Furthermore, a player gains at most 1 by deviating in the first period, but in this case the player then loses 2 because of the shift to (x, x) in the second period.

(d) Renegotiation can interfere with the equilibrium described in part (c). For example, if player 1 deviates by choosing y in the first period, he could then say the following to player 2: "Hey, I made a mistake. Let's not continue with the equilibrium we agreed to earlier, for it now specifies play of (x, x). It is strictly better for both of us to coordinate on (y, y)." Indeed, the players have a mutual interest in switching to (y, y). But anticipating that they would do this in the event of a first-period deviation, (z, z) is not sustainable in the first period. The theory of renegotiation is, by the way, an important topic at the frontier of current research in game theory.

EXERCISES

1. Consider the Bertrand oligopoly model, where n firms simultaneously and independently select their prices, $p_1, p_2, \ldots, p_n$, in a market. (These prices are greater than or equal to 0.) Consumers observe these prices and only purchase from the firm (or firms) with the lowest price $\underline{p}$, according to the demand curve $Q = 110 - \underline{p}$. ($\underline{p} = \min\{p_1, p_2, \ldots, p_n\}$.) That is, the firm with the lowest price gets all of the sales. If the lowest price is offered by more than one firm, then these firms equally share the quantity demanded. Assume that firms must supply the quantities demanded of them and that production takes place

at a constant cost of 10 per unit. (That is, the cost function for each firm is $c(q) = 10q$.) Determining the Nash equilibrium of this game was the subject of a previous exercise.

(a) Suppose that this game is infinitely repeated. (The firms play the game each period, for an infinite number of periods.) Define δ as the discount factor for the firms. Imagine that the firms wish to sustain a collusive arrangement in which they all select the monopoly price $p^{M} = 60$ each period. What strategies might support this behavior in equilibrium? (Do not solve for conditions under which equilibrium occurs. Just explain what the strategies are. Remember, this requires specifying how the firms punish each other. Use the Nash equilibrium price as punishment.)

(b) Derive a condition on n and δ that guarantees that collusion can be sustained.

(c) What does your answer to part (b) imply about the optimal size of cartels?

2. Examine the infinitely repeated tariff-setting game, where the stage game is the two-country tariff game in Chapter 10 (see also Exercise 3 in that chapter).

(a) Compute the Nash equilibrium of the stage game.

(b) Find conditions on the discount factor such that zero tariffs ($x_1 = x_2 = 0$) can be sustained each period by a subgame perfect equilibrium. Use the grim-trigger strategy profile.

(c) Find conditions on the discount factor such that a tariff level of $x_1 = x_2 = k$ can be sustained by a subgame perfect equilibrium, where k is some fixed number between 0 and 100.

3. Repeat the analysis of goodwill presented in this chapter for the following stage game:

1 \ 2	X	Y	Z
A	5, 5	0, 3	4, 8
B	0, 0	4, 4	0, 0

4. Consider an infinite-period repeated prisoners' dilemma game in which a long-run player 1 faces a sequence of short-run opponents. (You dealt with games like this in Exercise 8 of Chapter 22.) Formally, there is an infinite number of players—denoted $2^1, 2^2, 2^3, \ldots$—who play as player 2 in the stage game. In period t, player 1 plays the following prisoners' dilemma with player 2^t.

1 \ 2	C	D
C	2, 2	0, 3
D	3, 0	1, 1

Assume that all of the players observe the history of play. Let δ denote the discount factor of player 1.

(a) Show that the only subgame perfect equilibrium involves play of (D, D) each period.

(b) Next suppose that the players 2^t have the opportunity to buy and sell the right to play the game each period. Formally, suppose that, in order for player 2^t to play the stage game in period t, this player must purchase the right from player 2^{t-1}. Model the negotiation between successive players 2^{t-1} and 2^t as a joint decision. Further, assume the outcome of negotiation is consistent with the standard bargaining solution. Under what conditions on bargaining weights and the discount factor can (C, C) be supported over time?

5. Repeat the analysis of the Guided Exercise for the following stage game:

1 \ 2	x	y	z
x	0, 0	0, 0	9, 0
y	0, 0	6, 6	0, 0
z	0, 9	0, 0	8, 8

In your answer to part (d), discuss the role of bargaining weights and disagreement points in the renegotiation phase.

6. This exercise addresses the notion of goodwill between generations. Consider an "overlapping generations" environment, whereby an infinite-period game is played by successive generations of persons. Specifically, imagine a family comprising an infinite sequence of players, each of whom lives for two periods. At the beginning of period t, player t is born; he is young in period t, he is old in period $t + 1$, and he dies at the end of period $t + 1$. Thus, in any given period t, both player t and player $t - 1$ are alive. Assume that the game starts in period 1 with an old player 0 and a young player 1.

When each player is young, he starts the period with one unit of wealth. An old player begins the period with no wealth. Wealth cannot be saved across periods, but each player t can, while young, give a share of his wealth to the next older player $t-1$. Thus, consumption of the old players depends on gifts from the young players. Player t consumes whatever he does not give to player $t-1$. Let x_t denote the share of wealth that player t gives to player $t-1$. Player t's payoff is given by $(1-x_t)+2x_{t+1}$, meaning that players prefer to consume when they are old.

(a) Show that there is a subgame perfect equilibrium in which each player consumes all of his wealth when he is young; that is, $x_t = 0$ for each t.

(b) Show that there is a subgame perfect equilibrium in which the young give all of their wealth to the old. In this equilibrium, how is a player punished if he deviates?

(c) Compare the payoffs of the equilibria in parts (a) and (b).

7. Consider an infinitely repeated game with the following stage game:

1 \ 2	I	N
I	4, 4	−1, 8
N	6, −1	0, 0

Suppose that this repeated game is the underlying game in a contractual setting with external enforcement. Specifically, an external enforcer will compel transfer α from player 2 to player 1 in every period in which (N, I) is played, transfer β from player 2 to player 1 in every period in which (I, N) is played, and transfer γ from player 2 to player 1 in every period in which (N, N) is played. The numbers α, β, and γ are chosen by the players in their contract before the repeated interaction begins. These numbers are fixed for the entire repeated game. Assume that the players discount the future using discount factor δ.

(a) Suppose there is full verifiability, so that α, β, and γ can be different numbers. Under what conditions on δ is there a contract such that the players choose (I, I) in each period in equilibrium?

(b) Suppose there is limited verifiability, so that $\alpha = \beta = \gamma$ is required. Under what conditions on δ is there a contract such that the players choose (I, I) in each period in equilibrium? Consider the grim-trigger strategy profile.

(c) Continue to assume the conditions in part (b), with a value of δ that supports play of (I, I) in each period of the repeated game. Suppose that, when the players select α, β, and γ before the repeated game is played, they can also make an up-front monetary transfer m from player 2 to player 1. Assume that, if the players do not reach an agreement on α, β, γ, and m, then they will not play the repeated game and each player will get 0. If player 1's bargaining weight is $\pi_1 = 2/3$, what does the standard bargaining solution predict will be the value of m?

8. Consider the prisoners' dilemma stage game pictured here:

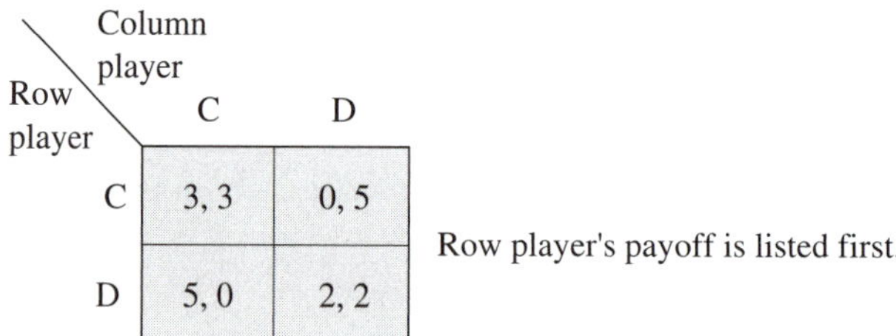

The following questions ask you to consider various discrete-time environments in which people meet and play the stage game. For each environment, you are to determine the extent to which cooperation can be sustained in a subgame perfect equilibrium.

(a) Suppose that two players interact in an infinitely repeated game, with the stage game pictured above, and that the players share the discount factor $\delta \in (0, 1)$. Under what conditions on δ is there a subgame perfect equilibrium that supports both players selecting C each period?

(b) Suppose that there is a society of k individuals. In each period, two people from the society are randomly selected to play the stage game. Individuals are equally likely to be selected. Furthermore, the random draw that determines who is selected in a given period is independent of the outcomes in previous periods. Thus, an individual has a probability $2/k$ of being selected to play the stage game in any given period. One of the people selected plays the role of Row player in the stage game, whereas the other person plays the role of Column player. (The stage game is symmetric, so the roles do not matter.) Those who are selected obtain the payoffs of the stage game, whereas the individuals who were not selected get 0 for this period. Everyone discounts the future using discount factor δ.

Assume that the individuals can identify each other. That is, individuals know the names and faces of everyone else in the society. Assume also that, before the stage game is played in a given period, everyone observes who is selected to play. Further, everyone observes the outcome of each period's stage game.

Is there a sense in which it is more difficult to sustain cooperation in this random-matching setting than was the case in part (a)? Explain why or why not. Calculate the cutoff value of δ under which cooperation can be sustained.

(c) Consider the same setting as in part (b) *except assume that the individuals only observe their own history of play*. As before, individuals can identify each other. However, an individual only knows when he is selected to play the stage game; if he is not selected to play in a given period, he observes neither who was selected to play nor the outcome of the stage game. When an individual is selected to play the stage game, he learns whom he is matched with; then the stage game is played, and the two matched individuals observe the outcome.

Explain why cooperation is more difficult to sustain in this setting than in the setting of part (b). If you can, calculate the cutoff value of δ under which cooperation can be sustained.

Information

Many strategic settings are interesting because players have different information at various junctures in a game. Most of the games presented in preceding chapters have some sort of information imperfection; that is, their extensive forms contain nontrivial information sets (those consisting of more than one node). The most interesting information problems feature *private information*, which exists when a player knows something that other players do not observe. This is also called *asymmetric information*.

The analysis presented in preceding chapters covers strategic settings in which there is asymmetric information only regarding players' *actions*. More broadly, there are important settings in which players have private information about other things as well. For example, suppose a buyer and a seller negotiate the price of a house. It may be that, in the negotiation process, actions are taken sequentially and in the open. Still, the buyer might know something that the seller does not know. The buyer may know his true valuation of the house—the highest price that he is willing to accept. The seller may have some belief about the buyer's reservation value (represented by a probability distribution), but the seller may not know for sure what it is.

The best way of modeling private information about intangible items, such as a person's value of a good, is to incorporate random events in the specification of a game—events that are out of the players' control. Game theorists call such random events *moves of nature*. In this sense, nature is a player in the game, whom we may call player 0. But nature is a nonstrategic player; nature's actions are defined by a fixed probability distribution and not by incentives. In the home-buying example, you can think of the buyer's reservation value as selected by nature at the start of the negotiation game. The buyer immediately observes this value; the seller only knows the distribution over nature's choices.

The expression *incomplete information* refers to games having moves of nature that generate asymmetric information between the players. Theorists use the term *type* to indicate the different moves of nature that a single player privately observes. Often, a player's private information has to do with his or her own personal attributes and tastes, to which the term *type* is particularly well suited. For example, we can speak of the different types of home buyer distinguished by different valuations of the house.

Incomplete information is present in a variety of trading environments: bargaining, competitive markets, contract environments, auctions, and so forth. In this part of the book, I introduce tools for analyzing games with moves of nature, and I survey some of the major applications of the theory. I start with games in which nature moves last, so there is no incomplete information between the players; this is the simplest class of games to study. I then turn to static games with incomplete information, followed by dynamic games with incomplete information.

RANDOM EVENTS AND INCOMPLETE INFORMATION 24

As already noted, it is useful to think of random events as acts of nature. In the extensive form, we model nature as "player 0," whose decisions are made according to a fixed probability distribution. Because nature is not a strategic player, no payoff numbers are associated with nature. Graphically, nature's decision nodes—also called *chance nodes*—are depicted by open circles, to distinguish them from the decision nodes of the strategic players.[1] Nature's "player 0" title is used very narrowly. Generally, when we speak of the "players in the game," we mean the strategic players. Thus, some researchers prefer to use the term "chance node" instead of "nature's decision node."

Figure 24.1 depicts a game of incomplete information. In this game, nature first determines the type of player 1, which is either Friend (with probability p) or Enemy (with probability $1 - p$). Player 1 then observes nature's move, so he knows his own identity. In this model, friends tend to keep desirable objects in their pockets, such as jewelry and new electronic gadgets. These objects are suitably gift wrapped. Enemies keep undesirable objects in their pockets, such as rocks and frogs. These things also are wrapped.

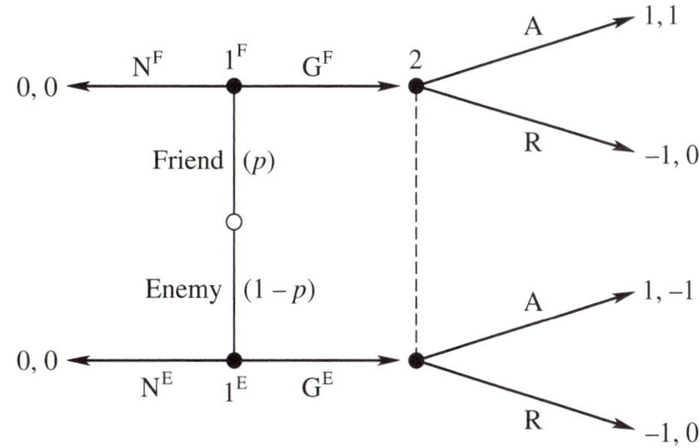

FIGURE 24.1
The gift game.

[1]The foundation for the analysis of incomplete information games is John Harsanyi's "Games with Incomplete Information Played by Bayesian Players," *Management Science* 14(1967–1968):159–182, 320–334, 486–502.

Player 1 then decides whether to offer a gift to player 2. If he chooses not to offer a gift (action N), then the game ends and both players receive zero. Offering a gift (action G) entails handing player 2 one of the handsomely wrapped objects from his pocket. In this case, player 2 must decide whether to accept (A) or reject (R) player 1's gift. Importantly, player 2 does not observe player 1's type (Friend or Enemy) directly. She knows only whether a gift has been offered, in which case she must make a decision.

The payoffs are meant to capture the following preferences: Player 1's favorite outcome occurs when he offers a gift and it is accepted. In other words, a friend enjoys seeing player 2 unwrap a piece of jewelry, whereas an enemy revels in the cruelty of insulting player 2 with a gift of a frog. Both types of player 1 prefer having not extended a gift to enduring the humiliation of a gift rejected. Player 2 prefers accepting a desirable gift to refusing a gift and prefers refusing a gift to discovering a frog inside the box. Apparently, player 2 does not mind associating with people whose pockets bulge.

Note that the game in Figure 24.1 is one of incomplete information in that player 1 has private information about nature's action. It is interesting that player 1's type determines player 2's value of accepting a gift. That is, player 1's personal attributes affect player 2's payoff. This is meant to illustrate that, although a player's private information often concerns his own payoffs, it sometimes has to do with the payoffs of another player.

Another thing you should recognize is that, in games of incomplete information, rational play will require a player who knows his own type to think about what he *would have done* had he been *another* type. For example, if you are a friend and you are considering whether to give a gift, you ought to imagine how you would behave if you were an enemy. The reason that putting yourself in another type's shoes is so important is that, although you know that you are a friend, player 2 does not know this. Your optimal decision depends on how player 2 will react to a gift, which, in turn, in part depends on whether player 2 thinks the enemy is duplicitous (gives gifts).

It can be helpful to study the normal-form version of a game with nature. Such a normal-form representation is called the *Bayesian normal form*. Translating the extensive form into the normal form requires just a bit more work in this case than it does in settings without moves of nature. But the procedure is the same. The key idea is that, because nature's moves are predetermined by a probability distribution, we can focus on the strategies of the strategic players and compute payoffs by averaging over the random events in the game. To illustrate, consider the gift game, where player 1 has four strategies and player 2 has two strategies. Figure 24.2 depicts the normal form. To obtain the payoffs in the normal form, one traces paths through the tree corresponding to strategy

FIGURE 24.2
The gift game in Bayesian normal form.

1 \ 2	A	R
G^FG^E	$1, 2p-1$	$-1, 0$
G^FN^E	p, p	$-p, 0$
N^FG^E	$1-p, p-1$	$p-1, 0$
N^FN^E	$0, 0$	$0, 0$

profiles. But contrary to the case without nature, where each strategy profile induces a single path, with nature there may be multiple paths. The payoffs of these paths are averaged according to nature's probability distribution, which yields a payoff vector for each cell of the matrix.

For example, to find the payoff of the strategy profile (G^FG^E, A), note that, with probability p, player 1's top node is reached. In this case, player 1 gives a gift that is accepted by player 2, leading to the payoff vector (1, 1). Likewise, with probability $1 - p$, player 1's lower decision node is reached, a gift is offered and accepted, and the payoff $(1, -1)$ is obtained. Thus, player 1's expected payoff from strategy profile (G^FG^E, A) is

$$p \cdot 1 + (1 - p) \cdot 1 = 1.$$

Player 2's expected payoff is

$$p \cdot 1 + (1 - p)(-1) = 2p - 1.$$

These numbers appear in the first cell of the matrix. To solidify your understanding of the normal form, verify the payoffs in the other cells.

Another example of a game with nature is depicted in Figure 24.3. In this strategic setting, players 1 and 2 play according to the matrix shown. However, player 1's payoff number x is private information. Player 2 knows only that $x = 12$ with probability 2/3 and $x = 0$ with probability 1/3. Note that

FIGURE 24.3
A game of incomplete information.

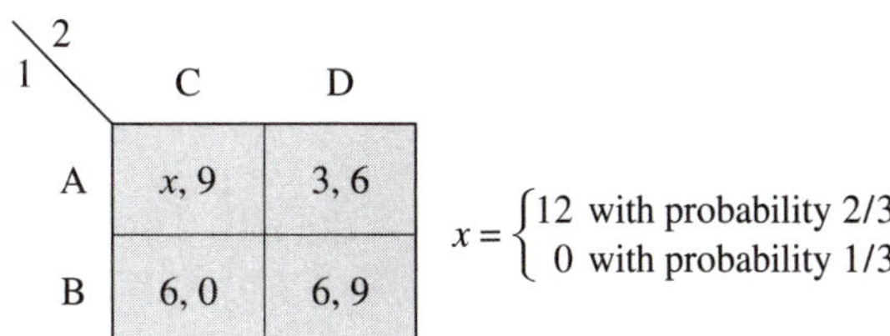

1 \ 2	C	D
A	x, 9	3, 6
B	6, 0	6, 9

$$x = \begin{cases} 12 \text{ with probability } 2/3 \\ 0 \text{ with probability } 1/3 \end{cases}$$

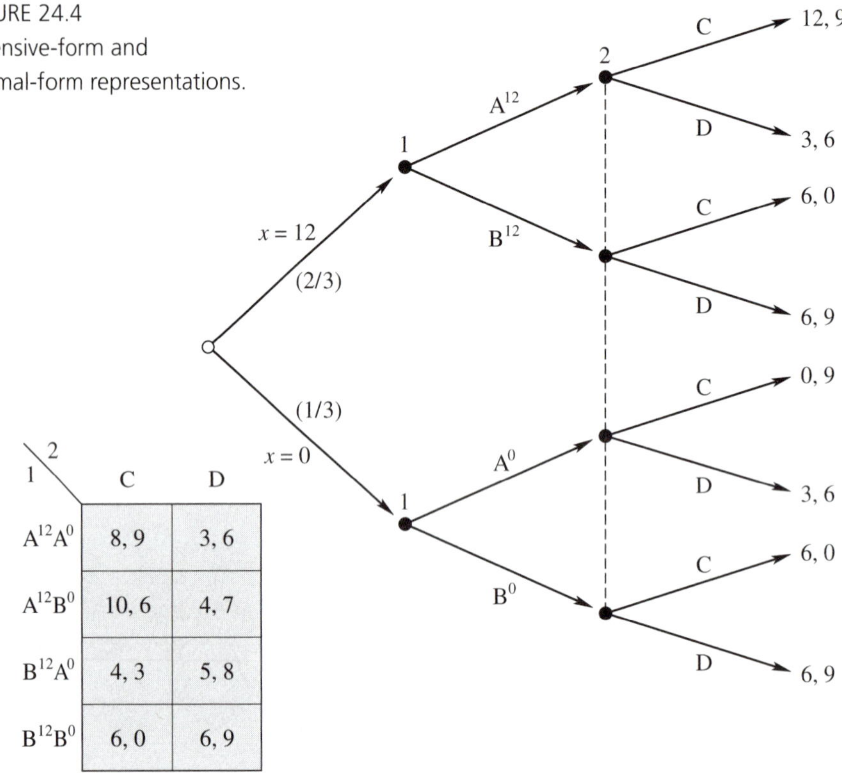

1 \ 2	C	D
$A^{12}A^0$	8, 9	3, 6
$A^{12}B^0$	10, 6	4, 7
$B^{12}A^0$	4, 3	5, 8
$B^{12}B^0$	6, 0	6, 9

FIGURE 24.4
Extensive-form and normal-form representations.

the matrix pictured is not the true normal form of the game because player 1 observes x before making his decision. The extensive- and normal-form representations of the game are pictured in Figure 24.4. As the extensive form shows, player 1 observes nature's action before selecting between A and B, yet player 2 must make his choice without observing player 1's type or action. Thus, player 1 has two decisions to make: (1) whether to select A or B after observing $x = 0$, and (2) whether to select A or B after observing $x = 12$. Using the abbreviations employed in the preceding example, we can write player 1's strategies as $A^{12}A^0$, $A^{12}B^0$, $B^{12}A^0$, and $B^{12}B^0$. The payoffs in the normal form of Figure 24.4 are the expected payoffs, given nature's probability distribution. For example, consider the strategy profile $(B^{12}A^0, D)$. When this profile is played, then, with probability 2/3, nature selects $x = 12$ and the payoff vector (6, 9) is obtained; with probability $1/3$, nature selects $x = 0$ and the payoff vector (3, 6) is obtained. Therefore, player 1's expected payoff is $6\cdot(2/3)+3\cdot(1/3) = 5$, and player 2's expected payoff is $9\cdot(2/3)+6\cdot(1/3) = 8$, as indicated in the normal-form matrix.

GUIDED EXERCISE

Problem: Consider the following market game. There are two firms: the "incumbent" and the "entrant." The incumbent firm has either high costs (H) or low costs (L); this is the incumbent's type, which is selected by nature at the beginning of the game. With probability q, the incumbent's type is H and, with probability $1 - q$, the incumbent's type is L. The incumbent observes its own type, but the entrant does not observe the incumbent's type. After observing its type, the incumbent selects either a high price ($\overline{p}$) or a low price ($\underline{p}$). The entrant observes the incumbent's price and then decides whether or not to enter the market (E or N). The incumbent's payoff is 0 if the entrant chooses E (regardless of the incumbent's type). If the entrant picks N and the incumbent's price is $\overline{p}$, then the high-type incumbent gets 2 and the low-type incumbent gets 4. If the entrant picks N and the incumbent's price is $\underline{p}$, then the high-type incumbent gets 0 and the low-type incumbent gets 2. The entrant obtains nothing if it does not enter. It obtains a payoff of 1 if it enters and faces the high-type incumbent. It obtains −1 if it enters and faces the low-type incumbent.

(a) Represent this game in the extensive form.

(b) Explain the relation between Nash equilibrium and subgame perfect equilibrium in this game.

(c) Represent this game in the Bayesian normal form.

Solution:

(a) Here is the extensive-form diagram:

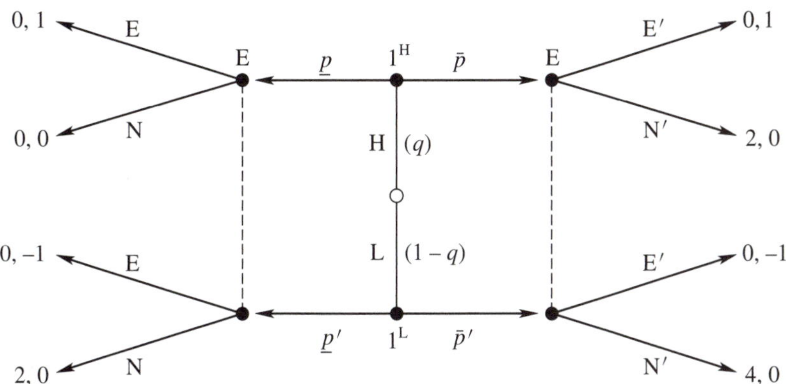

(b) Recall that a subgame perfect equilibrium involves specification of a Nash equilibrium on each subgame. Note that this game has *no* proper subgames, so the only subgame is the entire game. Thus, subgame perfection is the same as Nash equilibrium in this game.

(c) Here is the Bayesian normal-form matrix:

I \ E	EE′	EN′	NE′	NN′
$\overline{pp}'$	$0, 2q-1$	$4-2q, 0$	$0, 2q-1$	$4-2q, 0$
$\overline{p}\underline{p}'$	$0, 2q-1$	$2q, q-1$	$2-2q, q$	$2, 0$
$\underline{p}\overline{p}'$	$0, 2q-1$	$4-4q, q$	$0, q-1$	$4-4q, 0$
$\underline{pp}'$	$0, 2q-1$	$0, 2q-1$	$2-2q, 0$	$2-2q, 0$

To see how the payoffs are determined, consider as an example the strategy profile ($\overline{p}\underline{p}'$, EN′) and trace through the tree. If nature selects H, then player 1 chooses $\overline{p}$, player 2 chooses N′, and the payoff vector is (2, 0). If nature selects L, then player 1 chooses $\underline{p}'$, player 2 chooses E, and the payoff vector is (0, −1). Multiplying these, respectively, by nature's probabilities q and $1-q$ and summing yields the expected payoff vector $(2q, q-1)$, which is shown in the appropriate cell of the matrix.

EXERCISES

1. Here is a description of the simplest poker game. There are two players and only two cards in the deck, an Ace (A) and a King (K). First, the deck is shuffled and one of the two cards is dealt to player 1. That is, nature chooses the card for player 1. It is the Ace with probability 1/2 and the King with probability 1/2. *Player 2 does not receive a card.*

   Player 1 observes his card and then chooses whether to bid (B) or fold (F). If he folds, then the game ends with player 1 getting a payoff of −1 and player 2 getting a payoff of 1 (that is, player 1 loses his ante to player 2). If player 1 bids, then player 2 must decide whether to bid or fold. When player 2 makes this decision, she knows that player 1 bid, but she has not observed player 1's card. The game ends after player 2's action. If player 2 folds, then the payoff vector is (1, −1), meaning player 1 gets 1 and player 2 gets −1. If player 2 bids, then the payoff depends on player 1's card; if player 1 holds the Ace then the payoff vector is (2, −2); if player 1 holds the King, then the payoff vector is (−2, 2).

   Represent this game in the extensive form and in the Bayesian normal form.

2. Suppose Andy and Brian play a guessing game. There are two slips of paper; one is black and the other is white. Each player has one of the slips of paper pinned to his back. Neither of the players observes which of the two slips is pinned to his own back. (Assume that nature puts the black slip on each player's back with probability 1/2.) The players are arranged so that Andy can see the slip on Brian's back, but Brian sees neither his own slip nor Andy's slip.

   After nature's decision, the players interact as follows. First, Andy chooses between Y and N. If he selects Y, then the game ends; in this case, Brian gets a payoff of 0, and Andy obtains 10 if Andy's slip is black and −10 if his slip is white. If Andy selects N, then it becomes Brian's turn to move. Brian chooses between Y and N, ending the game. If Brian says Y and Brian's slip is black, then he obtains 10 and Andy obtains 0. If Brian chooses Y and the white slip is on his back, then he gets −10 and Andy gets 0. If Brian chooses N, then both players obtain 0.

   (a) Represent this game in the extensive form.

   (b) Draw the Bayesian normal-form matrix of this game.

3. Represent the following game in the Bayesian normal form.

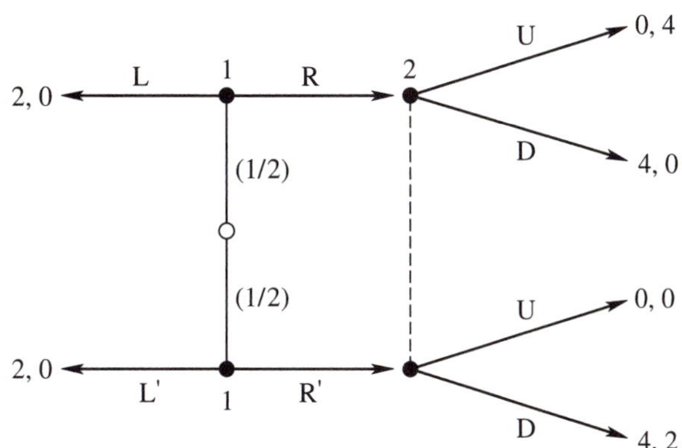

4. Draw the extensive-form representation of the following three-card poker game. There are three cards in the deck: an Ace, a King, and a Queen. The deck is shuffled, one card is dealt to player 1, and one card is dealt to player 2. Assume the deck is perfectly shuffled so that the six different ways in which the cards may be dealt are equally likely. Each player observes his own card but not the card dealt to the other player.

   Player 1 then decides whether to fold or bid. If he folds, then the game ends and the payoff vector is (−1, 1); that is, player 1 obtains −1 (he loses his ante), and player 2 obtains 1 (gaining player 1's ante). If player 1 bids, then player 2 must decide whether to fold or bid. If player 2 folds, then the game ends with payoff vector $(1, -1)$. If player 2 decides to bid, then the players reveal their cards. The player with the higher card wins, obtaining a payoff of 2; the player with the lower card loses, getting a payoff of -2. Note that, when both players bid, the stakes are raised. By the way, the Ace is considered the highest card, whereas the Queen is the lowest card.

RISK AND INCENTIVES IN CONTRACTING 25

The most approachable games with nature are those in which the moves of nature occur only at the end of the game. In other words, the players make decisions knowing that some random event will make payoffs uncertain, but during the game the players have the same information about nature. Such games can be analyzed by using the tools developed in part III of this book.

Payoff uncertainty is an important consideration in contractual relationships, because people usually care about whether they are exposed to risk. For example, the owner-manager of a retail shopping mall has a lease contract with each of his commercial tenants. The profits of the owner and of the tenants depend on the number of customers who visit the mall in any given month. The contracting parties face uncertainty because mall traffic is partly a function of random factors such as the strength of the local economy, demographics, fad and fashion, and whether the latest teen-idol or boy band agrees to perform on the mall's stage.

Owner–tenant contracts can be structured to distribute risk. A contract that requires the tenant to pay a fixed rent would allocate most of the risk to the tenant, because then the mall owner gets a guaranteed lease payment, whereas the tenant's revenue is subject to the random factors. Alternatively, the risk can be shifted to the mall owner by using a contract that makes the tenant's rent payment contingent on mall traffic.

The objective of distributing risk in any particular way can conflict with the goal of providing incentives. For example, if the mall owner gets a fixed lease payment, he may not be motivated to exert effort advertising the mall or running the kind of promotions that would increase mall traffic (and hence the profits of his tenants). The bottom line is that contracts sometimes have to balance the concerns of risk and incentives. In this chapter, I provide a model to demonstrate this trade-off.

RISK AVERSION

Before sketching the model of risk and incentives, I would like to elaborate on how payoffs represent players' preferences toward risk. I noted in Chapter 4

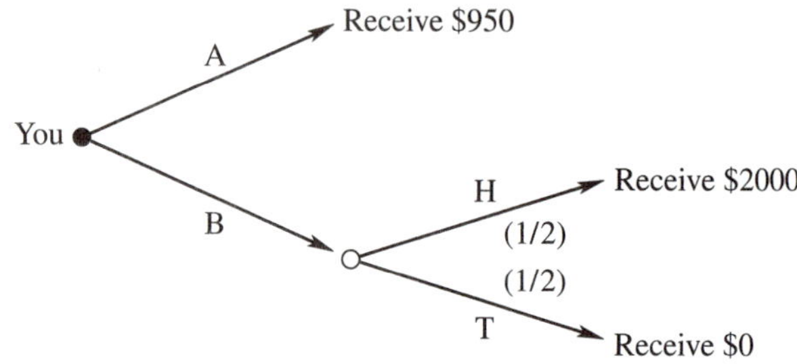

FIGURE 25.1
Lottery or sure thing.

that payoff functions measure more than just the players' rankings over certain outcomes; they also capture the players' preferences over *random* outcomes.[1] A simple thought exercise will demonstrate how this works. Suppose I offer you a choice between two alternatives, A and B. If you choose A, then I will give you $950. If you choose B, then I will flip a coin and give you $2000 if the coin toss yields "heads," $0 if it yields "tails." In other words, alternative B is a *lottery* that pays $2000 with probability $1/2$ and $0 with probability $1/2$. Figure 25.1 displays your choice as a game.

Note that the picture represents the outcome in words, rather than with utility or payoff numbers. As usual, we should convert the outcomes into payoffs to analyze the game. You might be inclined to use the dollar amounts as payoffs—as has been done many times so far in this book. But at this point I would like you to think more generally. Because the outcomes are all in monetary terms and because you ultimately care how much money you receive, we can imagine a function v that defines the relation between money and utility. That is, $v(x)$ is the utility of receiving x dollars. I use the term "utility" here because I want you to be thinking about preferences over random monetary payments, aside from any particular game. In the end, the utility function will indicate the payoffs in specific games.

Because you (presumably) prefer more money to less, we better assume that $v(0) < v(950) < v(2000)$. Lots of different functions satisfy these inequalities. One function that works is the identity function $v(x) = x$, which simply associates utility with the dollar amount. Other functions that are consistent with the "more money is better" assumption are $v(x) = x/1000$ and $v(x) = \sqrt{x}$. You should realize that *any increasing function* would do the job. But will just any increasing function accurately represent your preferences? In fact, no. Remember that random outcomes are evaluated by computing *expected utility*. Thus, your utility of selecting A is $v(950)$, whereas your ex-

[1] John von Neumann and Oscar Morgenstern, in their book *The Theory of Games and Economic Behavior* (1944, 1947), developed the theory of decision making under uncertainty now used by economists everywhere.

pected utility of selecting B is $(1/2)v(0) + (1/2)v(2000)$. If you strictly prefer alternative A, then your utility function has the property that

$$v(950) > \frac{v(0)}{2} + \frac{v(2000)}{2}.$$

If you prefer alternative B, then the opposite inequality holds. Your preference between A and B therefore further restricts the utility function.

What decision would you make? Most people who are given a choice such as this one opt for alternative A—the "sure thing" payoff of \$950. As a theorist, you might find this choice interesting because alternative B actually gives an expected monetary payment of $(1/2)(0) + (1/2)(2000) = 1000$, which is greater than the payment from picking A. However, there is a big difference between expected *monetary payments* and expected *utility,* the latter of which represents attitudes toward risk. Most people choose A over B because B is a lottery that subjects people to risk. Sure, B gives you an expected payment of \$1000, but you end up with \$0 half the time. Most people do not like to face such randomness and are willing to forego \$50 to get money for sure. This means that most people have utility functions satisfying $v(950) > v(0)/2 + v(2000)/2$. Note that the function $v(x) = x$ *does not* satisfy the inequality, because $950 < (1/2)(0) + (1/2)(2000)$. A utility function that *does* satisfy the inequality is $v(x) = \sqrt{x}$. You should verify that $\sqrt{950} > (1/2)(0) + (1/2)\sqrt{2000}$ (use a calculator).

Risk preferences are manifested in the way that a person's utility function curves. To interpret the curvature, compare the expected utility of any lottery with the utility of its expected monetary payment. For example, the lottery that pays 2000 with probability 1/2 has an expected dollar payment of $(1/2)(0) + (1/2)(2000) = 1000$. If a person strictly prefers to get \$1000 for sure, then his utility function has the property that

$$\frac{v(0)}{2} + \frac{v(2000)}{2} < v(1000).$$

Concave utility functions, such as $v(x) = \sqrt{x}$, have this property; linear functions, such as $v(x) = x$, do not (see Figure 25.2 on the next page). With the use of calculus, concavity is measured by the second derivative. If the second derivative is negative, then the function is concave (open side down). If the second derivative is equal to zero, then the function is linear (a straight line). For example, the second derivative of $\sqrt{x}$ is $-1/(4x^{3/2})$, which is negative (assuming $x > 0$; otherwise, the function is not defined).

A person is said to be *risk averse* if he strictly prefers to get a monetary payment for sure rather than playing a lottery that has the same expected pay-

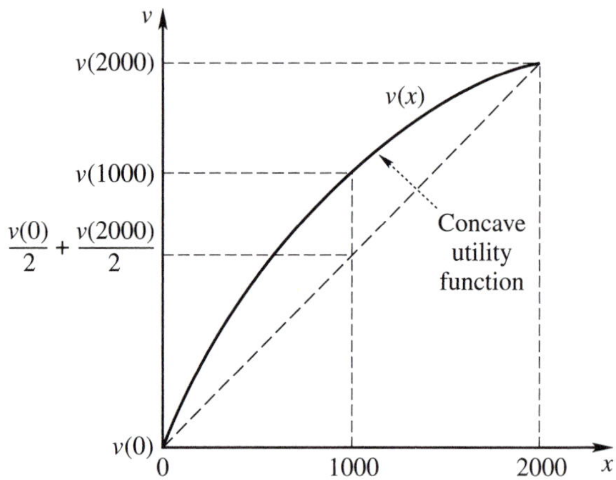

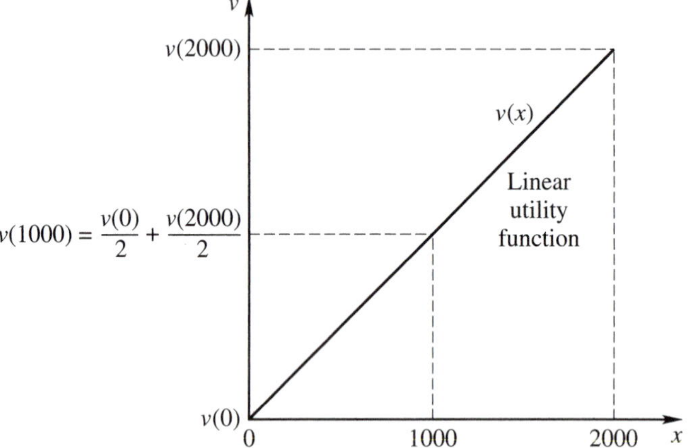

FIGURE 25.2
Utility functions and concavity.

ment. A person is *risk neutral* if he is indifferent between lotteries and their expected payments. As just shown, risk aversion is represented by a concave utility function. A higher degree of risk aversion implies a more concave utility function; it also implies a greater *risk premium,* which is the amount of expected monetary gain that a person is willing to give up to obtain a sure payment rather than a random one. In the preceding choice between alternatives A and B, if you choose A, then your risk premium is at least $50 because you are willing to give up this amount to get a sure payment.[2]

[2]A risk-loving person has a convex utility function and negative risk premiums. I do not address this kind of preference, although it does play a role for some people.

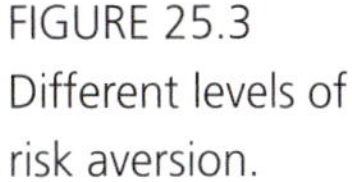
FIGURE 25.3
Different levels of
risk aversion.

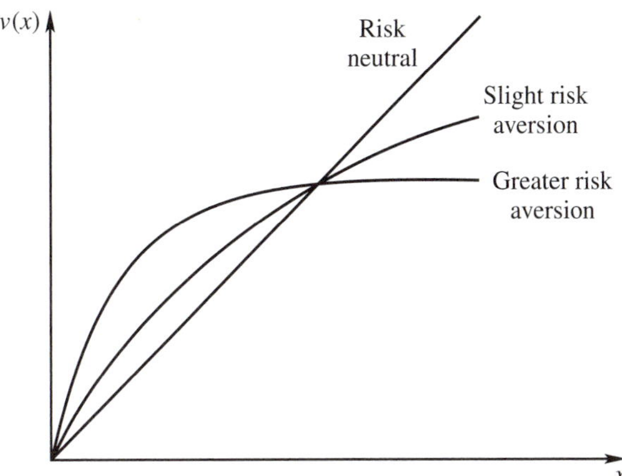

One measure of risk aversion—called the *Arrow-Pratt measure of relative risk aversion*—is given by $-xv''(x)/v'(x)$.[3] As an example, consider the utility function $v(x) = x^{\alpha}$, where α is a positive constant.[4] Taking the derivatives, we have $v'(x) = \alpha x^{\alpha-1}$ and $v''(x) = \alpha(\alpha - 1)x^{\alpha-2}$. The Arrow-Pratt measure for this utility function is thus $-x\alpha(\alpha - 1)x^{\alpha-2}/\alpha x^{\alpha-1}$, which equals $1 - \alpha$. Note that, if $\alpha = 1$, then the risk aversion measure is 0 and the utility function is a straight line. If α is less than 1, then the risk aversion measure is positive and the utility function is concave. As α gets closer to 0, the measure of risk aversion increases. This is pictured in Figure 25.3. Note that linear utility functions represent risk neutrality, where the agent is indifferent to risk and treats a lottery the same as its expected payment.

A PRINCIPAL–AGENT GAME

With the knowledge of how to represent attitudes toward risk, you can consider the management of risk and incentives in contracting. The basic model of risk and incentives is called the *Principal–Agent Model with Moral Hazard.* "Principal–agent" refers to a situation in which one party—the principal—hires another party—the agent—to work on a project on her behalf. "Moral hazard" stands for the setting in which the agent's effort is not verifiable, so

[3]Kenneth Arrow and John Pratt developed this and other measures in the following publications: J. W. Pratt, "Risk Aversion in the Small and in the Large," *Econometrica* 32(1964):122–136; and K. Arrow, *Essays in the Theory of Risk Bearing* (Chicago: Markham, 1970).
[4]Such a utility function is valid only for $x \geq 0$ because, when $\alpha \neq 1$, x^{α} is undefined for $x < 0$.

the parties cannot write an externally enforced contract specifying a transfer as a function of effort. Only the outcome of the project is verifiable. Furthermore, the project outcome depends not only on the agent's effort but also on random events.[5]

Suppose Pat manages a large computer software company and Allen is a talented program designer. Pat would like to hire Allen to develop a new software package. If Allen works for Pat, Allen must choose whether or not to expend high effort or low effort on the job. At the end of the work cycle (say, one year), Pat will learn whether the project is successful or not. A successful project yields revenue of 6 for the firm, whereas the revenue is 2 if the project is unsuccessful. (Imagine that these numbers are in hundreds of thousands of dollars.)

Success depends on Allen's high effort as well as on a random factor. Specifically, if Allen expends high effort, then success will be achieved with probability $1/2$; if Allen expends low effort, then the project is unsuccessful for sure. Assume that, to expend high effort, Allen must endure a personal cost of 1 in monetary terms. The parties can write a contract that specifies compensation for Allen conditional on whether the project is successful (but it cannot be conditioned directly on Allen's effort). Assume that Allen values money according to the utility function $v_A(x) = x^\alpha$. Assume Allen is risk averse, meaning that $0 < \alpha < 1$. Finally, suppose Pat is risk neutral (because she manages a large company that pools the risks of its various divisions) with utility function $v_P(x) = x$.

Imagine that the players interact as depicted in Figure 25.4. At the beginning of the game, Pat offers Allen a wage and bonus package. The wage w is

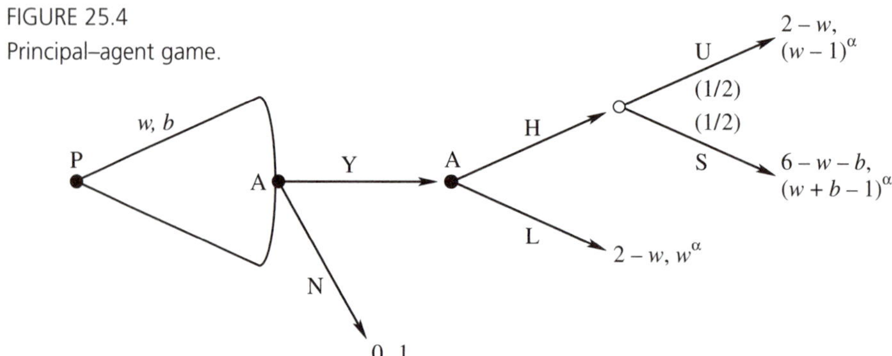

FIGURE 25.4
Principal–agent game.

[5]The example I present here is a special case of the model of S. Grossman and O. Hart, "An Analysis of the Principal–Agent Problem," *Econometrica* 51(1983):7–45.

FIGURE 25.5 Principal–agent game with expected payoffs.

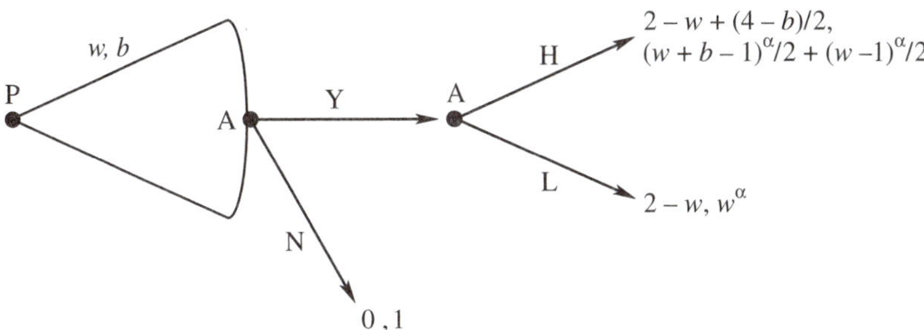

to be paid regardless of the project outcome, whereas the bonus b would be paid only if the project is successful. Then Allen decides whether or not to accept the contract. If he declines (N) then the game ends, with Pat getting 0 and Allen obtaining utility of 1 (corresponding to his outside opportunities).

If Allen accepts the contract (Y), then he decides whether to exert high (H) or low (L) effort. Low effort leads to an unsuccessful project, whereby Pat gets revenue of 2 minus the wage w and Allen gets his utility of the wage, w^{α}. High effort leads to a chance node, where nature picks whether the project is successful (with probability $1/2$) or not (probability $1/2$). An unsuccessful project implies the same payoffs as with low effort, except that, in this case, Allen also pays his effort cost of 1. A successful project raises Pat's revenue to 6 and triggers the bonus b paid to Allen in addition to the wage. Calculating the expected payoffs from the chance node, we can rewrite the game as shown in Figure 25.5.

To solve the game and learn about the interaction of risk and incentives, we use backward induction. Start by observing that Pat would certainly like Allen to exert high effort. In fact, it is *inefficient* for Allen to expend low effort. To see this, note that Pat can compensate Allen for exerting high effort by increasing his pay by 1 (offsetting Allen's effort cost). By doing so, Pat's expected revenue increases by 2. Another way of looking at this is that, if Allen's effort were verifiable, the parties would write a contract that induces high effort.

Given that high effort is desired, we must ask whether there is a contract that induces it. That is, can Pat find a wage and bonus package that motivates Allen to exert high effort and gives Pat a large payoff? Let us begin by checking what can be accomplished without a bonus (that is, $b = 0$). First note that $b = 0$ implies that Allen has *no* incentive to exert high effort; he gets w^{α} when he chooses L and less, $(w - 1)^{\alpha}$, when he chooses H. Knowing this, the best that Pat can do without a bonus is to offer $w = 1$, which Allen is willing to accept.[6] Thus, the best no-bonus contract ($w = 1$ and $b = 0$) yields the payoff vector (1, 1).

Next consider a bonus contract, designed to induce high effort. In order for him to be motivated to exert high effort, Allen's expected payoff from H must be at least as great as is his payoff from L:

$$\frac{1}{2}(w + b - 1)^{\alpha} + \frac{1}{2}(w - 1)^{\alpha} \geq w^{\alpha}. \tag{1}$$

In principal–agent models, this kind of inequality is usually called the **effort constraint** or **incentive compatibility condition**. In addition to the effort constraint, the contract must give Allen an expected payoff at least as great as the value of his outside opportunity; otherwise, Allen would reject the contract. Mathematically, this is

$$\frac{1}{2}(w + b - 1)^{\alpha} + \frac{1}{2}(w - 1)^{\alpha} \geq 1. \tag{2}$$

Theorists call this the **participation constraint**.

Assuming that she wants to motivate high effort, Pat will offer Allen the bonus contract satisfying inequalities 1 and 2 at terms most favorable to Pat (because she gets to make the contract offer). In fact, the *best* bonus contract satisfies inequalities 1 and 2 with equality (meaning that you can replace the "$\geq$" signs with "="). The reasoning goes as follows. First, if the participation constraint does not hold with equality, then there must be a way for Pat to lower both w and b such that both the effort constraint and the participation constraint remain satisfied. To maintain the effort constraint, w and b have to be adjusted in a related way, but the important thing is that they both are *lowered*. This would increase Pat's expected payoff (because she pays less to Allen in both the successful and the unsuccessful outcomes). We thus conclude that the best contract for Pat satisfies

$$\frac{1}{2}(w + b - 1)^{\alpha} + \frac{1}{2}(w - 1)^{\alpha} = 1. \tag{3}$$

Second, consider the effort constraint. If it holds strictly, then Allen is forced to face too much risk. We know that he is risk averse (because $0 < \alpha < 1$) and that a larger bonus increases his risk (because it increases the difference between what Allen gets in the successful and unsuccessful outcomes). Pat can lower the bonus and raise the wage slightly in a way that maintains equality 3 but lowers the risk premium that she has to pay Allen. This increases Pat's

[6] Allen will accept no less, because he would otherwise be better off rejecting the contract offer.

expected payoff while keeping Allen's unchanged.[7] Thus, the best contract for Pat satisfies inequality 1 with equality, meaning that

$$\frac{1}{2}(w+b-1)^{\alpha}+\frac{1}{2}(w-1)^{\alpha}=w^{\alpha}. \qquad (4)$$

Combining equations 3 and 4, we find that $w^{\alpha}=1$, so $w=1$. Substituting $w=1$ into equation (3), we get $b=2^{1/\alpha}$. In summary, Pat's best bonus contract specifies the wage $w=1$ and the bonus $b=2^{1/\alpha}$. Allen will accept this contract and expend high effort, yielding an expected payoff of

$$2-1+\frac{4-2^{1/\alpha}}{2}=3-2^{(1-\alpha)/\alpha}$$

for Pat and 1 for Allen. Note that if Allen were risk neutral—so that $\alpha=1$—then Pat's payoff would be $3-2^{0}=2$. The difference between this payoff and Pat's payoff when $\alpha<1$ is the risk premium that Allen requires to exert high effort. That is, the risk premium is

$$2-[3-2^{(1-\alpha)/\alpha}]=2^{(1-\alpha)/\alpha}-1.$$

If $\alpha=1$, then the risk premium is equal to 0, whereas the risk premium is positive when $\alpha<1$.

The final step in the analysis is to compare the optimal bonus and no-bonus contracts. Recall that Allen's payoff is 1 in either case. Pat's payoff is 1

[7] Here is a mathematical proof. Look at Allen's expected payoff from H and think about adjusting w and b in a way that does not change the expected payoff. Suppose that we want Allen to get an expected payoff of k. Let $f(w)$ be the function that gives the bonus b corresponding to wage w that yields expected payoff k. That is, $f(w)$ satisfies

$$\frac{1}{2}[w+f(w)-1]^{\alpha}+\frac{1}{2}(w-1)^{\alpha}=k.$$

Let us find the derivative of f, which tells us the amount that b has to be adjusted when w is increased, to keep Allen's expected payoff equal to k. Differentiating the preceding equation with respect to w yields

$$\frac{\alpha}{2}[w+f(w)-1]^{\alpha-1}[1+f'(w)]+\frac{\alpha}{2}(w-1)^{\alpha-1}=0.$$

Solving for $f'(w)$ and simplifying, we have

$$f'(w)=-1-\left(\frac{w+f(w)-1}{w-1}\right)^{1-\alpha},$$

which is always less than -2. In words, if w is increased by a small amount, then b can be decreased by more than two times this amount, such that Allen's expected payoff is held constant. Such an alteration of w and b is to Pat's benefit, because she has to pay the bonus with probability $1/2$. Shifting w up and b down tightens the effort constraint.

in the no-bonus case and $3 - 2^{(1-\alpha)/\alpha}$ in the bonus case. Pat will select a bonus contract if and only if

$$3 - 2^{(1-\alpha)/\alpha} \geq 1.$$

Rearranging this inequality yields $\alpha \geq 1/2$. That is, Pat opts for a bonus contract if and only if $\alpha \geq 1/2$. There is important intuition behind this statement. When α is close to 0—meaning that Allen is very risk averse—the risk premium that Pat would have to pay to motivate high effort is larger than is the expected revenue gain that Pat would get from Allen's high effort.

The moral of the model is that the amount a principal has to pay to motivate high effort increases with the agent's level of risk aversion. Therefore, with an agent who is highly risk averse, a principal will resort to a fixed-wage contract that induces low effort. With an agent who is closer to risk neutral, a principal will optimally motivate high effort by using a bonus contract. This result also indicates that the kind of contracts that one would expect in different industries in the world depends on the amount of risk in the production process. In fields where the productive outcome is influenced more by random factors than by an agent's behavior, we expect to see fixed-wage contracts. On the other hand, where an agent's effort is crucial to the success of a business project—such as in real estate transactions, automobile sales, and many other lines of work—bonus contracts are very useful.

GUIDED EXERCISE

Problem: Suppose I prefer \$20 for sure to a lottery that pays \$100 with probability 1/4 and \$0 with probability 3/4. Also suppose that I prefer lottery A to lottery B, where lottery A pays \$100 with probability 1/8, \$0 with probability 7/8, and lottery B pays \$20 with probability 1/2, \$0 with probability 1/2. Is there a utility function that is consistent with my preferences? If so, describe such a utility function. If not, explain why.

Solution: The first preference requires that

$$v(20) > \frac{1}{4}v(100) + \frac{3}{4}v(0).$$

One function that meets these requirements is $u(x) = \sqrt{x}$. That lottery A is preferred to lottery B implies

$$\frac{1}{8}v(100) + \frac{7}{8}v(0) > \frac{1}{2}v(20) + \frac{1}{2}v(0).$$

Subtracting $v(0)/2$ from each side and then multiplying by 2 yields

$$\frac{1}{4}v(100) + \frac{3}{4}v(0) > v(20),$$

which contradicts the first preference. Thus, there is no utility function that represents the stated preferences.

EXERCISES

1. Discuss some examples of jobs about which you know, highlighting the relation between risk, risk aversion, and the use of bonuses.

2. Repeat the analysis of the principal–agent game in this chapter, under the assumption that p is the probability of a successful project with high effort (rather than $1/2$ as in the basic model). How does the optimal contract depend on p?

3. A firm and worker interact as follows. First, the firm offers the worker a wage w and a job z; $z = 0$ denotes the "safe" job and $z = 1$ denotes the "risky" job. After observing the firm's contract offer (w, z), the worker accepts or rejects it. These are the only decisions made by the firm and worker in this game.

If the worker rejects the contract, then he gets a payoff of 100, which corresponds to his outside opportunities. If he accepts the job, then the worker cares about two things: his wage and his *status*. The worker's status depends on how he is rated by his peers, which is influenced by characteristics of his job as well as by random events. Specifically, his rating is given by x, which is either 1 (poor), 2 (good), or 3 (excellent). If the worker has the safe job, then $x = 2$ for sure. On the other hand, if the worker has the risky job, then $x = 3$ with probability q and $x = 1$ with probability $1 - q$. That is, with probability q, the worker's peers think of him as excellent.

When employed by this firm, the worker's payoff is $w + v(x)$, where $v(x)$ is the value of status x. Assume that $v(1) = 0$ and $v(3) = 100$, and let $y = v(2)$. The worker maximizes his expected payoff.

The firm obtains a return of $180 - w$ when the worker is employed in the safe job. The firm gets a return of $200 - w$ when the worker has the risky job. If the worker rejects the firm's offer, then the firm obtains 0.

Compute the subgame perfect equilibrium of this game by answering the following questions.

(a) How large must the wage offer be in order for the worker to rationally accept the safe job? What is the firm's maximal payoff in this case? The parameter y should be featured in your answer.

(b) How large must the wage offer be in order for the worker to rationally accept the risky job? What is the firm's maximal payoff in this case? The parameter q should be featured in your answer.

(c) What is the firm's optimal contract offer for the case in which $q = 1/2$? Your answer should include an inequality describing conditions under which $z = 1$ is optimal.

4. Most of this exercise does not concern players' preferences about risk, but it is a good example of a game with moves of nature that can be analyzed by using subgame perfection. Consider a T-period bargaining game like the alternating-offer game discussed in Chapter 19, except suppose that, in each period, nature chooses which player gets to make an offer. At the start of a period, nature selects player 1 with probability q_1 and player 2 with probability q_2, where $q_1 + q_2 = 1$. The selected player proposes a split of some surplus (of size 1). The other player then responds by accepting or rejecting the offer. Acceptance ends the game and yields the payoff associated with the proposal, discounted according to the period in which agreement is made. Rejection leads to the next period, except in period T, in which case the game ends with zero payoffs. Assume the players are risk neutral and have the same discount factor δ.

(a) Describe the extensive form of this game.

(b) Find and report the subgame perfect equilibrium.

(c) What are the equilibrium payoffs? Offer an interpretation in terms of the standard bargaining solution (bargaining weights and a disagreement point).

(d) Suppose $T = 2$ and that the players are not necessarily risk neutral. How would you expect the players' risk preferences to influence the equilibrium proposals that are made in the first period?

5. Consider a game inspired by the long-running television program *The Price Is Right*. Three players have to guess the price x of an object. It is common knowledge that x is a random variable that is distributed uniformly over the integers $1, 2, \ldots, 9$. That is, with probability $1/9$ the value of x is 1, with probability $1/9$ the value of x is 2, and so on. Thus, each player guesses a number from the set $\{1, 2, 3, 4, 5, 6, 7, 8, 9\}$.

The game runs as follows: First, player 1 chooses a number $n_1 \in \{1, 2, 3, 4, 5, 6, 7, 8, 9\}$, which the other players observe. Then player 2 chooses a number n_2 that is observed by the other players. Player 2 is *not* allowed to select the number already chosen by player 1; that is, n_2 must be different than n_1. Player 3 then selects a number n_3, where n_3 must be different from n_1 and n_2. Finally, the number x is drawn and the player whose guess is closest to x *without going over* wins \$1000; the other players get 0. That is, the winner must have guessed x correctly or have guessed a lower number that is closest to x. If all of the players' guesses are above x, then everyone gets 0.

(a) Suppose you are player 2 and you know that player 3 is sequentially rational. If player 1 selected $n_1 = 1$, what number should you pick to maximize your expected payoff?

(b) Again suppose you are player 2 and you know that player 3 is sequentially rational. If player 1 selected $n_1 = 5$, what number should you pick to maximize your expected payoff?

(c) Suppose you are player 1 and you know that the other two players are sequentially rational. What number should you pick to maximize your expected payoff?

26 BAYESIAN NASH EQUILIBRIUM AND RATIONALIZABILITY

The Nash equilibrium and rationalizability concepts can be applied directly to any game with random events. These static concepts are most valid for games in which the players' actions are taken simultaneously and independently. Many interesting applications fall into this class of games. For example, consider an auction, where bidders may have private information about their valuations of the item on the block. In the "sealed bid" auction format, the bidders choose their bids simultaneously. Similarly, interaction between oligopolist firms can often be modeled as a static game with incomplete information. In fact, for most static situations—from location games to partnership problems—it is worthwhile and interesting to study variants with incomplete information.

There are two methods of evaluating Bayesian games. The first method is to compute rationalizability and Nash equilibrium on the Bayesian normal form. This is recommended for any game that can be represented in matrix form. The second method entails treating the types of each player as separate players. For example, if player 1 is one of two types, A and B, then it may be helpful to think of 1A and 1B as distinct players. Treating types separately often simplifies the analysis of infinite games, where calculus may be required. Strictly speaking, the methods do not always yield identical solutions.[1] But no discrepancies arise in the applications studied here. By the way, when applied to a game with nature, solution concepts receive the "Bayesian" qualifier; thus, we have *Bayesian Nash equilibrium* and *Bayesian rationalizability*.

For an example of the first method, consider the game discussed at the end of Chapter 24 and pictured in Figures 24.3 and 24.4. The extensive-form and normal-form representations are reproduced here in Figure 26.1. Examining the normal form reveals that $A^{12}A^0$ and $B^{12}A^0$ are dominated strategies for player 1. With reference to the extensive form, this is confirmed by observing that player 1's action A^0 yields a lower payoff than does B^0, regardless of whether player 2 selects C or D. Continuing, note that player 2's strategy

[1] Differences arise in the rationalizability context and when there is a type of player that arises with zero probability. One of the exercises at the end of the chapter addresses rationalizability.

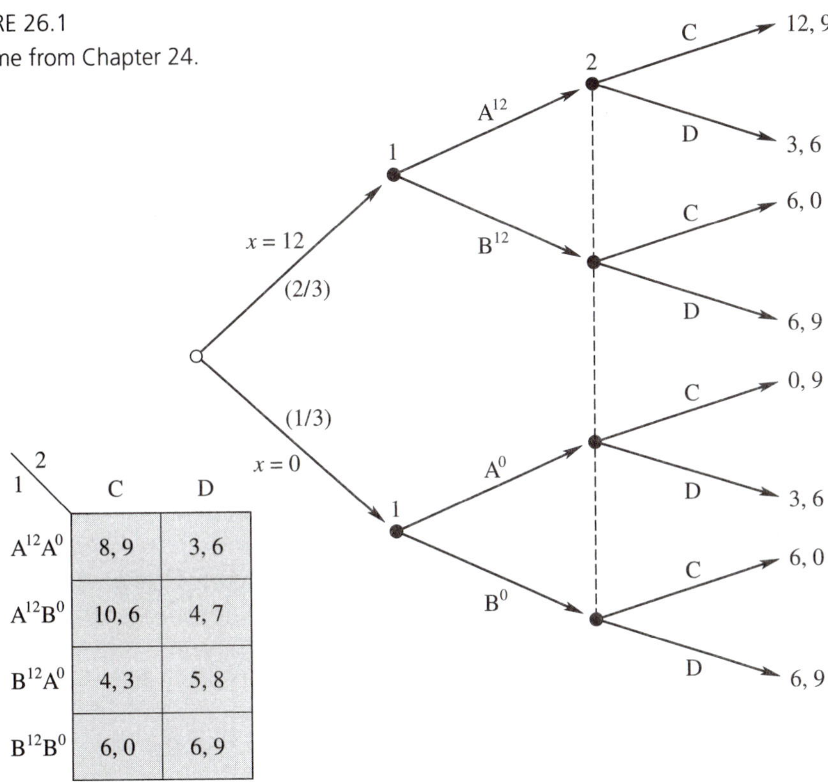

1 \ 2	C	D
A^{12}A^{0}	8, 9	3, 6
A^{12}B^{0}	10, 6	4, 7
B^{12}A^{0}	4, 3	5, 8
B^{12}B^{0}	6, 0	6, 9

FIGURE 26.1
A game from Chapter 24.

C is dominated in the next round of the iterated dominance procedure. The extensive form provides more details on this. Player 2 knows that player 1 of type 0 selects B^{0}; given that type 0 arises with probability $1/3$, player 2 strictly prefers D to C, regardless of his belief about the behavior of type 12. After C is removed, player 1's strategy A^{12}B^{0} also is discarded. That is, B^{12} is the only rational choice for type 12, given that player 2 selects D. The Bayesian rationalizable set is thus {(B^{12}B^{0}, D)}, which also identifies the unique Bayesian Nash equilibrium of the game.

As an example of the second approach to solving Bayesian games, consider a simple Cournot duopoly game with incomplete information. Suppose that demand is given by $p = 1 - Q$, where Q is the total quantity produced in the industry. Firm 1 selects a quantity q_1, which it produces at zero cost. Firm 2's cost of production is private information (selected by nature). With probability $1/2$, firm 2 produces at zero cost. With probability $1/2$, firm 2 produces with a marginal cost of $1/4$. Call the former type of firm 2 "L" and the latter type "H" (for low and high cost, respectively). Firm 2 knows its type, whereas firm 1 knows only the probability that L and H occur. Let q_2^{H} and q_2^{L}

denote the quantities selected by the two types of firm 2. Then when firm 2's type is L, its payoff is given by

$$u_2^L = (1 - q_1 - q_2^L)q_2^L.$$

When firm 2's type is H, its payoff is

$$u_2^H = (1 - q_1 - q_2^H)q_2^H - \frac{q_2^H}{4}.$$

As a function of the strategy profile $(q_1; q_2^L, q_2^H)$, firm 1's payoff is

$$u_1 = \frac{(1 - q_1 - q_2^L)q_1}{2} + \frac{(1 - q_1 - q_2^H)q_1}{2} = \left(1 - q_1 - \frac{q_2^L}{2} - \frac{q_2^H}{2}\right)q_1.$$

Note that firm 1's payoff is an expected payoff obtained by averaging the payoffs of facing the low and high types of firm 1, according to the probability of these types.

To find the Bayesian Nash equilibrium of this market game, consider the types of player 2 as separate players. Then find the best-response functions for the three player types and determine the strategy profile that solves them simultaneously. The best-response functions are calculated by evaluating the following derivative conditions:

$$\frac{\partial u_1}{\partial q_1} = 0, \quad \frac{\partial u_2^L}{\partial q_2^L} = 0, \quad \text{and} \quad \frac{\partial u_2^H}{\partial q_2^H} = 0.$$

This yields:

$$BR_1(q_2^L, q_2^H) = \frac{1}{2} - \frac{q_2^L}{4} - \frac{q_2^H}{4} \qquad \text{for player 1,}$$

$$BR_2^L(q_1) = \frac{1}{2} - \frac{q_1}{2} \qquad \text{for player-type 2L, and}$$

$$BR_2^H(q_1) = \frac{3}{8} - \frac{q_1}{2} \qquad \text{for player-type 2H.}$$

Solving the associated system of equalities,

$$q_1 = \frac{1}{2} - \frac{q_2^L}{4} - \frac{q_2^H}{4}, \quad q_2^L = \frac{1}{2} - \frac{q_1}{2}, \quad q_2^H = \frac{3}{8} - \frac{q_1}{2},$$

the Bayesian Nash equilibrium is found to be the profile $q_1 = 3/8$, $q_2^L = 5/16$, $q_2^H = 3/16$. In words, firm 1 produces $3/8$, whereas firm 2 produces $5/16$ if its cost is low and $3/16$ if its cost is high.

GUIDED EXERCISE

Problem: Consider the following game with nature:

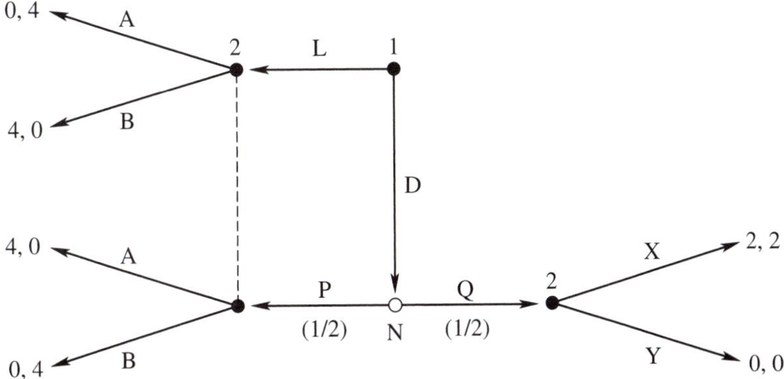

(a) Represent this game in the (Bayesian) normal form.

(b) Calculate the (Bayesian) rationalizable set. Does this game have any pure-strategy (Bayesian) Nash equilibria?

(c) Calculate the mixed-strategy Nash equilibrium.

Solution:

(a) Here is the normal-form matrix:

1 \ 2	AX	AY	BX	BY
L	0, 4	0, 4	4, 0	4, 0
D	3, 1	2, 0	1, 3	0, 2

(b) Regarding the rationalizable set, note that player 2's strategy BY is dominated by a mixed strategy that puts high probability on BX and a small probability on either AX or AY. After removing BY, nothing else is dominated for either player. Thus, the rationalizable set is

$$R = \{\text{L, D}\} \times \{\text{AX, AY, BX}\}.$$

You can easily verify that there is no pure-strategy Nash equilibrium.

(c) To find the mixed-strategy equilibrium, first note that it must involve player 1 randomizing between L and D, for if player 1 selects only L or only D then, given player 2's best response, player 1 would want to switch. Next observe that, because player 1 will choose D with positive probability, player 2 strictly prefers not to play AY. Thus, the mixed-strategy equilibrium will have player 1 mixing between L and D, and player 2 mixing between AX and BX. Suppose that player 1 puts probability p on L and $1 - p$ on D; suppose player 2 puts probability q on AX and $1 - q$ on BX. Then the required indifference conditions are

$$q \cdot 0 + (1 - q) \cdot 4 = q \cdot 3 + (1 - q) \cdot 1$$

and

$$p \cdot 4 + (1 - p) \cdot 1 = p \cdot 0 + (1 - p) \cdot 3.$$

Solving these equations, we learn that the mixed-strategy equilibrium specifies $p = 1/3$ and $q = 1/2$.

EXERCISES

1. Consider the following game. Nature selects A with probability $1/2$ and B with probability $1/2$. If nature selects A, then players 1 and 2 interact according to matrix "A." If nature selects B, then the players interact according to matrix "B." These matrices are pictured here.

1 \ 2	V	W
X	6, 0	4, 1
Y	0, 0	0, 1
Z	5, 1	3, 0

A

1 \ 2	V	W
X	0, 0	0, 1
Y	6, 0	4, 1
Z	5, 1	3, 0

B

(a) Suppose that, when the players choose their actions, the players do not know which matrix they are playing. That is, they think that with probability $1/2$ the payoffs are as in matrix A and that with probability $1/2$ the payoffs are as in matrix B. Write the normal-form matrix that describes this Bayesian game. (This matrix is the "average" of matrices A and B.) Using rationalizability, what is the strategy profile that is played?

(b) Now suppose that, before the players select their actions, player 1 observes nature's choice. (That is, player 1 knows which matrix is being played.) Player 2 does not observe nature's choice. Represent this game in the extensive form and in the Bayesian normal form. Using dominance, what is player 1's optimal strategy in this game? What is the set of rationalizable strategies in the game?

(c) In this example, is the statement "A player benefits from having more information" true or false?

2. Two players have to simultaneously and independently decide how much to contribute to a public good. If player 1 contributes x_1 and player 2 contributes x_2 then the value of the public good is $2(x_1+x_2+x_1x_2)$, which they each receive. Assume that x_1 and x_2 are positive numbers. Player 1 must pay a cost x_1^2 of contributing; thus, player 1's payoff in the game is $u_1 = 2(x_1+x_2+x_1x_2)-x_1^2$. Player 2 pays the cost tx_2^2 so that player 2's payoff is $u_2 = 2(x_1+x_2+x_1x_2) - tx_2^2$. The number t is private information to player 2; player 1 knows that t equals 2 with probability 1/2 and it equals 3 with probability 1/2. Compute the Bayesian Nash equilibrium of this game.

3. Suppose that nature selects A with probability 1/2 and B with probability 1/2. If nature selects A, then players 1 and 2 interact according to matrix "A." If nature selects B, then the players interact according to matrix "B." These matrices are pictured here. Suppose that, before the players select their actions, player 1 observes nature's choice. That is, player 1 knows from which matrix the payoffs are drawn and player 1 can condition his or her decision on this knowledge. Player 2 does not know which matrix is being played when he or she selects between L and R.

1 \ 2	L	R
U	2, 2	0, 0
D	0, 0	4, 4

A

1 \ 2	L	R
U'	0, 2	2, 0
D'	4, 0	0, 4

B

(a) Draw the extensive-form representation of this game. Also represent this game in Bayesian normal form. Compute the set of rationalizable strategies and find the Nash equilibria.

(b) Consider a three-player interpretation of this strategic setting in which each of player 1's types is modeled as a separate player. That is, the game is played by players 1A, 1B, and 2. Assume that player 1A's payoff is zero whenever nature chooses B; likewise, player 1B's payoff is zero whenever nature selects A. Depict this version of the game in the extensive form (remember that payoff vectors consist of three numbers) and in the normal form. Compute the set of rationalizable strategies and find the Nash equilibria.

(c) Explain why the predictions of parts (a) and (b) are the same in regard to equilibrium but different in regard to rationalizability. (Hint: The answer has to do with the scope of the players' beliefs.)

4. Demonstrate that, for the Cournot game discussed in this chapter, the only rationalizable strategy is the Bayesian Nash equilibrium.

5. Consider a differentiated duopoly market in which firms compete by selecting prices and produce to fill orders. Let p_1 be the price chosen by firm 1 and let p_2 be the price of firm 2. Let q_1 and q_2 denote the quantities demanded (and produced) by the two firms. Suppose that the demand for firm 1 is given by $q_1 = 22 - 2p_1 + p_2$, and the demand for firm 2 is given by $q_2 = 22 - 2p_2 + p_1$. Firm 1 produces at a constant marginal cost of 10 and no fixed cost. Firm 2 produces at a constant marginal cost of c and no fixed cost. The payoffs are the firms' individual profits.

(a) The firms' strategies are their prices. Represent the normal form by writing the firms' payoff functions.

(b) Calculate the firms' best-response functions.

(c) Suppose that $c = 10$ so the firms are identical (the game is symmetric). Calculate the Nash equilibrium prices.

(d) Now suppose that firm 1 does not know firm 2's marginal cost c. With probability 1/2 nature picks $c = 14$ and with probability 1/2 nature picks $c = 6$. Firm 2 knows its own cost (that is, it observes nature's move), but firm 1 only knows that firm 2's marginal cost is either 6 or 14 (with equal probabilities). Calculate the best-response functions of player 1 and the two types ($c = 6$ and $c = 14$) of player 2 and calculate the Bayesian Nash equilibrium quantities.

6. Find the Bayesian Nash equilibrium of the game pictured here. Note that Exercise 3 of Chapter 24 asked you to convert this into the normal form.

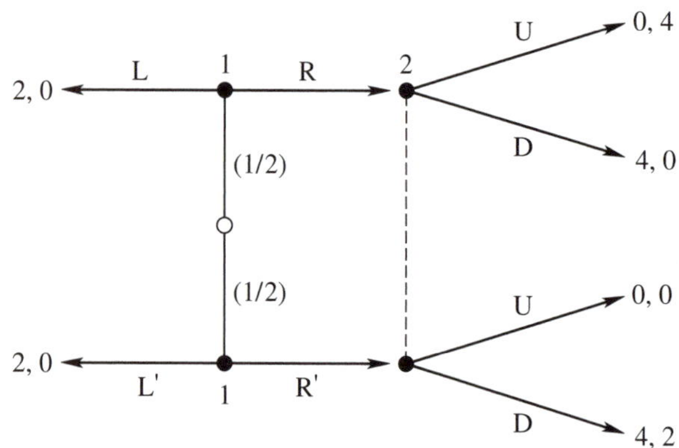

7. Consider the following static game of incomplete information. Nature selects the type (c) of player 1, where $c = 2$ with probability 2/3 and $c = 0$ with probability 1/3. Player 1 observes c (he knows his own type), but player 2 does not observe c. Then the players make simultaneous and independent choices and receive payoffs as described by the following matrix.

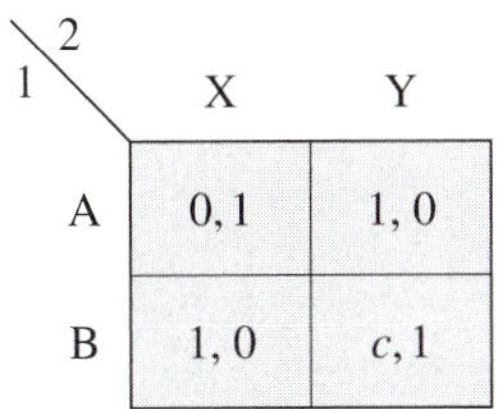

1 \ 2	X	Y
A	0, 1	1, 0
B	1, 0	c, 1

(a) Draw the normal-form matrix of this game.

(b) Compute the Bayesian Nash equilibrium.

8. Consider a simple simultaneous-bid poker game. First, nature selects numbers x_1 and x_2. Assume that these numbers are independently and uniformly distributed between 0 and 1. Player 1 observes x_1 and player 2 observes x_2, but neither player observes the number given to the other player. Simultaneously and independently, the players choose either to fold or to bid. If both players fold, then they both get the payoff -1. If only one player folds, then he obtains -1 while the other player gets 1. If both players elected to bid, then each player receives 2 if his number is at least as large as the other player's number; otherwise, he gets -2. Compute the Bayesian Nash equilibrium of this game. (Hint: Look for a symmetric equilibrium in which a player bids if and only if his number is greater than some constant α. Your analysis will reveal the equilibrium value of α.)

9. Consider the simple poker game described in Exercise 1 of Chapter 24, where there are just two cards in the deck and one card is dealt to player 1. This game has a single Nash equilibrium (perhaps in mixed strategies). Calculate and report the equilibrium strategy profile. Explain whether bluffing occurs in equilibrium.

27 LEMONS, AUCTIONS, AND INFORMATION AGGREGATION

In this chapter, I present some examples of how incomplete information affects trade and the aggregation of information between two or more economic agents. Each of the settings is modeled as a static game.

MARKETS AND LEMONS

You may have had the experience of buying or selling a used automobile. If so, you know something about markets with incomplete information. In the used-car market, sellers generally have more information about their cars than do prospective buyers. A typical seller knows whether his car has a shrouded engine problem—something that the buyer might not notice but that would likely require a costly repair before long. The seller knows whether the car tends to overheat in the summer months. The seller knows the myriad idiosyncracies that the car has developed since he acquired it. Prospective buyers may know only what they can gather from a cursory inspection of the vehicle. Thus, buyers are at an informational disadvantage. You would expect that, as a result, the buyers would not fare well in the market. Nonetheless, sellers might lose if the market failed owing to justifiably cautious buyers.[1]

To illustrate, suppose Jerry is in the market for a used car. One day he meets a shifty looking man named Freddie, who offers an attractive fifteen-year-old sedan for sale. Jerry likes the car's appearance. He imagines himself at the wheel, cruising up and down Broadway, taking in the flirtatious glances of many a woman through his sunglasses. Then he imagines the engine exploding, followed by an embarrassing scene in which he watches from the curb as a firefighter dowses his vehicle with water. Jerry says to Freddie, "The car looks good, but how do I know it isn't a lemon?" Freddie rejoins, "You have my word; this car is a peach; it's in great shape!" Jerry insists, "Galimatias! Let's put aside the hep-talk. Since I will not likely see you again, your word means

[1]The trading example that I present here is inspired by G. Akerlof, "The Market for Lemons: Qualitative Uncertainty and the Market Mechanism," *Quarterly Journal of Economics* 84(1970):488–500. For his work, George Akerlof was awarded the 2001 Nobel Prize in Economics, with corecipients A. Michael Spence and Joseph Stiglitz.

FIGURE 27.1 The lemons problem.

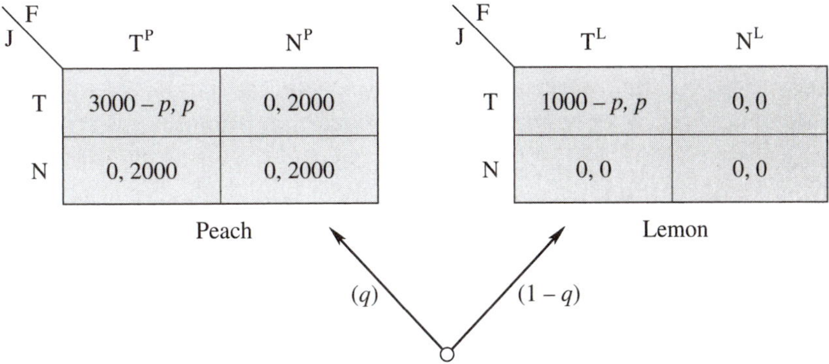

nothing to me. I know game theory and, as Professor Watson clearly explains, you have no way of, or interest in, establishing a reputation with someone you will never see again." Freddie nods and says, "Okay, let's talk turkey. The *Blue Book* tells us the market price for this car.[2] You can look the car over as you please. In the end, you will have to decide whether you are willing to pay the *Blue Book* price for the car, as I must decide whether to offer the car at this price."

Interaction between Jerry and Freddie may be quite elaborate, but I would like to abstract from this complexity and simply focus on the bottom line. Suppose there is some fixed market price for fifteen-year-old sedans of the type that Freddie is selling. Call this exogenously given price p. Assume that Jerry and Freddie play the game depicted in Figure 27.1. Nature first chooses whether the car is a peach or a lemon. If the car is a peach, then it is worth \$3000 to Jerry and \$2000 to Freddie. If the car is a lemon, then it is worth \$1000 to Jerry and \$0 to Freddie. Note that, in both cases, Jerry values the car more than does Freddie, so efficiency requires that the car be traded and the surplus (in each case \$1000) be divided between them. But there is incomplete information; Freddie observes nature's choice, whereas Jerry knows only that the car is a peach with probability q. Then the players simultaneously and independently decide whether to trade (T) or not (N) at the market price p. If both elect to trade, then the trade takes place. Otherwise, Freddie keeps the car.

Two kinds of equilibria are possible in this game. In the first kind, only the lemon is traded. Let us check whether there are values of p for which only the lemon is traded in equilibrium. That is, Jerry selects T and Freddie plays the

[2]The *Kelley Blue Book* is a publication in the United States that establishes price guidelines for used automobiles.

strategy N^PT^L (no trade if peach; trade if lemon). In order for Freddie's strategy to be optimal, it must be that $0 \leq p$ and $p \leq 2000$ (otherwise Freddie would either not want to trade a lemon or want to trade a peach). In order for Jerry's strategy to be optimal, it must be that Jerry is willing to trade, *conditional on knowing that Freddie only offers the lemon for sale*. Jerry obtains an expected payoff of $q \cdot 0 + (1 - q)(1000 - p)$ if he chooses T, 0 if he chooses N. Jerry is willing to trade if and only if

$$(1 - q)(1000 - p) \geq 0,$$

which simplifies to $p \leq 1000$. Putting the incentive conditions together, we see that, if $p \in (0, 1000)$, then there is an equilibrium in which only the lemon is traded. Intuitively, if the market price is below \$1000, Freddie would want to bring only the lemon to market. Anticipating that only a lemon will be for sale, Jerry is willing to pay no more than \$1000.

The second kind of equilibrium features trade of both the lemon and the peach. That is, Jerry selects T and Freddie plays the strategy T^PT^L. In order for this equilibrium to exist, the market price must be high enough so that Freddie is willing to sell the peach; specifically, $p \geq 2000$. In addition, Jerry's expected value of owning the car must be at least as great as the price. That is, it must be that

$$3000q + 1000(1 - q) \geq p,$$

which simplifies to $1000 + 2000q \geq p$. Thus, there is an equilibrium in which both types of car are traded as long as

$$1000 + 2000q \geq p \geq 2000.$$

Note that there is a price p that works if and only if

$$1000 + 2000q \geq 2000,$$

which simplifies to $q \geq 1/2$. In words, unless the probability of a peach is sufficiently high (there are not too many lemons in the world), there is no equilibrium in which the peach is traded.

If $q < 1/2$, then only lemons are traded in equilibrium. Recall that this outcome is *inefficient,* because trading the peach creates value. Thus, the model demonstrates that asymmetric information sometimes causes markets to malfunction.

AUCTIONS

In the lemons example, the seller has private information about his and the buyers' valuations of the good to be traded. Other markets have different information structures. In many instances, prospective *buyers* have private information about their valuations for the good. Furthermore, different prospective buyers have different tastes, needs, and abilities, leading to variations in people's willingness to pay for things such as houses, artwork, and productive inputs.

The seller of a good naturally wants to trade at the highest price that she can obtain. When the seller has one object to sell and there are multiple potential buyers, the seller would like to find the buyer with the highest willingness to pay for the object and then consummate a deal with this buyer at a price close to the buyer's valuation of the good. Unfortunately for the seller, she may not know the prospective buyers' valuations. One way for the seller to encourage competition between prospective buyers and to identify the highest valuation is to hold an auction.

Auctions are quite common in reality. All sorts of merchandise is sold at formal and informal auction houses, often over the Internet. Many different auction formats are in use as well. There are sealed-bid auctions, where bidders simultaneously and independently submit offers; sealed bids are often used in home sales and for government procurement. There are dynamic oral auctions, where an auctioneer suggests prices in a sequence and the prospective buyers signal or call out their bids. Several versions of these auction forms are in prominent use.

To give you a taste of auction theory and its elemental intuition, I shall present an analysis of two examples of sealed-bid auctions. Imagine that a person—the *seller*—has a painting that is worth nothing to her personally. She hopes to make some money by selling the art. There are two potential buyers, whom I call bidders 1 and 2. Let v_1 and v_2 denote the valuations of the two bidders. If bidder i wins the painting and has to pay x for it, then bidder i's payoff is $v_i - x$. Suppose that v_1 and v_2 are chosen independently by nature and that each is uniformly distributed between 0 and 1000. Technically speaking, the probability that $v_i < y$ is $y/1000$. The bidders observe their own valuations before engaging in the auction. However, they do not observe each other's valuations, so each must bid knowing only his own valuation and that the other's valuation is distributed uniformly between 0 and 1000. The seller does not observe the bidders' valuations; she knows only their distributions.

In general, the seller designs the auction in which the bidders participate. The auction implies a game between bidders 1 and 2, whom we can therefore call players 1 and 2. The auction game specifies which bidder gets the painting (or neither) and what monetary payments each bidder must make, as a function of the bidding behavior.

Consider a second-price auction.[3] In this game, players simultaneously and independently submit bids b_1 and b_2. The painting is awarded to the highest bidder at a price equal to the *second*-highest bid. For example, if player 1 bids 420 and player 2 bids 380, then player 1 gets the painting and is required to pay $380. In this outcome, player 1's payoff would be $v_1 - 380$ and player 2's payoff would be 0 (because player 2 neither gets the painting nor has to pay anything).[4]

Finding the Bayesian Nash equilibrium of the second-price auction is not too difficult, once you notice that bidding one's valuation is a *weakly dominant strategy* for each player. Roughly speaking, a weakly dominant strategy (if one exists) is a strategy that is a best response to every strategy of the other player. Mathematically, with the use of standard notation, strategy s_i is weakly dominant for player i only if

$$u_i(s_i, s_{-i}) \geq u_i(s_i', s_{-i})$$

for all $s_i' \in S_i$ and all $s_{-i} \in S_{-i}$.[5] (Compare this with the definition of dominance.) When all players have a weakly dominant strategy, then this strategy profile is an equilibrium in the game.

Here's how you can prove that bidding one's valuation is a weakly dominant strategy in the second-price auction. Suppose that player i has valuation v_i and that he is considering whether to bid $b_i = v_i$ or to bid some other amount $b_i = x$. I will show that bidding v_i always gives him a weakly higher payoff than does bidding x, regardless of the other player's (player j's) bid. Further, bidding v_i yields a strictly higher payoff for some bids of player j.

Suppose, for instance, that $x > v_i$ and consider the following three possibilities. First, it may be that player j's bid b_j is at least as large as x. In

[3]The second-price auction, as well as related mechanisms, is associated with William Vickrey, who, with James Mirrlees, won the 1996 Nobel Prize in Economics. These men made fundamental contributions to the theory of incentives in environments of incomplete information. Vickrey's auction analysis is contained in his "Counterspeculation, Auctions and Competitive Sealed Tenders," *Journal of Finance* 16(1961):8–37.

[4]The second-price auction is growing in use and is also closely related to the dynamic ascending bid auction form.

[5]In a comparison of two strategies s_i and s_i', the standard meaning of "s_i weakly dominates s_i'" is that $u_i(s_i, s_{-i}) \geq u_i(s_i', s_{-i})$ for every s_{-i} and that $u_i(s_i, s_{-i}') > u_i(s_i', s_{-i}')$ for at least one strategy s_{-i}' of the other players.

this case, player i will lose the auction regardless of whether he bids x or v_i.[6] Second, it may be that b_j is between v_i and x. In this case, player i actually does *worse* by bidding x than by bidding v_i. If he bids v_i, then he will lose the auction and get a payoff of 0; on the other hand, if he bids x, then he wins the auction and has to pay b_j, giving him the negative payoff $v_i - b_j$. Third, consider the case in which b_j is less than v_i. In this case, bidding v_i and bidding x yield the same payoff to player i; either way, player i wins the painting and has to pay b_j, for a payoff of $v_i - b_j$.

In summary, player i is better off bidding his value v_i rather than bidding any other amount x, regardless of what player j does. You can check that the same result is reached when $x < v_i$ (Exercise 3 at the end of this chapter asks you to do this). Thus, bidding one's valuation is a weakly dominant strategy and there is a Bayesian equilibrium in which both players use this strategy.

The equilibrium of the second-price auction is efficient because the painting goes to the player with the highest valuation. Therefore, by running an auction, the seller locates the highest valuation bidder. But the seller is not able to appropriate all of the surplus of the trade, because the winning bidder pays only the second-highest bid. The seller's expected revenue equals the expected second-highest valuation, which is 1000/3.[7]

Next consider a first-price, sealed-bid auction. The players simultaneously and independently submit bids b_1 and b_2 as before. The painting is awarded to the highest bidder, who must pay his bid. For example, if player 1 bids 290 and player 2 bids 310, then player 2 obtains the painting for the price of \$310. Player 2's payoff would then be $v_2 - 310$, whereas player 1's payoff would be 0. First-price auctions such as this one are very common.

Computing the Bayesian Nash equilibrium of the first-price auction is more tricky than is the analysis for the second-price auction. Fortunately, some intuition and a guess can help. Note that a player has no reason to bid more than his valuation, because he would then get a negative payoff in the event

---

[6] I have not described what the auction specifies (who wins the painting or with what probability) in the case in which the bids are equal ($b_1 = b_2$). In fact, it will not matter what is specified for this contingency, so I ignore it at this point.

[7] To see how this is computed, note that the expected second-highest valuation, conditional on knowing the value v_1, is

$$\text{Prob}[v_2 \geq v_1](v_1) + \text{Prob}[v_2 < v_1]E[v_2 \mid v_2 < v_1].$$

The first term represents the case in which v_1 is the second-highest bid (occurring when $v_2 \geq v_1$), whereas the second term represents the case in which v_2 is the second-highest bid. $E[v_2 \mid v_2 < v_1]$ denotes the expected value of v_2 conditional on v_2 being less than v_1. Because v_2 is distributed uniformly, we know that, with v_1 fixed, $\text{Prob}[v_2 < v_1] = v_1/1000$, $\text{Prob}[v_2 \geq v_1] = 1 - v_1/1000$, and $E[v_2 \mid v_2 < v_1] = v_1/2$. If we substitute these expressions and simplify, the expected second-highest valuation, conditional on knowing v_1, is $v_1 - v_1^2/2000$. Taking the expectation with respect to v_1 (integrating with the density function $1/1000$) yields $1000/3$.

that he wins the auction. In fact, each player ought to bid *less* than his valuation so that he can obtain a positive payoff if he wins. Let us start by conjecturing the *form* of the players' equilibrium bidding strategies and then, by analyzing incentives, determine whether the form is correct. Presume that each player adopts a strategy in which he bids a fraction a of his valuation. That is, when player i's valuation is v_i, he bids $b_i = av_i$. We can compute whether this form of symmetric strategy profile is an equilibrium and, if so, we will be able to calculate what the parameter a must be.

Let us compute player i's optimal strategy under the assumption that bidder j uses the strategy $b_j = av_j$. Suppose that player i's valuation is v_i and that he is considering a bid of x. If player i wins the auction, then his payoff will be $v_i - x$. Of course, contingent on winning, player i prefers x to be small. But lowering x reduces the chance that he wins the auction. In particular, player i wins if and only if player j's bid falls below x. Because player j bids according to the function $b_j = av_j$, a bid of x would be made by the type of player j whose valuation is x/a; types of player j with valuations below x/a will bid less than x. Because v_j is distributed uniformly between 0 and 1000, the probability that $v_j < x/a$ is $x/1000a$. Thus, if player i bids x, then he can expect to win the auction with probability $x/1000a$.

By bidding x, player i's expected payoff is equal to the probability of winning times the surplus that he gets if he wins:

$$\frac{(v_i - x)x}{1000a}.$$

Note that this payoff is a concave parabola as a function of x. To find player i's optimal bid, take the derivative with respect to x and set it equal to 0. Solving for x yields $x = v_i/2$. In words, player i's best response to j's strategy is to bid exactly half his valuation. That is, $b_i = v_i/2$ is player i's optimal strategy. Because this strategy is of the form that we assumed at the beginning, we know that we have found a Bayesian Nash equilibrium and that the bidding parameter is $a = 1/2$. That is, $b_1(v_1) = v_1/2$ and $b_2(v_2) = v_2/2$ constitute a Bayesian Nash equilibrium.

As with the second-price auction, the equilibrium of the first-price auction is efficient; the player with the highest valuation wins the auction. Furthermore, the winner pays one-half his valuation, which bears a resemblance to the outcome of the second-price auction. Specifically, recall that the winner of the second-price auction pays the second-highest bid, which (as you can check) happens to be one-half of the winner's valuation on average. Thus, the first-price and second-price auctions yield the *same* expected revenue for the seller, 1000/3. This fact about "revenue equivalence" between different auction forms actually holds in many auction environments.[8]

What can you learn from all this, besides the fact that auction theory is complicated? First, in standard first-price auctions, there is a trade-off between the probability of winning and the surplus obtained by winning. Second, under first-price auction rules, it is optimal to bid less than one's valuation. Third, auctions can be designed to induce "truthful" bidding (as is the case with the second-price form). Fourth, competitive bidding produces information about the bidders' valuations and can allow the seller to extract surplus from the trade. The next time you have an item to sell, why not try an auction?

INFORMATION AGGREGATION

The lemon/peach and auction examples just described are cases in which players must make a joint decision (whether and how to trade) based on private information. More generally, there are many situations in which a group, perhaps all of society, must make a collective decision in an environment of incomplete information. For instance, a city may be considering whether to extend a trolley line and, to make the socially optimal choice, it will depend on its individual citizens to accurately report what their own personal costs and benefits from the project would be. Another example is society's decision as to whether to reelect a president, again optimally a function of the citizens' individual and privately known preferences. A third example is the question of whether to convict an accused felon, which may be best decided by collecting the various pieces of information that members of a jury have about the defendant and the nature of the crime.

These examples involve information aggregation—that is, combining information from individual players to inform a single decision that affects them all. Such settings are typically called **social-choice problems**. A key issue in social choice is whether a communication and decision-making rule (known as a **mechanism** in the technical literature) can be found to achieve social objectives. One component of the analysis is determining the incentives of the players under any given mechanism. Although a formal analysis of general social-choice problems is not a topic for this book, it is worthwhile to exam-

[8]For more on the revenue equivalence result, and auction theory in general, see the following articles: P. McAfee and J. McMillan, "Auctions and Bidding," *Journal of Economic Literature* 25(1987):699–738; P. Milgrom, "Auction Theory," pp. 1–32, in *Advances in Economic Theory,* Fifth World Congress, ed. T. Bewley (Cambridge, UK: Cambridge University Press, 1987); and P. Milgrom and R. Weber, "A Theory of Auctions and Competitive Bidding," *Econometrica* 50(1982):1089–1122.

ine an applied example to help you recognize nuances in the rational reporting of private information. In plainer language, you will see that players do not always have the incentive to "vote their information."

Consider a setting in which a jury must decide whether to acquit or convict a defendant who is on trial.[9] For simplicity, suppose that the jury consists of two people who are the players in the game. The defendant's identity is either guilty or innocent and, from the player's perspective, this is determined by a move of nature that assigns equal probability to each possibility. During the trial, each juror obtains a signal of the defendant's identity. Because the jurors have different spheres of expertise and are alert at different times in the trial, the signal that juror/player 1 obtains is independent of the signal that juror/player 2 obtains. Suppose that each signal is either I or G, with the following distribution: Conditional on an innocent defendant, player i's signal is I with probability $3/4$ and G with probability $1/4$. Conditional on a guilty defendant, player i's signal is I with probability $1/4$ and G with probability $3/4$. Thus, I is an indication of innocence and G is an indication of guilt, although neither is absolute.

For example, note that, conditional on the defendant being guilty, the probability of two G signals is

$$\text{Prob}[\text{GG} \mid \text{guilty}] = \frac{3}{4} \cdot \frac{3}{4} = \frac{9}{16}.$$

Conditional on the defendant being innocent, the probability of two G signals is

$$\text{Prob}[\text{GG} \mid \text{innocent}] = \frac{1}{4} \cdot \frac{1}{4} = \frac{1}{16}.$$

In these mathematical expressions, the symbol "|" stands for "conditional on." The calculations show that getting two G signals is nine times as likely with the guilty defendant as with the innocent defendant.

We can use *Bayes' rule* to calculate the conditional probability of the defendant being guilty given the signals. That is, we can ask a question such as: Given that we initially thought that the defendant was guilty with probability $1/2$, and now having learned that both players received signal G, what should we believe is the probability that the defendant is guilty? The initial belief ($1/2$ here) is also known as the **prior belief**. The assessment of the likelihood

[9] What follows is an example along the lines of D. Austen-Smith and J. Banks, "Information Aggregation, Rationality, and the Condorcet Jury Theorem," *American Political Science Review* 90(1996):34–45. For a more general analysis, see T. J. Feddersen and W. Pesendorfer, "Voting Behavior and Information Aggregation in Elections with Private Information," *Econometrica* 65(1997):1029–1058.

of a guilty defendant conditional on the signals is called the **updated** or **posterior belief**. Bayes' rule tells us that the updated probability of guilt equals the probability of the particular signals conditional on a guilty defendant, times the prior probability that the defendant is guilty, divided by the total probability that the particular signals occur (irrespective of the defendant's type):

$$\text{Prob[guilty} \mid \text{GG]} = \frac{\text{Prob[GG} \mid \text{guilty] Prob[guilty]}}{\text{Prob[GG]}}.$$

Note that the overall probability of GG in the denominator is a sum of probabilities over the events of guilty and innocent defendants; that is,

$$\text{Prob[GG]} = \text{Prob[GG} \mid \text{guilty] Prob[guilty]} + \text{Prob[GG} \mid \text{innocent] Prob[innocent]}.$$

Plugging in the numbers, the Bayes' rule formula yields

$$\text{Prob[guilty} \mid \text{GG]} = \frac{(9/16)(1/2)}{(9/16)(1/2) + (1/16)(1/2)} = \frac{9}{10}.$$

This means that observing two G signals would cause us to believe that the defendant is guilty with 90 percent probability. Incidentally, if this is your first exposure to Bayes' rule, or you could use a general refresher on the basic definitions of probability, please read Appendix A. You can test your understanding by performing the calculations necessary to show that if one player gets the I signal and the other player gets the G signal, then the updated probability of a guilty defendant is

$$\text{Prob[guilty} \mid \text{IG]} = \frac{\text{Prob[IG} \mid \text{guilty] Prob[guilty]}}{\text{Prob[IG]}} = \frac{1}{2}.$$

At the end of the trial, the court (representing society) must declare the defendant either acquitted or convicted. Suppose that the jurors' preferences are identical and depend only on the defendant's identity and on the court's ruling. Each juror gets a payoff of 0 if the defendant is acquitted, regardless of the defendant's identity. If the defendant is guilty and is convicted, then each juror gets a payoff of 3; if the defendant is innocent and is convicted, then each juror gets a payoff of -2. Thus, jurors want to convict the guilty and acquit the innocent.

The court's ruling is issued on the advice of the jurors. Suppose that the players are not allowed to converse about the case and that, at the end of the trial, they are each asked to vote for acquittal or conviction. Furthermore, consider a *unanimity voting rule*, in which the court declares the defendant convicted if and only if both jurors vote to convict. The social objective is for each player to vote for conviction if and only if his signal is G, so that the defendant is convicted only when both signals are G.

How do rational jurors behave in this game? In particular, does each player want to pass along his information by voting to convict if and only if he gets the G signal? You might find it curious that the answer is "no." To be precise, voting in this way does *not* constitute a Bayesian Nash equilibrium of the game. To see this, put yourself in player 1's shoes under the assumption that player 2 will vote for conviction if and only if player 2 gets the G signal. Let us then calculate whether behaving the same way is a best response.

Note that your vote affects the court's ruling only in the situation in which player 2 is voting for conviction. You see, if player 2 votes for acquittal then, under the unanimity rule, the defendant will be acquitted regardless of what you do. On the other hand, if player 2 votes for conviction, then the defendant's fate is in your hands. In this case, if you vote to convict, then the defendant will be convicted. If you vote to acquit, then the defendant will be acquitted. In the jargon of research on voting behavior, your vote is *pivotal* to the outcome only if player 2 votes for conviction. Because our working assumption is that player 2 votes for conviction just when his signal is G, you know that your vote is pivotal precisely when player 2 gets the G signal; your vote makes no difference to the outcome when player 2 gets the I signal.

Also note that because you, as player 1, care only about the defendant's identity and the court's ruling, your action affects your payoff only when your vote is pivotal. Thus, when deciding how to vote, you can limit your attention to the case in which player 2's signal is G. Let us consider your incentives on how to vote in the situation in which your signal is I. Conditional on these signals, your updated probability of the guilty defendant is 1/2. Thus, if you vote to convict (in which case the defendant is convicted), your payoff is

$$\text{Prob[guilty} \mid \text{IG]} \cdot 3 + \text{Prob[innocent} \mid \text{IG]} \cdot (-2) = \frac{1}{2} \cdot 3 + \frac{1}{2} \cdot (-2) = \frac{1}{2}.$$

If you vote to acquit, then the defendant is acquitted and you get 0. The bottom line is that you strictly prefer voting for conviction rather than acquittal when your signal is I, which is contrary to the desired social policy.

If "voting one's information" is not an equilibrium, what is an equilibrium of this game? In fact, there are several equilibria. Here is a simple pure-strategy equilibrium: Player 2 always votes for conviction, regardless of player 2's signal, whereas player 1 votes to convict if and only if his signal is G. To see that this is an equilibrium, let us evaluate the incentives of both players. Given player 2's strategy, player 1 knows that his vote is pivotal regardless of player 2's signal; it is as though there is a jury of one. If player 1 gets signal

I then, using Bayes' rule, player 1 believes that the defendant is guilty with probability

$$\text{Prob}[\text{guilty} \mid \text{I}] = \frac{\text{Prob}[\text{I} \mid \text{guilty}]\,\text{Prob}[\text{guilty}]}{\text{Prob}[\text{I}]}$$
$$= \frac{(1/4)(1/2)}{(1/4)(1/2) + (3/4)(1/2)} = \frac{1}{4}.$$

In this case, convicting the defendant yields an expected payoff of $(1/4) \cdot 3 + (3/4) \cdot (-2) = -3/4$, which is lower than the payoff of acquittal, and so voting to acquit (as presumed) is better. If player 1 gets signal G, then, using Bayes' rule, player 1 believes that the defendant is guilty with probability

$$\text{Prob}[\text{guilty} \mid \text{G}] = \frac{\text{Prob}[\text{G} \mid \text{guilty}]\,\text{Prob}[\text{guilty}]}{\text{Prob}[\text{G}]}$$
$$= \frac{(3/4)(1/2)}{(3/4)(1/2) + (1/4)(1/2)} = \frac{3}{4},$$

which, as you can check, makes conviction preferred. As for player 2, his preferences are exactly as analyzed in the previous paragraphs (under the assumption that player 1 votes to convict on signal G only), and so player 2 rationally always votes to convict.

In this equilibrium, only player 1's signal is used to determine the court's ruling, which is less precise than is the stated social objective. There is another pure-strategy equilibrium in which the roles are reversed, with player 1 always voting for conviction. There is also a mixed-strategy equilibrium in which players vote for conviction when they receive the I signal and randomize when they get the G signal.

In all of the equilibria, less information is transmitted from the jury to the court than society prefers. The main reason for this is that I have assumed in the design of the game that the players actually have different preferences than does society. To see this, recall that I suggested a social objective of convicting the defendant only when both signals are G. Thus, society wants to acquit the defendant if one player gets the I signal while the other gets the G signal. In this contingency, the updated probability of the defendant being guilty is $1/2$. But the players have a strict incentive to convict the defendant because, with equal probabilities of innocence and guilt, the benefit of conviction in the case of a guilty defendant (a payoff of 3) outweighs the loss of convicting an innocent defendant (a payoff of -2).

If the payoffs were changed so that the players had the same preferences as the greater society, then there would be an efficient equilibrium, which can be shown. Thus, the potential for inefficiency in the voting game rests on a discrepancy between the preferences of the jurors and society, or perhaps on a discrepancy between the preferences of different jurors.[10] Another issue is whether inefficiencies disappear as the number of jurors increases; the answer to this question is generally "yes." One might also wonder if society would gain by allowing the jurors to talk to one another and share information before voting, as is the case in actual jury deliberations. Again, the answer is "yes." Designing and managing the optimal mechanism would be challenging, however, if the players had different preferences; the scientific literature has left room for analysis in this direction.

GUIDED EXERCISE

Problem: Consider the auction environment discussed in this chapter, where the bidders' values are independently drawn and distributed uniformly on the interval [0, 1000], but now suppose that there are n players. Compute the equilibrium of the first-price auction. (Hint: Presume that the players use a bidding function of the form $b_i = av_i$; note that, for a particular group of $n-1$ bidders, the probability that all of these bidders' valuations are simultaneously below x is $(x/1000)^{n-1}$.)

Solution: Assume that the equilibrium strategies take the form $b_i = av_i$. Then, given that the other players are using this bidding strategy (for some constant a), player i's expected payoff of bidding x is

$$(v_i - x)\left[\frac{x}{1000a}\right]^{n-1}.$$

In this expression, the fraction $x/1000a$ represents the probability that an individual player j (someone other than player i) bids less than x, given that player j is using strategy $b_j = av_j$ and that j's valuation is uniformly distributed between 0 and 1000. Raising this fraction to the power $n-1$ yields the probability that the bids of *all* $n-1$ *other players* are below x, in which case player i wins the auction. The first-order condition for player i's best response is

$$(n-1)(v_i - x)x^{n-2} - x^{n-1} = 0.$$

[10] On this, see A. Costinot and N. Kartik, "Information Aggregation, Strategic Voting, and Institutional Reform," unpublished manuscript, UC San Diego Department of Economics, 2006.

Solving for x yields $x = v_i(n-1)/n$, which identifies the parameter a to be $(n-1)/n$. That is, in equilibrium each of the players uses the bidding function $b_i = (n-1)v_i/n$. Note that, as $n \to \infty$, a approaches 1, and the players bid approximately their valuations (good for the seller).

EXERCISES

1. Regarding the trade game played by Jerry and Freddie that was analyzed in this chapter, are there values of p such that no equilibrium exists? Are there values of p such that the equilibrium entails *no* trade whatsoever?

2. Suppose you and one other bidder are competing in a private-value auction. The auction format is sealed bid, first price. Let v and b denote your valuation and bid, respectively, and let $\hat{v}$ and $\hat{b}$ denote the valuation and bid of your opponent. Your payoff is $v - b$ if it is the case that $b \geq \hat{b}$. Your payoff is 0 otherwise. Although you do not observe $\hat{v}$, you know that $\hat{v}$ is uniformly distributed over the interval between 0 and 1. That is, v' is the probability that $\hat{v} < v'$. You also know that your opponent bids according to the function $\hat{b}(\hat{v}) = \hat{v}^2$. Suppose your value is 3/5. What is your optimal bid?

3. Complete the analysis of the second-price auction by showing that bidding one's valuation v_i is weakly preferred to bidding any $x < v_i$.

4. Suppose that a person (the "seller") wishes to sell a single desk. Ten people are interested in buying the desk: Ann, Bill, Colin, Dave, Ellen, Frank, Gale, Hal, Irwin, and Jim. Each of the potential buyers would derive some utility from owning the desk, and this utility is measured in dollar terms by the buyer's "value." The valuations of the 10 potential buyers are shown in the following table.

Ann	Bill	Colin	Dave	Ellen	Frank	Gale	Hal	Irwin	Jim
45	53	92	61	26	78	82	70	65	56

Each bidder knows his or her own valuation of owning the desk. Using the appropriate concepts of rationality, answer these questions:

(a) If the seller holds a second-price, sealed-bid auction, who will win the auction and how much will this person pay?

(b) Suppose that the bidders' valuations are common knowledge among them. That is, it is common knowledge that each bidder knows the valuations of all of the other bidders. Suppose that the seller does not observe the bidders' valuations directly and knows only that they are all between 0 and 100. If the seller holds a first-price, sealed-bid auction, who will win the desk and how much will he or she have to pay? (Think about the Nash equilibrium in the bidding game. The analysis of this game is a bit different from the [more complicated] analysis of the first-price auction in this chapter, because here the bidders know one another's valuations.)

(c) Now suppose that the seller knows that the buyers' valuations are 45, 53, 92, 61, 26, 78, 82, 70, 65, and 56, but the seller does not know exactly which buyer has which valuation. The buyers know their own valuations but not one another's valuations. Suppose that the seller runs the following auction: She first announces a *reserve price* $\underline{p}$. Then simultaneously and independently the players select their bids; if a player bids below $\underline{p}$, then this player is disqualified from the auction and therefore cannot win. The highest bidder wins the desk and has to pay the amount of his or her bid. This is called a "sealed-bid, first-price auction with a reserve price." What is the optimal reserve price $\underline{p}$ for the seller to announce? Who will win the auction? And what will the winning bid be?

5. Consider the two-bidder auction environment discussed in this chapter, where the bidders' values are independently drawn and distributed uniformly on the interval [0, 1000]. Demonstrate that, by setting a reserve price, the auctioneer can obtain an expected revenue that exceeds what he would expect from the standard first- and second-price auctions. You do not have to compute the optimal reserve price or the actual equilibrium strategies to answer this question.

6. Consider an "all-pay auction" with two players (the bidders). Player 1's valuation v_1 for the object being auctioned is uniformly distributed between 0 and 1. That is, for any $x \in [0, 1]$, player 1's valuation is below x with probability x. Player 2's valuation is also uniformly distributed between 0 and 1, so the game is symmetric. After nature chooses the players' valuations, each player observes his/her own valuation but not that of the other player. Simultaneously and independently, the players submit bids. The player who bids higher wins the object, but *both* players must pay their bids. That is, if player i bids b_i, then his/her payoff is $-b_i$ if he/she does not win the auction; his/her payoff is $v_i - b_i$ if he/she wins the auction. Calculate the Bayesian Nash equilibrium strategies (bidding functions). (Hint: The equilibrium bidding function for player i is of the form $b_i(v_i) = kv_i^2$ for some number k.)

7. Suppose that you and two other people are competing in a *third-price,* sealed-bid auction. In this auction, players simultaneously and independently submit bids. The highest bidder wins the object but only has to pay the bid of the third-highest bidder. Sure, this is a silly type of auction, but it makes for a nice exercise. Suppose that your value of the object is 20. You do not know the values of the other two bidders. Demonstrate that, in contrast with a second-price auction, it may be strictly optimal for you to bid 25 instead of 20. Show this by finding a belief about the other players' bids under which 25 is a best-response action, yet 20 is *not* a best-response action.

8. Suppose John owns a share of stock in Columbus Research, a computer software firm. Jessica is interested in investing in the company. John and Jessica each receive a signal of the stock's value v, which is the dollar amount that the owner will receive in the future. John observes x_1, whereas Jessica observes x_2. It is common knowledge that x_1 and x_2 are independent random variables with probability distributions F_1 and F_2, respectively. These numbers are between 100 and 1000. The value of the stock is equal to $v = (x_1 + x_2)/2$.

Imagine a game in which John and Jessica may agree to trade the share of stock at price p, which is exogenously given. (Perhaps some third party or some external market mechanism sets the price.) Simultaneously and independently, John and Jessica select either "trade" or "not." A party that chooses "trade" must pay a trading cost of \$1. If they both choose to trade, then the stock is traded at price p; otherwise, the stock is not traded. Thus, if both announce "trade," then John's payoff is $p - 1$ and Jessica's payoff is $v - p - 1$. If John chooses "trade" and Jessica chooses "not," then John receives $v - 1$ and Jessica gets 0. If Jessica announces "trade" and John picks "not," then John receives v and Jessica receives -1. Finally, if they both say "not," John's payoff is v and Jessica's payoff is 0.

(a) Suppose that the probability distributions F_1 and F_2 are the same and that they each assign positive probability to just two numbers. Specifically, $x_i = 200$ with probability $1/2$, and $x_i = 1000$ with probability $1/2$, for $i = 1, 2$. In other words, there are two types of player John (200 and 1000) and there are two types of player Jessica (200 and 1000). Remember that p is a fixed constant, which you can assume is between 100 and 1000. Compute the Bayesian Nash equilibria of the trading game, given these assumptions on F_1 and F_2.

(b) Does trade occur in equilibrium with positive probability? Do the predictions of this model conform to your view of the stock market? In what ways would the model have to be altered to better explain market phenomena?

(c) Prove, if you can, that the result of part (a) holds more generally—for any p, F_1, and F_2. To complete this proof, you will need to evaluate some double integrals.

9. Consider a "common-value auction" with two players, where the value of the object being auctioned is the *same* for both players. Call this value Y and suppose that $Y = y_1 + y_2$, where y_1 and y_2 are both uniformly distributed between 0 and 10. That is, for any $x \in [0, 10]$, we have $y_i < x$ with probability $x/10$. Suppose that player 1 observes y_1 (but does not observe y_2) and that player 2 observes y_2 (but does not observe y_1). The players simultaneously submit bids,

b_1 and b_2. If player i bids higher than does player j, then player i wins the auction and gets the payoff $Y - b_i$ whereas player j gets 0.

(a) Suppose player 2 uses the bidding strategy $b_2(y_2) = 3 + y_2$. What is player 1's best-response bidding strategy?

(b) Suppose each player will not select a strategy that would surely give him/her a negative payoff conditional on winning the auction. Suppose also that this fact is common knowledge between the players. What can you conclude about how high the players are willing to bid?

PERFECT BAYESIAN EQUILIBRIUM 28

The games with moves of nature that were surveyed in preceding chapters are those to which standard concepts of equilibrium, subgame perfection, and rationalizability can be applied. Two classes of games were distinguished: those in which nature moves last and those in which all of the players' actions are taken simultaneously. This chapter examines a third class of games, where players have private information and they move sequentially. Many interesting economic settings have these features, from some types of auctions to cases of signaling one's quality to a prospective business partner. To study games with sequential moves and incomplete information, the appropriate notion of equilibrium goes beyond subgame perfection to allow sequential rationality to be applied to all information sets.

The gift game described in Chapter 24 illustrates the need for a modification of the subgame perfect equilibrium concept. Consider the version of the game pictured in Figure 28.1. Relative to the game in Figure 24.1, in the variant player 2 enjoys opening gifts from *both* types of player 1. Suppose that you are interested in applying equilibrium theory to this game. Because the gift game has sequential decisions, it seems appropriate to look for subgame perfect equilibria. But the game has *no* proper subgames, so every Nash equilibrium is subgame perfect. In particular, $(N^F N^E, R)$ is a subgame perfect

FIGURE 28.1
A variant of the gift game.

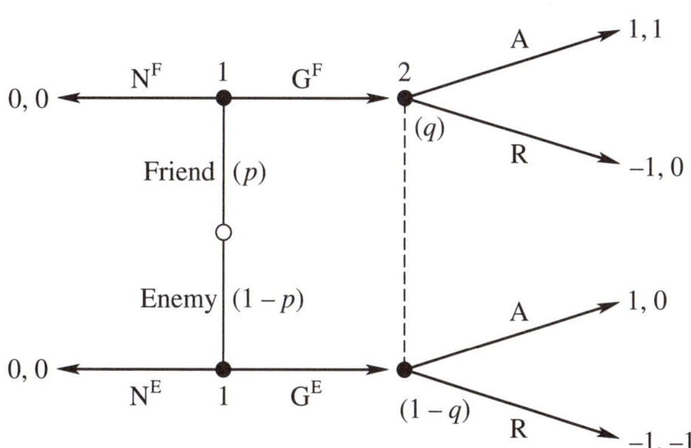

equilibrium of the game; you should verify this through examination of the Bayesian normal form. In this equilibrium, both types of player 1 choose not to give a gift and player 2 plans to refuse gifts.

One problem with the profile ($N^F N^E$, R) is that it prescribes behavior for player 2 that is clearly irrational conditional on the game reaching his information set. Regardless of player 1's type, player 2 prefers to accept *any* gift offered. This preference is not incorporated into the subgame perfect equilibrium because (1) player 2's information set is not reached on the path induced by ($N^F N^E$, R), and (2) player 2's information set does not represent the start of a subgame. As this example shows, not all information sets are necessarily evaluated in a subgame perfect equilibrium. In other words, the concept of subgame perfection does not sufficiently capture *sequential rationality* (that players maximize their payoffs from each of their information sets).

To better address sequential rationality, we must employ an equilibrium concept that isolates *every* information set for examination. *Perfect Bayesian equilibrium* does just that. The key to this equilibrium concept is that it combines a strategy profile with a description of beliefs that the players have at each of their information sets. The beliefs represent the players' assessments about each other's types, conditional on reaching different points in the game.

CONDITIONAL BELIEFS ABOUT TYPES

The gift game in Figure 28.1 illustrates the idea of a conditional belief. Recall that, in this game, player 2 does not observe nature's decision. Therefore, at the beginning of the game, player 2 knows only that player 1 is the friend type with probability p and the enemy type with probability $1 - p$. This belief p is called player 2's *initial belief about player 1's type*. Keep in mind that this is a belief about player 1's *type*, not a belief about player 1's *strategy* (which is the sort of belief with which we were dealing in Parts I through III of this book).

Although player 2 does not observe nature's decision, player 2 does observe whether player 1 decided to give a gift. Furthermore, player 2 might learn something about player 1's type by observing player 1's action. As a result, player 2 will have an *updated belief* about player 1's type. For example, suppose that you are player 2 in the gift game and suppose that player 1 behaves according to strategy $N^F G^E$; thus, you expect only to receive a gift from the enemy type. What should you conclude, then, if player 1 actually gives you a gift? Given player 1's strategy, you should conclude that player 1 is an enemy. In reference to Figure 28.1, when your information set is reached, you believe that you are playing at the lower of the two nodes in the information set.

In general, player 2 has an updated belief about player 1's type, conditional on arriving at player 2's information set (that is, conditional on receiving a gift). Note that player 2's updated belief about player 1's type can be put in terms of a probability distribution over the nodes in player 2's information set. In Figure 28.1, this probability distribution is described by the numbers q and $1 - q$ that appear beside the nodes. Literally, q is the probability that player 2 believes he is at the top node when his information set is reached. Thus, q is the probability that player 2 believes player 1 is the friend type, conditional on receiving a gift.

SEQUENTIAL RATIONALITY

Taking account of conditional beliefs allows us to evaluate rational behavior at all information sets, even those that may not be reached in equilibrium play. Consider, again, the gift game pictured in Figure 28.1. Regardless of player 1's strategy, player 2 will have *some* updated belief q at his information set. This number has meaning even if player 2 believes that player 1 adopts the strategy $N^F N^E$ (where neither type gives a gift). In this case, q represents player 2's belief about the type of player 1 when the "surprise" of a gift occurs. Given the belief q, we can determine player 2's optimal action at his information set. You can readily confirm that action A is best for player 2, whatever is q. Thus, sequential rationality requires that player 2 select A.

For another example, consider the gift game pictured in Figure 28.2. This is the same game discussed in Chapter 24. Note that, regardless of the probability q, player 2 receives a payoff of 0 if he selects R at his information set.

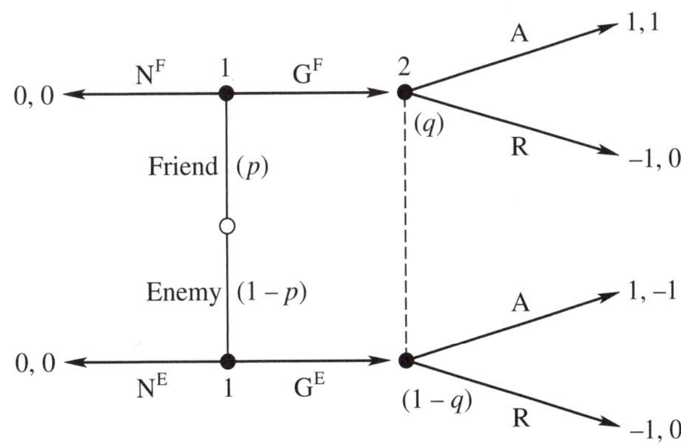

FIGURE 28.2
The gift game.

On the other hand, if player 2 chooses A, then he gets a payoff of 1 with probability q (the probability that his decision is taken from the top node in his information set) and he gets a payoff of -1 with probability $1 - q$. Player 2's expected payoff of selecting A is therefore

$$q + (-1)(1 - q) = 2q - 1.$$

Player 2 will select A if $q > 1/2$, he will select R if $q < 1/2$, and he will be indifferent between A and R if $q = 1/2$.

CONSISTENCY OF BELIEFS

In an equilibrium, player 2's updated belief should be consistent with nature's probability distribution and player 1's strategy. For example, as noted earlier, if player 2 knows that player 1 adopts strategy $N^F G^E$, then player 2's updated belief should specify $q = 0$; that is, conditional on receiving a gift, player 2 believes that player 1 is the enemy type. In general, consistency between nature's probability distribution, player 1's strategy, and player 2's updated belief can be evaluated by using *Bayes' rule*. Recall that Bayes' rule was discussed in the context of information aggregation in the previous chapter. If you did not read about it there, or if you could use a brief review, please read Appendix A.

The Bayes' rule calculation is quite simple and intuitive. Here is the general form for the gift game in Figure 28.2. At player 2's information set, his updated belief gives the relative likelihood that player 2 thinks his top and bottom nodes have been reached. Let r^F and r^E be the probabilities of arriving at player 2's top and bottom nodes, respectively. That is, r^F is the probability that nature selects F and then player 1 selects G^F. Likewise, r^E is the probability that nature selects E and then player 1 chooses G^E. As an example, suppose that $r^F = 1/8$ and $r^E = 1/16$. In this case, player 2's information set is reached with probability $1/8 + 1/16 = 3/16$, which is not a very likely event. But note that the top node is twice as likely as is the bottom node. Thus, conditional on player 2's information set actually being reached, player 2 ought to believe that it is twice as likely that he is at the top node than at the bottom node. Because the probabilities must sum to 1, this updated belief is represented by a probability of 2/3 on the top node and 1/3 on the bottom node.

In general, the relation between r^F, r^E, and q is given by

$$q = \frac{r^F}{r^F + r^E}.$$

In words, q is the probability of reaching the top node divided by the total probability of reaching the top and the bottom nodes (the latter of which is the probability of reaching the information set). Numbers r^{F} and r^{E} can be calculated from nature's probability distribution and player 1's strategy. Specifically, let α^{F} and α^{E} denote the probabilities that the friend and enemy types of player 1 choose to give a gift. Then player 2's top node is reached with probability $r^{\mathrm{F}} = p\alpha^{\mathrm{F}}$, whereas player 2's bottom node is reached with probability $r^{\mathrm{E}} = (1 - p)\alpha^{\mathrm{E}}$. Therefore,

$$q \equiv \frac{p\alpha^{\mathrm{F}}}{p\alpha^{\mathrm{F}} + (1 - p)\alpha^{\mathrm{E}}}.$$

This fraction can be represented in a more intuitive way. Let Prob[G] denote the overall probability that player 1 gives a gift, which is the denominator of the fraction. The numerator is the probability that nature selects the friend type, Prob[F], times the probability that the friend gives a gift, Prob[G | F]. The number q is the probability that player 1 is a friend conditional on player 1 giving a gift. Substituting for the terms in the preceding fraction, we have

$$q = \mathrm{Prob}[\mathrm{F} \mid \mathrm{G}] \equiv \frac{\mathrm{Prob}[\mathrm{G} \mid \mathrm{F}]\ \mathrm{Prob}[\mathrm{F}]}{\mathrm{Prob}[\mathrm{G}]},$$

which is the familiar Bayes' rule expression.

Note that Bayes' rule cannot be applied if player 2's information set is reached with 0 probability, which is the case when player 1 employs strategy $\mathrm{N^F N^E}$. In this situation, q is still meaningful—it is the belief of player 2 when he is surprised to learn that player 1 has given a gift—but q is not restricted to be any particular number. In other words, *any* updated belief is feasible after a surprise event.

EQUILIBRIUM DEFINITION

Perfect Bayesian equilibrium is a concept that incorporates sequential rationality and consistency of beliefs. In essence, a perfect Bayesian equilibrium is a coherent story that describes beliefs and behavior in a game. The beliefs must be consistent with the players' strategy profile, while the strategy profile must specify rational behavior at all information sets, given the players' beliefs. In more formal language:

> Consider a strategy profile for the players, as well as beliefs over the nodes at all information sets. These are called a **perfect Bayesian equilibrium (PBE)** if: (1) each player's strategy specifies optimal

actions, given his beliefs and the strategies of the other players, and (2) the beliefs are consistent with Bayes' rule wherever possible.[1]

Two additional terms are useful in categorizing the classes of potential equilibria. Specifically, we call an equilibrium **separating** if the types of a player behave differently. On the other hand, in a **pooling** equilibrium, the types behave the same.

To determine the set of PBE for a game, you can use the following procedure.

Steps for calculating perfect Bayesian equilibria:

1. Start with a strategy for player 1 (pooling or separating).
2. If possible, calculate updated beliefs (q in the example) by using Bayes' rule. In the event that Bayes' rule cannot be used, you must arbitrarily select an updated belief; here you will generally have to check different potential values for the updated belief with the next steps of the procedure.
3. Given the updated beliefs, calculate player 2's optimal action.
4. Check whether player 1's strategy is a best response to player 2's strategy. If so, you have found a PBE.

To solidify your understanding of the PBE concept, follow along with the computation of equilibria in the gift game of Figure 28.2. Let us focus on pure-strategy equilibria, as usual. Then there are four potential equilibria: (1) a separating equilibrium featuring strategy $N^F G^E$, (2) a separating equilibrium featuring strategy $G^F N^E$, (3) a pooling equilibrium with strategy $G^F G^E$, and (4) a pooling equilibrium with $N^F N^E$.

Here is the procedure in action:

Separating with $N^F G^E$: Given this strategy for player 1, it must be that $q = 0$. Thus, player 2's optimal strategy is R. But then player 1 would strictly prefer not to play G^E when of the enemy type. Therefore, there is no PBE in which $N^F G^E$ is played.

Separating with $G^F N^E$: Given this strategy for player 1, it must be that $q = 1$. Thus, player 2's optimal strategy is A. But then the enemy type of player 1 would strictly prefer to play G^E rather than N^E. Therefore, there is no PBE in which $G^F N^E$ is played.

---

[1]The progression of ideas leading to the standard, formal definition of perfect Bayesian equilibrium is well represented by the following articles: R. Selten, "Reexamination of the Perfectness Concept for Equilibrium Points in Extensive Games," *International Journal of Game Theory* 4(1975):25–55; D. M. Kreps and R. Wilson, "Sequential Equilibria," *Econometrica* 50(1982):863–894; D. Fudenberg and J. Tirole, "Perfect Bayesian and Sequential Equilibrium," *Journal of Economic Theory* 53(1991):236–260; and P. Battigalli, "Strategic Independence and Perfect Bayesian Equilibria," *Journal of Economic Theory* 70(1996):201–234.

Pooling with $G^F G^E$: Here Bayes' rule requires $q = p$, so player 2 optimally selects A if and only if $p \geq 1/2$. In the event that $p < 1/2$, player 2 must select R, in which case neither type of player 1 wishes to play G in the first place. Thus, there is no PBE of this type when $p < 1/2$. When $p \geq 1/2$, there is a PBE in which $q = p$ and ($G^F G^E$, A) is played.

Pooling with $N^F N^E$: In this case, Bayes' rule does not determine q. But notice that the types of player 1 prefer not giving gifts only if player 2 selects R. In order for R to be chosen, player 2 must have a sufficiently pessimistic belief regarding the type of player 1 after the "surprise" event in which a gift is given. Strategy R is optimal as long as $q \leq 1/2$. Thus, for every $q \leq 1/2$, there is a PBE in which player 2's belief is q and the strategy profile ($N^F N^E$, R) is played.

The example shows that, because the types of player 1 have the same preferences over outcomes, there is no separating equilibrium. There is always a pooling equilibrium in which no gift is given. In this equilibrium, player 2 believes that a gift signals the presence of the enemy. You probably know a misanthrope who has beliefs such as this. Finally, if there is a great enough chance of encountering a friend (so that $p \geq 1/2$), then there is a pooling equilibrium in which gifts are given by both types. In this equilibrium, a sanguine player 2 gladly accepts gifts.

GUIDED EXERCISE

Problem: Consider a game between two friends, Amy and Brenda. Amy wants Brenda to give her a ride to the mall. Brenda has no interest in going to the mall unless her favorite shoes are on sale (S) at the large department store there. Amy likes these shoes as well, but she wants to go to the mall even if the shoes are not on sale (N). Only Amy subscribes to the newspaper, which carries a daily advertisement of the department store. The advertisement lists all items that are on sale, so Amy learns whether or not the shoes are on sale. Amy can prove whether or not the shoes are on sale by showing the newspaper to Brenda. But this is costly for Amy, because she will have to take the newspaper away from her sister, who will yell at her later for doing so.

In this game, nature first decides whether or not the shoes are on sale, and this information is made known to Amy. (Amy observes whether nature chose S or N.) Nature chooses S with probability p and N with probability $1 - p$. Then Amy decides whether or not to take the newspaper to Brenda (T or D). If she takes the newspaper to Brenda, then it reveals to Brenda whether the shoes are on sale. In any case, Brenda must then decide whether to take Amy to the mall (Y) or to forget it (F). If the shoes are on sale, then going to the mall is worth 1 unit of utility to Brenda and 3 to Amy. If the shoes are not on sale, then traveling to the mall is worth 1 to Amy and -1 to Brenda. Both players obtain 0 utility when they do not go to the mall. Amy's personal cost of taking the newspaper to Brenda is 2 units of utility, which is subtracted from her other utility amounts.

(a) Draw the extensive form of this game.

(b) Does this game have a *separating* perfect Bayesian equilibrium? If so, fully describe it.

(c) Does this game have a *pooling* perfect Bayesian equilibrium? If so, fully describe it.

Solution:

(a) Here is the extensive-form diagram, with Amy's payoffs given first:

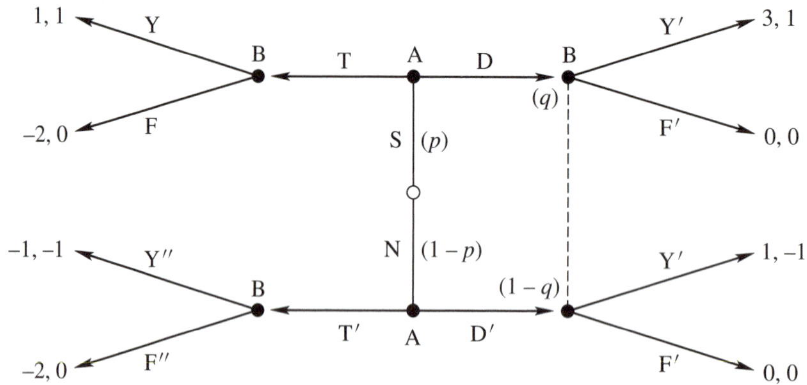

Note that the newspaper provides verifiable information of whether the shoes are on sale. Thus, Brenda can distinguish between her top-left node or her bottom-left node. Therefore, in any PBE, Brenda must use a strategy that selects Y and F″, respectively, at these two information sets.

(b) This game does have a separating PBE. As shown in the extensive form, let q denote Brenda's updated probability that the shoes are on sale, conditional on Amy not showing her the newspaper. Clearly, Amy has no incentive to choose T′ because Brenda selects F″. Thus, the only candidate for Amy's strategy in a separating equilibrium is TD′, which implies $q = 0$. Given this value of q, Brenda's optimal action at her right information set is F′, which makes Amy's prescribed strategy a best response. In summary, the separating equilibrium is (TD′, YF′F″) with $q = 0$.

(c) This game has a pooling equilibrium if and only if $p \geq 1/2$. As observed for part (b), Amy must choose D′ in any equilibrium, so the only candidate for Amy's strategy in a pooling equilibrium is DD′, which implies that Brenda's updated belief is $q = p$. If $p \geq 1/2$, so that $q \geq 1/2$ as well, then Brenda prefers action Y′ at her right information set. This, in turn, justifies Amy selecting the prescribed strategy. On the other hand, if $p < 1/2$ then Brenda strictly prefers F′ and it would not be rational for Amy to select D. To summarize, if $p \geq 1/2$, then there is a pooling equilibrium with strategy profile (DD′, YY′F″) and $q = 0$. If $p < 1/2$, then there is no pooling equilibrium and Amy is forced to bring the newspaper to Brenda in the event of a shoe sale.

EXERCISES

1. Consider the market game in the Guided Exercise of Chapter 24.

   (a) Find a separating perfect Bayesian equilibrium for this game. (Report strategies and beliefs.)

   (b) Find a pooling perfect Bayesian equilibrium for this game. Under what values of q does it exist?

2. Consider the following game with nature:

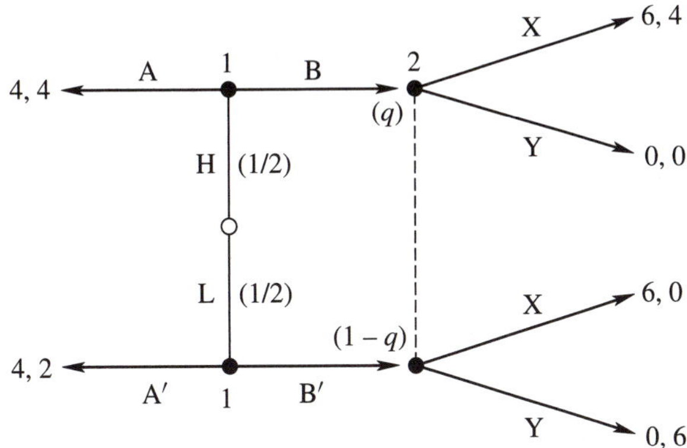

   (a) Does this game have any *separating* perfect Bayesian equilibrium? Show your analysis and, if there is such an equilibrium, report it.

   (b) Does this game have any *pooling* perfect Bayesian equilibrium? Show your analysis and, if there is such an equilibrium, report it.

   (c) Draw the normal-form matrix for this game.

3. Consider the following game of incomplete information.

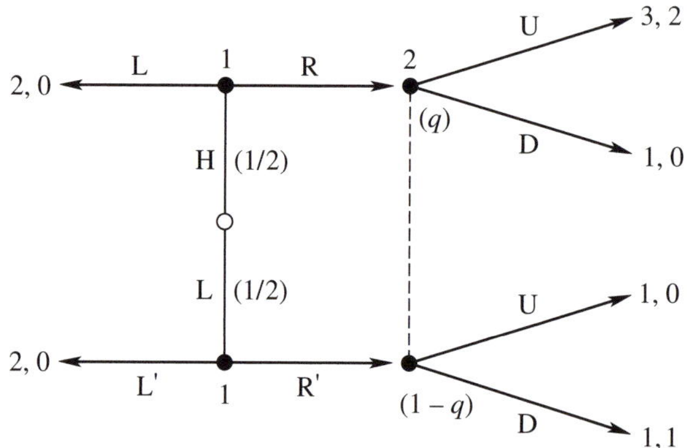

(a) Does this game have a *separating* perfect Bayesian equilibrium? If so, fully describe it.

(b) Does this game have a *pooling* perfect Bayesian equilibrium? If so, fully describe it.

4. Consider an extensive-form game in which player 1 is one of two types: A and B. Suppose that types A and B have *exactly* the same preferences; the difference between these types has something to do with the payoff of another player. Is it possible for such a game to have a separating PBE, where A and B behave differently?

5. A defendant in a court case appears before the judge. Suppose the actual harm to the plaintiff caused by the defendant is equal to $1000x$ dollars, where either $x = 0$ or $x = 1$. That is, the defendant is either innocent ($x = 0$) or guilty of 1000 dollars of damage ($x = 1$). The defendant knows x and has evidence to prove it. The judge does not observe x directly; she only knows that $x = 1$ with probability $1/2$ and $x = 0$ with probability $1/2$.

The judge and defendant interact as follows: First, the defendant has an opportunity to provide his evidence of x. He freely chooses whether or not to provide the evidence; the court cannot force him to do it. Providing evidence to the court costs the defendant one dollar (for photocopying). If the defendant chooses to provide the evidence, then it reveals x to the judge. Whether or not evidence is provided, the judge then decides the level of damages y (in thousands of dollars) that the defendant must pay. The judge prefers to select y "fairly"; she would like y to be as close as possible to x. The defendant wishes to minimize his monetary loss. These preferences and the players' interaction are summarized by the extensive-form diagram shown here. Note that "E" stands for "provide evidence" and N stands for "do not provide evidence."

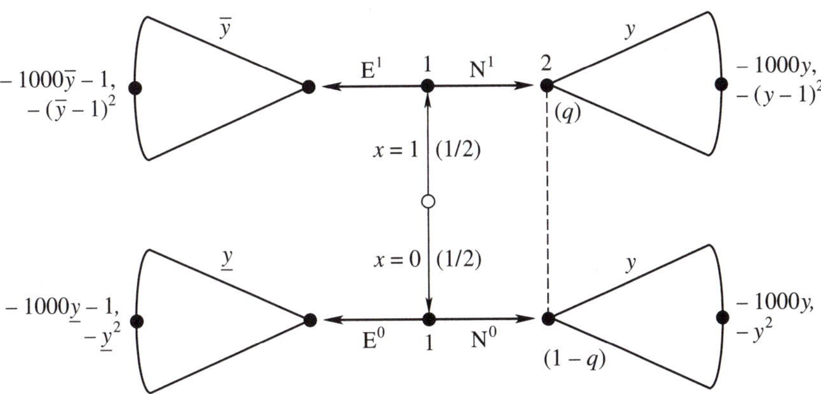

(a) This game has a unique perfect Bayesian equilibrium. Find and report it. (Hint: Start by showing that it is optimal for the judge to set y equal to the expected value of x, given her belief.)

(b) In one or two sentences, explain why the result of part (a) is interesting from an economic standpoint.

(c) Consider a version of the game in which x is an integer between 0 and K, inclusive, with each of these values equally likely. Compute the perfect Bayesian equilibrium of this game. (Hint: Use your intuition from part (a).)

6. In the classic Rob Reiner movie *The Princess Bride,* there is a scene at the end where Wesley (the protagonist) confronts the evil prince Humperdinck. The interaction can be modeled as the following game: Wesley is one of two types: weak or strong. Wesley knows whether he is weak or strong, but the prince only knows that he is weak with probability $1/2$ and strong with probability $1/2$. Wesley is lying in a bed in the prince's castle when the prince enters the room. Wesley decides whether to get out of bed (O) or stay in bed (B). The prince observes Wesley's action but does not observe Wesley's type. The prince then decides whether to fight (F) or surrender (S) to Wesley. The payoffs are such that the prince prefers to fight only with the weak Wesley, because otherwise the prince is an inferior swordsman. Also, the weak Wesley must pay a cost c to get out of bed. The extensive-form representation of the game is shown at the top of the next page.

(a) What conditions on c guarantee the existence of a separating PBE? Fully describe such an equilibrium.

(b) For what values of c is there a pooling equilibrium in which both strong and weak Wesleys get out of bed? Fully describe such an equilibrium.

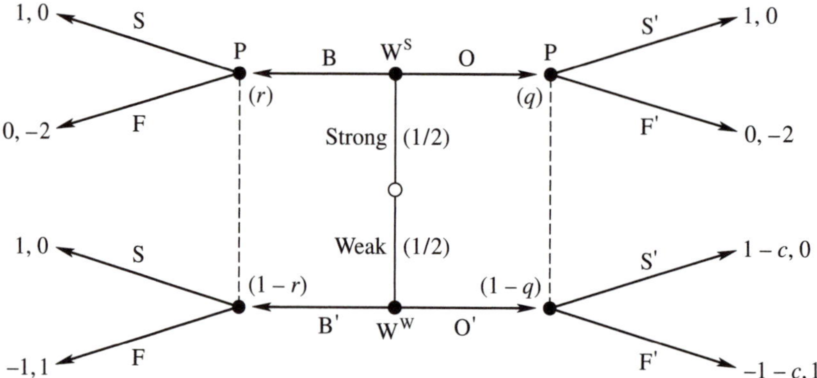

7. Consider the "worker status" model of Exercise 3 in Chapter 25. Suppose there are two possible types of worker, H and L. The types differ in the parameters y and q. Specifically, $y = 75$ and $q = 3/5$ for the H type, whereas $y = 65$ and $q = 2/5$ for the L type.

   (a) Using your answer from Exercise 3 of Chapter 25, what contract would the firm offer if it knew that the worker's type is L? What if it knew that the worker's type is H?

   (b) Consider the game of incomplete information in which the worker knows his own values of y and q, but the firm only knows that the worker is L with probability p and H with probability $1 - p$. Suppose the firm can offer two contracts to the worker, which we can write as $(w^0, 0)$ and $(w^1, 1)$. The interpretation is that the firm is willing to pay w^0 for the safe job and w^1 for the risky job. After observing the firm's offer, the worker decides whether to accept a job and, if so, which contract to take. Suppose that, because of market pressure, the firm is constrained to set $w^0 = 35$. (Other firms are offering job $z = 0$ at wage 35.) Compute the firm's optimal choice of w^1. Explain the steps you take to solve this problem.

   (c) Try to find the optimal contract offers for the firm when w^0 is not constrained to equal 35.

8. Compute the PBE of the three-card poker game described in Exercise 4 of Chapter 24. (Hint: Start by determining whether there are any information sets at which a player has an optimal action that is independent of his belief about his opponent's strategy.)

JOB-MARKET SIGNALING AND REPUTATION 29

In this chapter, I present two examples of dynamic games with incomplete information. The examples illustrate two important economic ideas: (1) that a worker can signal his ability by engaging in costly schooling, and (2) that a person may have an incentive to establish a reputation for being someone she is not.

JOBS AND SCHOOL

Have you ever wondered why you are working so hard to complete an academic degree? Is it because schooling helps you develop valuable skills, learn facts, and recognize important insights? Will academic wisdom serve you in the workplace? Will an understanding of the causes of the Peloponnesian War help you to be a better nurse, lawyer, or salesperson? Do you think your potential future employers care about what you learn in college? Why do universities set such formal standards for degrees?

I think it would be insane to suggest that academic training has no intrinsic value. Writing essays about Peloponnesia helps you develop a critical mind and good writing skills. Studying game theory helps you to develop a logical mind and understand social interaction. Surely these facilities will enhance your productivity later in life, particularly on the job. Prospective employers ought to pay a premium for the labor of a well-trained, intelligent person like yourself. Education adds value.

Yet, formal education has another—perhaps equally important—role in the marketplace. An academic degree is a sign of quality to the extent that highly productive people are more likely than less-productive people to get degrees.[1] People obtain degrees to prove their quality to prospective employers, colleagues, friends, and spouses. In this respect, degrees serve as signaling mechanisms. Perhaps rather than helping people *become* smart, colleges exist

[1]This is a gross generalization. There are many unjust barriers that keep some of the brightest people from going to college. Nonetheless, the impression in our society is that academic degrees are associated with intelligence and skills.

FIGURE 29.1 Job-market signaling.

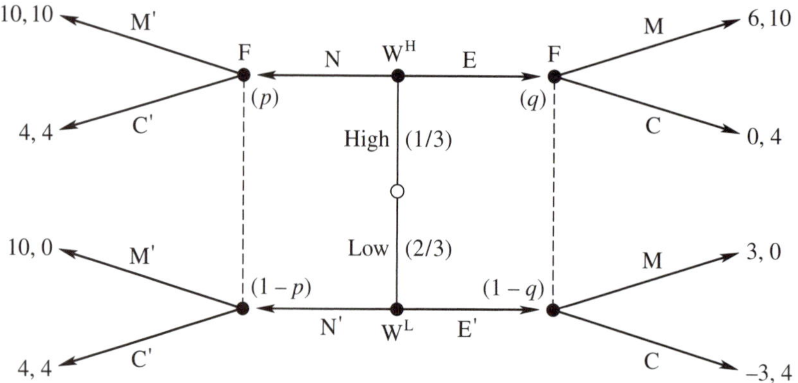

merely to help people who are already smart *prove* that they are smart. It is an extreme view—offensive to some—but not without some validity.

Here is a simple model demonstrating the signaling role of education.[2] Imagine a game of incomplete information played by a worker (W) and a firm (F); see Figure 29.1. The worker has private information about her level of ability. With probability 1/3 she is a high type (H) and with probability 2/3 she is a low type (L). After observing her own type, the worker decides whether to obtain a costly education (E) or not (N); think of E as getting a degree. The firm observes the worker's education (which is described in her resumé), but the firm does not observe the worker's quality type. The firm then decides whether to employ the worker in an important managerial job (M) or in a much less important clerical job (C). In equilibrium, the firm may deduce the worker's quality level on the basis of the worker's education.

The worker's payoff is listed first at the terminal nodes of the extensive form. As the payoff numbers indicate, both types of worker would like to have the managerial job. In particular, the managerial job yields a benefit of 10 units of utility to the worker, irrespective of type. The clerical job yields a benefit of 4 to both types of worker. On the other hand, the high and low types have *different* education costs; to obtain an education, the high type pays 4 units of utility, whereas the low type pays 7 units. The education cost is subtracted from the job benefit in the event that the worker obtains an education. The interpretation of differential education costs is that low-ability workers have a more difficult time getting through college; they have to work harder, hire tutors, and so on.

---

[2]This model is due to M. Spence, "Job Market Signaling," *Quarterly Journal of Economics* 87(1973): 355–374.

The firm would like to put the high-ability worker in the management position; this yields the firm a profit of 10. In contrast, giving the management job to the low-ability worker would be a disaster for the firm, leading to a profit of 0. When assigned to the clerical job, both types of worker produce a profit of 4 for the firm. Importantly, education is of no direct value to the firm; the firm's payoff does not depend on whether the worker gets an education. Thus, education is inefficient in this model. However, it can serve to separate the two types of worker.

To compute the perfect Bayesian equilibria of this game, divide the analysis into two cases: pooling equilibria and separating equilibria. Regarding pooling equilibria, first note that there is no pooling equilibrium in which the worker plays strategy EE′. This is because with EE′ the only consistent belief at the firm's right information set is $q = 1/3$ (because both types get an education, the firm learns nothing at this information set). But under the belief $q = 1/3$, C is the firm's optimal action. This gives the low type a negative payoff, which is worse than the least she could expect by playing N′.

There is a pooling equilibrium in which the worker adopts strategy NN′. The firm's belief at its left information set is $p = 1/3$ and there it selects action C′. In this equilibrium, the high-type worker is deterred from selecting E by anticipating the firm's selection of C at its right information set. The firm's choice of C is justified by a belief q that is less than 2/5.[3] In summary, there is a pooling equilibrium with strategies NN′ and CC′ and beliefs given by $p = 1/3$ and any $q \leq 2/5$.

The game has a single separating equilibrium, which captures the idea of education as a signaling device. To find the separating equilibrium, first realize that NE′ cannot be part of a PBE. If this strategy were played by the worker, then the only consistent beliefs would be $p = 1$ and $q = 0$ and the firm would select CM′; facing CM′, E′ is obviously not rational for the low-type worker. Thus, the separating equilibrium must feature strategy EN′ by the worker, implying beliefs $p = 0$ and $q = 1$ for the firm. The firm's equilibrium strategy is MC′.

You should take two insights away from this example. First, the *only* way for the high-type worker to get the job that she deserves is to signal her type by getting an education. Otherwise, the firm judges the worker to be a low type. The connection between education and perceived ability level gibes with reality. Second, the value of education as a signaling device depends on the types' differential education costs, not on any skill enhancement that education delivers. This insight forces us to think more deeply about career preparation

---

[3] Because the right information set is not reached, given the worker's strategy, any q is consistent with Bayes' rule.

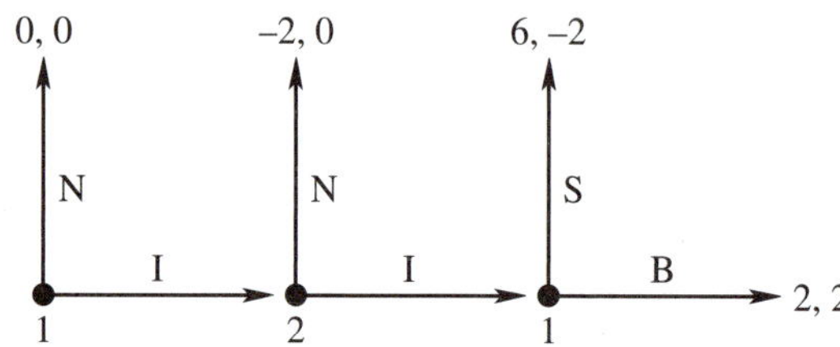

FIGURE 29.2
An investment game.

in the real world; at the least, we should try to separate between the signaling and the human capital aspects of education.

REPUTATION AND INCOMPLETE INFORMATION

Here is the *last* example of this textbook—are you relieved or dejected to have reached the terminus? I hope the latter. Examine the extensive-form game pictured in Figure 29.2. This is an investment game played between two people. Player 1 owns an asset that can be put to productive use only if both players make an investment. For example, the asset might be a motorcycle that is in need of repair. Player 1 might be an expert in electrical systems, so his investment would be to perform the electrical repairs on the bike. Player 2 might be a mechanical specialist, whose investment would be to repair the engine mechanics.

At the beginning of the game, player 1 decides whether to invest in the asset (perform the electrical repair). Player 1's choice is observed by player 2. If player 1 decides not to invest (N), then the game ends with zero payoffs. If player 1 invests (I), then player 2 must decide whether to invest (repair the engine). If he fails to invest (N), then the asset is of no productive use; in this case, the game ends and player 1 gets a negative payoff owing to his wasted investment. If player 2 invests (I), then the asset is made productive, creating a net value of 4. That is, investment by both players puts the motorcycle in operating condition so that it can be enjoyed at the local park for off-road vehicles. But because player 1 owns the asset, he determines how it will be used. He can decide to be benevolent (B) by sharing the asset with player 2 (that is, allowing player 2 to ride the bike) or he can be selfish (S) and hoard the asset.

The game is easily solved by backward induction. Given the payoffs, player 1 would select S at the last decision node in the game. Anticipating this selection from the middle decision node, player 2 would choose N. That is, player 2 would have no reason to invest, because she would expect that player 1 would not share the fruits of investment with her. Expecting this, player 1 would select N at the beginning of the game. Thus, the unique sub-

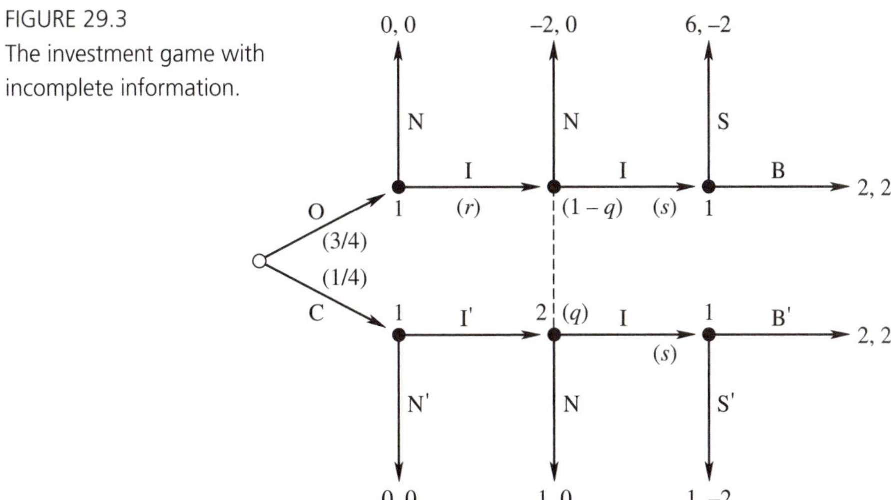

FIGURE 29.3
The investment game with incomplete information.

game perfect equilibrium is (NS, N). Investments are not made and the asset sits idle. It is a sorry situation.

Next imagine the slight, possibly more realistic, variation of the investment game pictured in Figure 29.3. There are two types of player 1. The "ordinary" type (O) has payoffs as specified in the original game. You can see this by confirming that the top half of the extensive form in Figure 29.3 is identical with the extensive form in Figure 29.2. The "cooperative" type (C) is more altruistic; this type likes to share and likes to invest regardless of what the other player does. Nature picks the ordinary type with probability $3/4$ and the cooperative type with probability $1/4$. Player 1 knows whether he is cooperative or ordinary; player 2 does not observe player 1's type.[4]

To compute the perfect Bayesian equilibria of this game of incomplete information, we need to find strategies for the players as well as the belief of player 2 at her information set (designated q in Figure 29.3). We should begin the analysis by observing that the cooperative player 1 has a strict incentive to invest and take the benevolent action, regardless of player 2's behavior. In other words, player 1's optimal choice at his lower-right decision node is B′ and his optimal choice at the lower-left decision node is I′. Furthermore, as with the original game, player 1's optimal choice at his top-right decision node is S. Thus, every PBE specifies S, I′, and B′. Regarding player 1, only the behavior

[4]Games similar to this investment game were first analyzed by theorists David Kreps, Paul Milgrom, John Roberts, and Robert Wilson. An example of their work is "Rational Cooperation in the Finitely Repeated Prisoners Dilemma," *Journal of Economic Theory* 27(1982):245–252. There is a decent chance that, by the time you read this book, one or more of these researchers will have been awarded the Nobel Prize in Economics.

of the ordinary type at the beginning of the game has yet to be determined. This depends on what player 2 does.

Examine player 2's information set, where q is the probability she attaches to the bottom node (the probability of the C type conditional on investment by player 1). Anticipating rational play from player 1, when player 2 selects I, she expects a payoff of 2 if player 1 is cooperative and -2 in the event that player 1 is ordinary. Given her belief q, selecting I will give player 2 an expected payoff of

$$2q - 2(1 - q) = 4q - 2.$$

Player 2 has the incentive to choose I if and only if $4q - 2 \geq 0$, which simplifies to $q \geq 1/2$. Player 2 is indifferent between I and N if $q = 1/2$.

With an understanding of the relation between player 2's belief and optimal strategy, we can return to evaluate player 1's action at the top-left node. Can the ordinary type of player 1 select N in a PBE? In fact, no. Here is why. If the O type chooses N and the C type chooses I′, then, in the event that player 2's information set is reached, she can deduce that player 1's type is C. In mathematical terms, Bayes' rule implies that $q = 1$. Thus, player 2 must select I. But anticipating that player 2 will invest, the ordinary player 1 will want to choose I, contradicting what we presumed.

Can type O choose I in a PBE? Again, no. If O chooses I, then—because C chooses I as well—player 2 learns nothing about player 1's type upon reaching her information set. Mathematically, here Bayes' rule implies that $q = 1/4$. Therefore, player 2 must choose N, making type O's choice of I a bad one indeed.

You might think that, having checked both actions I and N for the ordinary player 1, the analysis is complete. In neither case did we find a PBE, so you might be inclined to conclude that no PBE exists in this game. *Finish the book already,* you say. Not so fast. In fact, so far we have examined only *pure* strategies. What if the ordinary player 1 selects a *mixed* strategy? What if player 2 chooses a mixed strategy as well? In Figure 29.3, r designates the probability that the O type of player 1 chooses I; s designates the probability that player 2 selects I.

In order for the O type to be indifferent between I and N, it must be that his expected payoff from I is exactly zero. This occurs only when player 2 mixes between I and N, putting probability 1/4 on I. Thus, it must be that $s = 1/4$. Further, player 2 is willing to randomize only if her expected payoff from I is exactly zero (because she obtains zero by choosing N, regardless of player 1's type). For this to be the case, player 2's belief must put equal weight on the two types of player 1; that is, $q = 1/2$. However, player 2's belief must be *justified* in that it is consistent with Bayes' rule and player 1's strategy. Bayes' rule requires

$$q = \frac{\frac{1}{4}}{\frac{1}{4} + \frac{3}{4}r}.$$

The numerator is the probability that player 2's lower decision node is reached, whereas the denominator is the probability that her information set is reached (the sum of the probabilities of reaching the two nodes in her information set). Plugging in $1/2$ for q and solving for r yields $r = 1/3$. That is, if the O type picks I with probability $1/3$, then, upon reaching her information set, player 2 will believe that player 1's type is C with probability $1/2$.

In summary, there is a unique PBE in this game of incomplete information. Player 1 uses the strategy specifying S, I$'$, and B$'$ with certainty and action I with probability $1/3$. Player 2 holds belief $q = 1/2$ and she chooses I with probability $1/4$. Player 2 and the ordinary player 1 get payoffs of zero, but the cooperative player 1 obtains a *positive* payoff of $5/4$. Thus, with some probability, the players invest in the asset and it creates value. Interestingly, in equilibrium, the behavior of the ordinary type of player 1 causes investment to serve as a signal that player 1 is the cooperative type. On observing that player 1 has invested, player 2 upgrades her assessment that player 1 is the cooperative type from a probability of $1/4$ to a probability of $1/2$. When the ordinary player 1 invests, he is pretending to be cooperative. In other words, the presence of the cooperative type helps the ordinary type establish a reputation for cooperative behavior.

GUIDED EXERCISE

Problem: Consider the following bargaining game with incomplete information: Player 1 owns a television that he does not use; thus, the value of the television to him is zero. Player 2 would like to have the television; his value of owning the television is v, which is private information to player 2. (Assume $v > 0$.) These players engage in a two-period, alternating-offer bargaining game to establish whether player 1 will trade the television to player 2 for a price. The players discount the second period according to the discount factor δ. In the first period, player 1 makes a price offer and player 2 responds with "yes" or "no." If player 2 rejects player 1's offer, then player 2 makes the offer in the second period. Agreement in the first period yields the payoff p_1 to player 1 and $v - p_1$ to player 2. Agreement in the second period yields δp_2 to player 1 and $\delta(v - p_2)$ to player 2. (Note the discounting.)

(a) Suppose that nature selects $v = 2$ with probability r and $v = 1$ with probability $1 - r$. Player 2 observes nature's choice; player 1 knows only these probabilities. Compute the perfect Bayesian equilibrium of this game and note how it depends on δ and r. (Assume that, if a player is indifferent between accepting and rejecting an offer, then he or she accepts it.)

(b) Now suppose that nature selects v according to a uniform distribution on the interval [0, 1]. That is, for any number a between 0 and 1, the probability that nature selects $v < a$ is equal to a. Compute the perfect Bayesian equilibrium of this game and note how it depends on δ. (Hint: Use the knowledge that you have gained from part (a) and do not try to draw the extensive form—it is a bit messy.)

Solution:

(a) The perfect Bayesian equilibrium is easy to calculate in this example because there are no difficult updated beliefs to analyze. In fact, we can basically solve the game using backward induction. It is easy to see that the outcome of negotiation in the second period is trade at price $p_2 = 0$, because it is the lowest price that player 1 will accept. Thus, if the game reaches the second period, then player 2's payoff will be δv and player 1 will get nothing. An implication is that, in the first period, player 2 will accept any price p_1 that satisfies

$$v - p_1 \geq \delta v,$$

which simplifies to $p_1 \leq (1 - \delta)v$. Note that the high-type buyer is willing to accept a higher price than is the low-type buyer. Because player 1 does not observe v, this leaves player 1 with a simple choice: (1) offer a price that only the high-type buyer will accept, or (2) offer a lower price that both types will accept. With alternative (1), the best price to offer is $2(1-\delta)$ because it is the highest price that the high-type buyer will accept. At this price, player 1's expected payoff would be $2(1-\delta)r$. With alternative (2), the best price to offer is $1 - \delta$, which yields to player 1 the payoff $1-\delta$. Alternative (1) yields a higher expected payoff than does alternative (2) if $r \geq 1/2$. Thus, in the case of $r \geq 1/2$, player 1 offers $p_1 = 2(1-\delta)$; in the case of $r < 1/2$, player 1 offers $p_1 = 1 - \delta$.

Player 1's offer is decreasing in the discount factor; that is, a more patient player 2 implies a lower p_1. For low values of r, where player 1 is reasonably sure that player 2 is the low type with $v = 1$, player 1 makes the "safe" price offer $1 - \delta$ and agreement is reached without delay. On the other hand, for high values of r, where player 1 thinks it more likely that player 2 is the high type, player 1 offers a higher price and agreement is delayed (and inefficiency results) with probability $1 - r$.

(b) Behavior in the second period is exactly as described in part (a). Also, as before, in the first period type v of player 2 accepts any price $p_1 \leq (1-\delta)v$. Rearranging this inequality yields

$$v \geq \frac{p_1}{1-\delta},$$

which is to say that all types above $p_1/(1-\delta)$ will accept the offer p_1. Because v is uniformly distributed on $[0, 1]$, the probability that v exceeds $p_1/(1-\delta)$ is

$$1 - \frac{p_1}{1-\delta}.$$

Thus, player 1's expected payoff of offering price p_1 is

$$p_1\left(1 - \frac{p_1}{1-\delta}\right).$$

To find player 1's optimal offer, we take the derivative of this expression, set it equal to 0, and solve for p_1. This yields $p^* = (1-\delta)/2$.

As in part (a), player 1's offer is decreasing in the discount factor. Also, the equilibrium is inefficient because agreement is delayed in the event that v is less than $1/2$.

EXERCISES

1. Consider the job-market signaling model analyzed in this chapter. Would an education be a useful signal of the worker's type if the types had the same education cost? Explain your answer.

2. Compute the PBE of the job-market signaling model under the assumption that the worker is a high type with probability $1/2$ and a low type with probability $1/2$ (rather than probabilities of $1/3$ and $2/3$ as assumed in the text).

3. Consider the extensive-form game of incomplete information pictured on the next page. There is a firm and a worker. In this game, nature first chooses the "type" of the firm (player 1). With probability p, the firm is of high quality (H) and, with probability $1 - p$, the firm is of low quality (L). The firm chooses either to offer a job to the worker (O) or not to offer a job (N). If no job is offered, the game ends and both parties receive 0. If the firm offers a job, then

the worker either accepts (A) or rejects (R) the offer. The worker's effort on the job brings the firm a profit of 2. If the worker rejects an offer of employment, then the firm gets a payoff of −1 (associated with being jilted). Rejecting an offer yields a payoff of 0 to the worker. Accepting an offer yields the worker a payoff of 2 if the firm is of high quality and −1 if the firm is of low quality. The worker does not observe the quality of the firm directly.

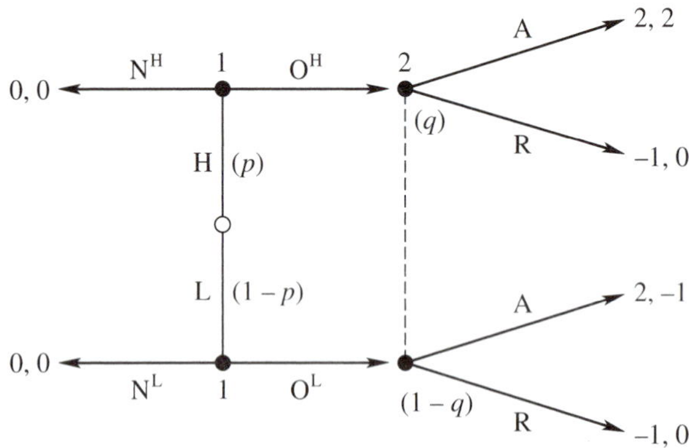

(a) Is there a separating PBE in this game? If so, specify the equilibrium and explain under what conditions it exists. If not, briefly demonstrate why.

(b) Is there a pooling PBE in which both types of firms offer a job? If so, specify the equilibrium and explain under what conditions it exists. If not, briefly demonstrate why.

(c) Is there a pooling PBE in which neither type of firm offers a job? If so, specify the equilibrium and explain under what conditions it exists. If not, briefly demonstrate why.

4. What is the relation between the PBE of the investment–reputation game in this chapter and the Bayesian Nash equilibria of its normal-form representation? You do not need to draw the normal-form representation to determine the answer, but it will help. Explain why you find this relation.

5. Compute the PBE for the investment–reputation model discussed in this chapter, under the assumption that nature chooses the cooperative type with probability p (rather than $1/4$ as assumed in the text). What happens if $p > 1/2$?

6. Recall the bargaining game from part (b) of this chapter's Guided Exercise, where player 1 is interested in selling a television to player 2. Player 2's value of owning the television, v, is privately known to player 2. Player 1 only knows that v is uniformly distributed between 0 and 1. Consider a variation of the game in which player 1 makes the offer in both periods, rather than having alternating offers as in the Guided Exercise. That is, in the first period, player 1 offers a price p_1. If player 2 rejects this offer, then play proceeds to the second period, where player 1 makes another offer, p_2. Assume, as before, that the players discount second-period payoffs by the factor δ.

Calculate the perfect Bayesian equilibrium of this game. Here are some hints to help you with the analysis. In the perfect Bayesian equilibrium, player 2 uses a cutoff rule in the first period. The cutoff rule is characterized by a function $c : [0, 1] \rightarrow [0, 1]$, whereby player 2 accepts an offer p_1 if and only if $v \geq c(p_1)$. Thus, whatever player 1 offers in the first period, if the offer is rejected, then player 1's updated belief about player 2's type is that it is uniformly distributed between 0 and $c(p_1)$. In other words, because player 1 knows that any type $v \geq c(p_1)$ would have accepted p_1, that the offer was rejected tells player 1 that player 2's valuation is between 0 and $c(p_1)$. (This is Bayes' rule in action.)

Begin your analysis by determining player 2's optimal behavior in the second period and then calculating player 1's optimal second-period price offer when facing the types in some interval $[0, x]$. Then try to determine the function c. In this regard, the key insight is that type $c(p_1)$ will be indifferent between accepting p_1 and waiting for the offer in period 2.

7. Suppose that two people (person 1 and person 2) are considering whether to form a partnership firm. Person 2's productivity (type) is unknown to person 1 at the time at which these people must decide whether to create a firm, but person 1 knows that, with probability p, person 2's productivity is high (H) and, with probability $1 - p$, person 2's productivity is low (L). Person 2 knows her own productivity. If these two people do not form a firm, then they each receive a payoff of 0. If they form a firm, then their payoffs are as follows: If person 2's type is H, then each person receives 10. If person 2's type is L, then person 2 receives 5 and person 1 receives -4 (which has to do with person 1 having to work very hard in the future to salvage the firm).

(a) Consider the game in which person 1 chooses between forming the firm (F) or not forming the firm (O). Draw the extensive form of this game (using a move of nature at the beginning to select person 2's type). Note that only one person has a move in this game. What is the Bayesian Nash equilibrium of the game? (It depends on p.)

(b) Now suppose that, before person 1 decides whether to form the firm, person 2 chooses whether or not to give person 1 a gift (such as a dinner). Player 1 observes person 2's choice between G and N (gift or no gift) before selecting F or O. The gift entails a cost of g units of utility for person 2; this cost is subtracted from person 2's payoff designated earlier. The gift, if given, adds w to person 1's utility. If person 2 does not give the gift, then it costs her nothing.

Assume that w and g are positive numbers. Draw the extensive form of this new game.

(c) Under what conditions (values of g and w) does a (separating) perfect Bayesian equilibrium exist in which the low type of person 2 does not give the gift, the high type gives the gift, and person 1 forms the firm if and only if a gift is given? Completely specify such an equilibrium.

(d) Is there a pooling equilibrium in this game? Fully describe it and note how it depends on p.

8. Consider the same setting as that in Exercise 7, except now consider that both players have private information. Player 1's type is either high or low as well. The players' types are independently selected by nature with the probability p. (Thus, p^2 is the probability that both players are the high type, $p(1 - p)$ is the probability that player 1's type is H and player 2's type is L, and so on.) Suppose that, if the firm is not formed, both players receive 0. If the firm is formed, then the payoffs are as follows: If both players have high productivity, then they each receive 10. If both have low productivity, then they each receive 0. If one of the players has high productivity and the other has low productivity, then the high-type player gets -4 and the low-type player gets 5.

(a) Consider the game in which the players simultaneously and independently select Y or O. The firm is formed if they both select Y. Otherwise, the firm is not formed. Note that a strategy for each player specifies what the player should do conditional on being the high type and conditional on being the low type. Demonstrate that (O, O; O, O) is a Bayesian Nash equilibrium in this game, regardless of p.

(b) Under what values of p is (Y, Y; Y, Y) an equilibrium? (In this equilibrium, both types of both players choose Y.)

(c) Now suppose that, before the players choose between Y and N, they have an opportunity to give each other gifts. The players simultaneously and independently choose between giving (G) and not giving (N) gifts, with the cost and benefit of gifts as specified in Exercise 7. Under what conditions is there a perfect Bayesian equilibrium in which the high types give gifts and the firm is formed if and only if both players give gifts? Specify the equilibrium as best you can.

9. A manager and a worker interact as follows. The manager would like the worker to exert some effort on a project. Let e denote the worker's effort. Each unit of effort produces a unit of revenue for the firm; that is, revenue is e. The worker bears a cost of effort given by αe^2, where α is a positive constant. The manager can pay the worker some money, which enters their payoffs in an additive way. Thus, if the worker picks effort level e and the manager pays the worker x, then the manager's payoff is $e - x$ and the worker's payoff is $x - \alpha e^2$. Assume that effort is verifiable and externally enforceable, meaning that the parties can commit to a payment and effort level.

Imagine that the parties interact as follows: First, the manager makes a contract offer to the worker. The contract is a specification of effort $\hat{e}$ and a wage $\hat{x}$. Then the worker accepts or rejects the offer. If she rejects, then the game ends and both parties obtain payoffs of 0. If she accepts, then the contract is enforced (effort $\hat{e}$ is taken and $\hat{x}$ is paid). Because the contract is externally enforced, you do not have to concern yourself with the worker's incentive to exert effort.

(a) Solve the game under the assumption that α is common knowledge between the worker and the manager. That is, find the manager's optimal contract offer. Is the outcome efficient? (Does the contract maximize the sum of the players' payoffs?) Note how the equilibrium contract depends on α.

(b) Let $\underline{e}$ and $\underline{x}$ denote the equilibrium contract in the case in which $\alpha = 1/8$ and let $\overline{e}$ and $\overline{x}$ denote the equilibrium contract in the case in which $\alpha = 3/8$. Calculate these values. Let us call $\alpha = 3/8$ the *high-type* worker and $\alpha = 1/8$ the *low-type* worker.

(c) Suppose that α is private information to the worker. The manager knows only that $\alpha = 1/8$ with probability $1/2$ and $\alpha = 3/8$ with probability $1/2$. Suppose that the manager offers the worker a choice between contracts $(\underline{e}, \underline{x})$ and $(\overline{e}, \overline{x})$—that is, the manager offers a *menu* of contracts—in the hope that the high type will choose $(\overline{e}, \overline{x})$ and the low type will choose $(\underline{e}, \underline{x})$. Will each type pick the contract intended for him? If not, what will happen and why?

(d) Suppose that the manager offers a menu of two contracts (e_L, x_L) and (e_H, x_H), where he hopes that the first contract will be accepted by the low type and the second will be accepted by the high type. Under what conditions will each type accept the contract intended for him? Your answer should consist of four inequalities. The first two—called the *incentive compatibility conditions*—reflect that each type prefers the contract intended for him rather than the one intended for the other type. The last two inequalities—called the *participation constraints*—reflect that, by selecting the contract meant for him, each type gets a payoff that is at least 0.

(e) Compute the manager's optimal menu (e_L, x_L) and (e_H, x_H). Note that the manager wants to maximize his expected payoff

$$(1/2)[e_H - x_H] + (1/2)[e_L - x_L],$$

subject to the worker's incentive compatibility conditions and participation constraints. (Hint: Only two of the inequalities of part (d) actually *bind*—that is, hold as equalities. The ones that bind are the high type's participation constraint and the low type's incentive compatibility condition. Using these two equations to substitute for x_L and x_H, write the manager's payoff as a function of e_L and e_H. Solve the problem by taking the partial derivatives with respect to these two variables and setting the derivatives equal to 0.)

(f) Comment on the relation between the solution to part (e) and the efficient contracts identified in part (b). How does the optimal menu under asymmetric information distort away from efficiency?

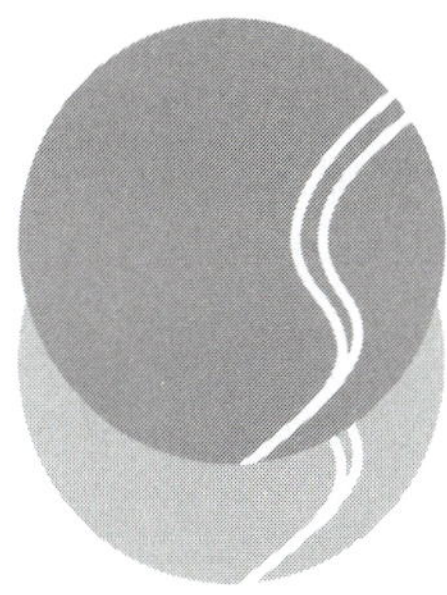

Appendices

These appendices contain some additional technical material that you might find handy as you read the textbook. Appendix A reviews some basic mathematical concepts and Appendix B elaborates on the definitions of rationalizability and Nash equilibrium.

REVIEW OF MATHEMATICS A

In this appendix, I define and discuss a few basic mathematical concepts used in the book. The coverage here is in no way general; it is not intended to stand alone as an introduction to mathematical analysis. Rather, it is designed to help you refresh your memory and solidify your understanding of the concepts by reviewing some elementary definitions and examples. For a serious treatment of mathematical analysis, you should consult calculus and analysis textbooks.[1]

SETS

A *set* is any collection of distinct items. For example, the "days of the week" is a set comprising the following seven things: Monday, Tuesday, Wednesday, Thursday, Friday, Saturday, Sunday. Each member of a set is called an *element* of the set; the term *point* has the same meaning. One way of mathematically describing a set is to list its elements, separated by commas and surrounded by brackets, as such:

$$\{\text{M, Tu, W, Th, F, Sa, Su}\}.$$

Note that, for convenience, I use M to stand for Monday, Tu for Tuesday, and so on. Abbreviations such as these are often helpful.

Another example of a set is "integers from 1 to 100," which can be represented as

$$\{1, 2, 3, \ldots, 100\}.$$

Note how an ellipsis (...) is used to signify the continuation of a pattern. The set of "positive odd integers" has no upper bound, so we represent it as

$$\{1, 3, 5, \ldots\}.$$

[1]For analysis, I recommend K. Binmore, *Mathematical Analysis,* 2nd ed. (New York: Cambridge University Press, 1982). For probability theory, you could try S. Ross, *Introduction to Probability Models,* 6th ed. (San Diego: Academic Press, 1997).

Sometimes sets are most easily defined by mathematical conditions, such as "the set of numbers that are the squares of positive integers":

$$\{x \mid x = n^2,\ \text{for}\ n = 1, 2, 3, \ldots\}.$$

That is, x is an element of this set if there is a positive integer n such that $x = n^2$. *Finite* sets contain only a finite number of elements; "the days of the week" is an example of a finite set. *Infinite* sets contain an infinite number of elements; the set of positive odd integers is an infinite set.

An *interval* is a set of numbers between two *endpoints*. For example, the set of numbers between 0 and 1, including these two numbers, is an interval. It can be written as

$$\{x \mid 0 \leq x \leq 1\}.$$

It is also standard to represent intervals by enclosing the endpoints in parentheses or square brackets. For example, $[a, b]$ denotes the set of points between a and b, including the endpoints—that is, a and b are included in this set; (a, b) describes the same interval except without the endpoints included. Finally, $(a, b]$ and $[a, b)$ denote the intervals with only one of the endpoints included (b in the first case, a in the second).

Often capital letters, such as X and S are used to name sets. For example, we could have

$$X \equiv \{\text{M, Tu, W, Th, F, Sa, Su}\},$$

where "$\equiv$" denotes "is defined as." Then we can write X instead of {M, Tu, W, Th, F, Sa, Su} in our calculations. The symbol "$\in$" stands for "is an element of." Thus, we write $x \in X$ to say that x is an element of X. In general expressions, sometimes lowercase letters refer to elements and uppercase letters denote sets. A letter that represents individual elements of a set is called a *variable*. For example, if X is the set of days of the week and I write $x \in X$, then I mean that the variable x can stand for any of the individual days of the week. I could "plug in" Tu for x, or M for x, and so forth.

The symbol "$\subset$" means "is a subset of." For two sets X and Y, $X \subset Y$ means that every element of X is also an element of Y; that is, $x \in X$ implies $x \in Y$. For example, if we define

$$X \equiv \{1, 3, 5, \ldots\},$$

and

$$Y \equiv \{1, 2, 3, \ldots\},$$

then $X \subset Y$. The *union* of two sets consists of the elements that are in at least one of the sets. The *intersection* of two sets consists of the elements that are in *both* of the sets. Formally, for sets X and Y, the union is

$$X \cup Y \equiv \{x \mid x \in X \text{ or } x \in Y \text{ or both}\}$$

and the intersection is

$$X \cap Y \equiv \{x \mid x \in X \text{ and } x \in Y\}.$$

The symbol "∅" represents the set with no elements; we call this the empty set. The symbol "**R**" represents the set of real numbers (all of the usual numbers that you can think of). We sometimes use the term *space* for a large set that contains all of the elements relevant for a particular mathematical problem, in which case all of the sets that we define are subsets of the space.

A *vector* is what you get when you put elements of different sets together in a specific order. For example, you might want to combine a day of the week with an hour of the morning, such as Tuesday and eight o'clock. Mathematically, this vector would be written (Tu, 8). Other vectors of this type are (M, 9) and (Sa, 11). Note that the first component of these vectors is an element of the set $D \equiv \{\text{M, Tu, W, Th, F, Sa, Su}\}$, whereas the second component is an element of $T \equiv \{1, 2, \ldots, 11\}$. The vectors themselves are elements of a larger set called the *Cartesian product* of the individual sets. The product set is written $D \times T$ and it comprises all of the vectors of the form (d, t) where $d \in D$ and $t \in T$. That is,

$$D \times T = \{(\text{M}, 1), (\text{M}, 2), \ldots, (\text{M}, 11), (\text{Tu}, 1), (\text{Tu}, 2), \ldots, (\text{Su}, 10), (\text{Su}, 11)\}.$$

Note that $D \times T$ contains 77 elements (7 elements in D times 11 elements in T). For a simpler example of a product set, note that the product of {A, B} and {1, 2} is

$$\{(\text{A}, 1), (\text{A}, 2), (\text{B}, 1), (\text{B}, 2)\}.$$

In general, if we have n sets $X_1, X_2, \ldots, X_n$, then the product set X is given by

$$X \equiv \{(x_1, x_2, \ldots, x_n) \mid x_1 \in X_1, x_2 \in X_2, \ldots, x_n \in X_n\}.$$

FUNCTIONS AND CALCULUS

A *function* describes a way of associating the elements of one set X with elements of another set Y. For *each* point $x \in X$, the function names a *single*

FIGURE A.1
The graph of $f(x) = x^2$.

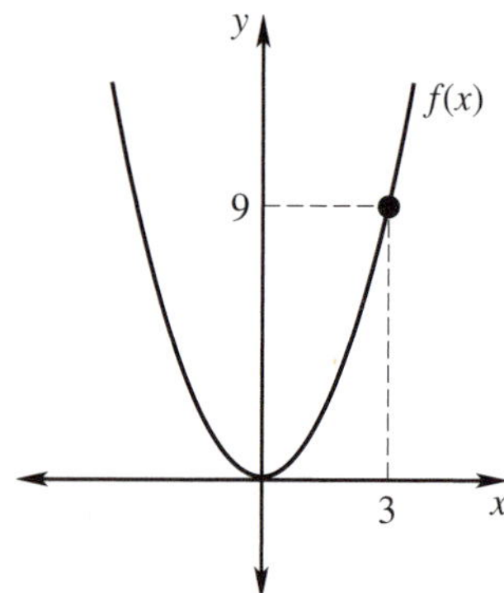

point $y \in Y$; the point y associated with x is denoted by $y = f(x)$. For example, suppose $X = \{a, b\}$ and $Y = \{3, 4\}$. A function f might be defined by $f(a) = 4$ and $f(b) = 3$. Another function g might be defined by $g(a) = 3$ and $g(b) = 3$. (A single y value can be associated with more than one x value.) To make clear between which sets a function relates, we usually write the expression $f : X \rightarrow Y$, which means "f maps X into Y." Here X is called the *domain* and Y is called the *codomain*. The set of points that f can return, which is defined by

$$\{f(x) \mid x \in X\},$$

is called the *range* of f. Many functions of interest map the real numbers into the same set. For example, we could define the function $f : \mathbf{R} \rightarrow \mathbf{R}$ by $f(x) = x^2$. In this case, if $x = 3$ then $y = f(3) = 9$, if $x = 5$ then $y = 25$, and so forth.

Real functions (those with real domains and ranges) can be graphed in the $\mathbf{R} \times \mathbf{R}$ product space—this is the usual x/y plane. The graph plots the set of vectors of the form (x, y), where $x \in \mathbf{R}$ and $y = f(x)$ is the *value at* x. The graph of $f(x) = x^2$ is depicted in Figure A.1. To determine the graph of a function, start by plugging in some numbers for x and plotting the resulting (x, y) vectors. Also find the x- and y-intercepts, which are the vectors that correspond to $x = 0$ in the first case and $y = 0$ in the second case. Another useful exercise is to check what happens to y as x converges to positive and negative infinity. For example, as x gets either very high or very low (large negative), x^2 becomes a very large positive number. Thus, the graph of $f(x) = x^2$ points upward on the right and the left.

The *slope* of a function measures the degree to which the value of the function increases or decreases as x goes up. The slope may be different at various points in the function's domain, and it depends on the amount that x

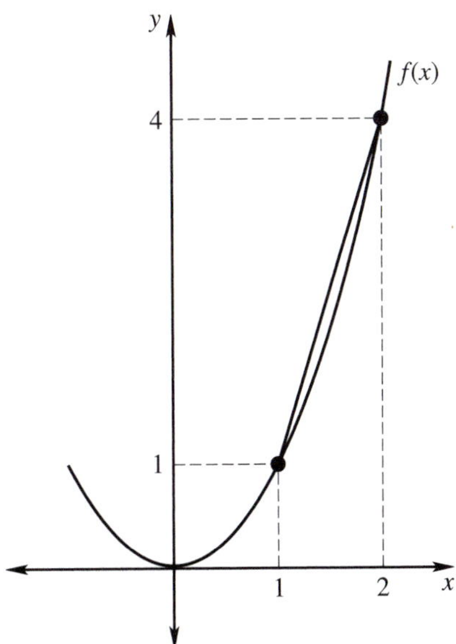

FIGURE A.2
Magnified graph of $f(x) = x^2$.

is changed. Technically, the slope equals the *rise* (the difference in y) divided by the *run* (the difference in x). For example, once again look at the function $f(x) = x^2$, whose graph near $x = 1$ is magnified in Figure A.2. To find the slope of f between the points $x = 1$ and $x = 2$, we simply compute the rise and run between these two points. The rise is

$$f(2) - f(1) = 2^2 - 1^2 = 4 - 1 = 3,$$

whereas the run is $2 - 1 = 1$. Thus, the slope is $3/1 = 3$. In Figure A.2, the slope corresponds to the straight line segment drawn between vectors $(1, 1)$ and $(2, 4)$.

The *derivative* of a function is defined as the limit of the slope (where the slope goes) as the change in x (that is, the run) becomes small. In other words, the derivative is the instantaneous slope at a point x. In formal terms, the derivative of f at x is defined as

$$\lim_{\varepsilon \to 0} \frac{f(x+\varepsilon) - f(x)}{\varepsilon}.$$

The expression "$\varepsilon \to 0$" means that the fraction is evaluated for a number ε that is taken to be very close to zero. Note that, in this definition, the run is ε because this is the distance from the starting point x that we are measuring. We denote the derivative of f at x by $f'(x)$; alternatively, where $y = f(x)$, we also write the derivative as dy/dx.

FIGURE A.3
Functions of the form $f(x) = ax^2 + bx + c$.

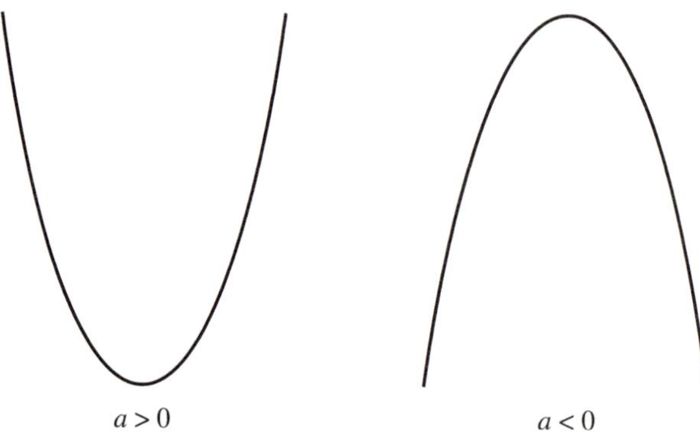

Let us compute the derivative of the function $f(x) = x^2$ at an arbitrary point x. In accord with the definition in the preceding paragraph, this is the limit of

$$\frac{(x+\varepsilon)^2 - x^2}{\varepsilon}$$

as ε gets close to zero. Expanding the term in parentheses, we get

$$\frac{x^2 + 2\varepsilon x + \varepsilon^2 - x^2}{\varepsilon} = 2x + \varepsilon.$$

As ε approaches zero, clearly this value converges to $2x$. Thus, $f'(x) = 2x$. You can check that multiplying a function by a fixed number a implies that the derivative also is multiplied by a. Thus, if $f(x) = ax^2$, then $f'(x) = 2ax$. You can also compute that, for any constant b and the function $f(x) = bx$, the derivative is $f'(x) = b$. Finally, if $f(x) = c$ for some constant c, then $f'(x) = 0$. Putting these things together, it is no surprise that, for any function of the form $f(x) = ax^2 + bx + c$, we have $f'(x) = 2ax + b$. Such a function f is called a *quadratic function*.[2]

The graph of any quadratic function is a parabola, such as that pictured in Figures A.1 and A.2. If the constant a is positive, then the ends of the parabola point upward; if the constant is negative, then the ends point downward. This is shown in Figure A.3. In the case in which $a < 0$, the maximum value attained by f occurs where the slope is equal to zero. Thus, to find the value of x that maximizes $f(x)$, simply solve the equation $f'(x^*) = 0$ for x^*. In the case in which $a > 0$, the function is minimized where the slope equals zero. In general, for any function with a well-defined derivative, you can find the maxima and minima (at least the "local" ones) by calculating where the

[2]In general, for a function of the form $f(x) = ax^k$, we have $f'(x) = akx^{k-1}$.

derivative equals zero. For most of the exercises in this textbook that require calculus, computing where the derivative equals zero is all you have to do. To reassure yourself about whether you are finding a maximum or a minimum, it is helpful to sketch the graph of the function and check whether it looks like one of the cases shown in Figure A.3.[3]

Some functions are defined on several variables. For example, the function $f(x, y) = xy + 2y$ associates a number $z = f(x, y)$ with any two numbers x and y. In this case, we write $f : \mathbf{R} \times \mathbf{R} \rightarrow \mathbf{R}$, because with this function we plug in two real numbers and get out one real number. With *multivariate* functions such as this, we sometimes need to evaluate slopes on just one of the dimensions of the domain (x or y). We can do so by simply treating the other variables as fixed constants. For example, if $f(x, y) = xy + 2y$, then the derivative of f with respect to x is defined as the change of z divided by the change of x, holding y fixed. This defines the *partial derivative* of f with respect to x, which is the standard derivative calculation with y held constant. We write $\partial z/\partial x$ for the partial derivative of f with respect to x. With the function $z = f(x, y) = xy + 2y$, the partial derivatives are $\partial z/\partial x = y$ and $\partial z/\partial y = x + 2$. In general, suppose x is an n-dimensional vector, so that $x = (x_1, x_2, \ldots, x_n)$. If $y = f(x)$, then we can compute the n partial derivatives, $\partial y/\partial x_1, \partial y/\partial x_2, \ldots, \partial y/\partial x_n$.

PROBABILITY

A coin toss produces a random event, where the outcome can be either "heads" or "tails." For a "fair" coin, we assess that these outcomes are equally likely. Sometimes we use the phrase "fifty-fifty" to describe the prospects. In more formal terms, we would say that the *probability* of heads is $1/2$ and the probability of tails is $1/2$. Using the concept of probability, we have an organized and logical way of considering random events.

There are all sorts of situations that have random components. The weather is random, machines fail randomly, and sometimes people behave randomly. To study the random component, it is useful to consider a *state space* that describes all of the possible resolutions of the random forces. For example, if you are interested in a particular horse race, the state space may consist of all of the different orderings of the horses (ways in which they could finish in the race). If you are a poker player, the state space may comprise the different ways in which the cards can be dealt. If you like flipping coins, the state space

[3]You can also use second-derivative conditions. If the second derivative (the derivative of the derivative) is negative, then the function roughly has a downward parabola shape, meaning that $f'(x^*) = 0$ identifies a maximum. If the second derivative is positive, then $f'(x^*) = 0$ identifies a minimum.

is {heads, tails}. Each of the states in the state space is assumed to describe an outcome that is mutually exclusive of what other states describe. Further, the states collectively exhaust all of the possibilities. In other words, one and only one state actually occurs.

To describe the relative likelihood of the different individual states, we can posit a *probability distribution* over the state space. For example, suppose the state space is $\{A, B, C\}$. Think of this as the possible outcomes of a horse race, where all you care about is which horse wins (horse A, horse B, or horse C). A probability distribution for this state space implies a function $p : \{A, B, C\} \to [0, 1]$ from which we get three numbers, $p(A)$, $p(B)$, and $p(C)$. The number $p(A)$ is the "probability that A occurs," $p(B)$ is the probability that B occurs, and $p(C)$ is the probability that C occurs. Each of these numbers is assumed to be between 0 and 1 (as you can see from the codomain designation of p) and the numbers sum to 1. For example, if horse A is twice as likely to win as are horses B and C individually and if B and C are equally likely to win, then $p(A) = 1/2$, $p(B) = 1/4$, and $p(C) = 1/4$.

In general, for a finite state space X, a probability distribution is represented by a function $p : X \to [0, 1]$ such that $\sum_{x \in X} p(x) = 1$. Note that "$\sum_{x \in X}$" means adding the numbers that you get by plugging in all of the different states, where a generic state is represented by x. A *random variable* is a variable that describes the actual state that occurs. For example, using the horse-race state space, we could let y denote the winning horse. Then y is called a random variable; it equals A with probability $p(A)$, B with probability $p(B)$, and C with probability $p(C)$. More generally, a random variable is any function of the state space.

In many cases, a random variable gives a real number—your dollar winnings at the racetrack, for instance. We can then speak of the *expected value* of the random variable, which is its average value weighted by the probability distribution. For example, suppose that you will get a payment of $60 if horse A wins the race and $0 if horse A does not win. Suppose you believe that A will win the race with probability 1/2. Then your expected payment is 30, which is the average of 60 and 0. If A will win only with probability 1/4, then your expected payment is 15. This is because there is a one-quarter chance that you will get $60 (in other words, "one-quarter of the time"), so the expected value is 1/4 times the payment of $60. In general, expected value is the sum of the possible payment amounts, each weighted by its probability. With the use of E to denote expectation, the expected value of random variable y is defined as

$$E[y] = \sum_{x \in X} y(x)p(x).$$

In this expression, $y(x)$ is the value of the random variable (your payment) in state x. As an example, suppose you get \$60 in state A, \$20 in state B, and \$0 in state C. Further suppose that $p(A) = 1/2$, $p(B) = 1/4$, and $p(C) = 1/4$, as before. Then your expected payment is

$$(1/2)(60) + (1/4)(20) + (1/4)(0) = 35.$$

Probability distributions attribute probabilities not only to single states but also to sets of states. In the horse-race example, imagine that horses A and B are from Kentucky, whereas horse C is from Tennessee. If you have a bet with your friend regarding whether a horse from Kentucky wins, then you care about the set $\{A, B\}$. Such a set of states is called an *event*. In formal terms, an event is any subset of states—including a singleton set containing only one state. Because all states are mutually exclusive (only one occurs), the probability of an event is just the sum of the probabilities of the states that compose the event. For instance, the probability that a horse from Kentucky wins is $p(A) + p(B)$. In general, if the state space X is finite, an event is any set $D \subset X$. The probability of event D is defined as

$$\text{Prob}[D] \equiv \sum_{x \in D} p(x).$$

Note that the probability of the entire state space is 1; that is, $\text{Prob}[X] = 1$.

There are two ways of thinking about probability. The *frequentist* way associates probability numbers with an objective test: the fraction of times that a state will occur if the underlying situation were repeated over and over again. For example, if you flipped a coin 1,000 times and discovered that it came up heads 504 times and tails 496 times, then you would say that the probabilities of heads and tails are about $1/2$ and $1/2$. The second way of thinking about probabilities is the *Bayesian* approach, where the probability numbers are not derived from repetitive testing but are subjectively assessed. I can look at a coin and surmise that heads and tails are equally likely results of a coin flip. I can size up a horse at the racetrack and conclude that he will win the race with probability $1/3$, although it is not even possible to verify this number through testing (which would require repeating the race an arbitrary number of times under identical conditions). In fact, my probability assessment may be *wrong* in the sense that I ignored some information that would have led me to a different assessment—the horse may have been scratched from the race, but I did not realize it. However, my assessment is still valid for me because it is the basis of my decision making.

Game theory utilizes both the frequentist and the Bayesian interpretations of probability to some degree. However, game theory leans heavily toward the

Bayesian approach, especially regarding players' beliefs about each other's actions. Probability theory is essential for game-theoretic reasoning because it is the foundation for dealing with uncertainty in a logically consistent and practical way.

For an infinite state space, probability distributions are more complicated. In fact, it may be that *no* state has positive probability by itself. Although the infinite setting requires a general, more sophisticated theory of probability measure, you can often deal with infinite state spaces without delving far into the mathematics. In the most common case of an infinite state space, the state is a real number. For example, the state x might be a number between a and b, where $a < b$, in which case the state space is $[a, b]$. A probability distribution is then represented by a *cumulative probability function* $F : [a, b] \rightarrow [0, 1]$, which gives the probability that x is less than or equal to any given number. That is, $F(z)$ is the probability of the *event* that $x \leq z$. It is proper, therefore, to write

$$F(z) = \text{Prob}[x \leq z].$$

The cumulative probability function satisfies two main conditions. First, because the state is between a and b, the probability that $x \leq b$ must be 1; that is, $F(b) = 1$. Second, F is weakly increasing, meaning that $F(z) \leq F(z')$ for any numbers z and z' such that $z < z'$. The greater is z, the higher is the probability that $x \leq z$.

The *uniform distribution* is a particularly simple and useful one. Roughly, in a uniform distribution, all numbers between a and b are equally likely. Because there are infinite numbers between a and b, this means that each individual number has probability 0. However, some standard events can be easily assessed. For example, the probability that $x < (a + b)/2$ is $1/2$, because $(a+b)/2$ is the midpoint between a and b. In other words, $(a+b)/2$ is the point that divides the state space in half; the probability that x lies below $(a + b)/2$ is the same as the probability that x lies above $(a + b)/2$. More generally, if we look at any point between a and b, the probability that x lies below this point is equal to the proportion that this point represents on the way to b from a. That is, the cumulative probability function for the uniform distribution on $[a, b]$ is given by

$$F(z) = \text{Prob}[x \leq z] = \frac{z - a}{b - a},$$

where $z \in [a, b]$. The expected value of a uniformly distributed random variable is simply the midpoint of the distribution, $(a + b)/2$.

Sometimes people start with a state space and probability distribution and then *update* their assessment on the basis of new information. When a probability distribution arises because of new information, it is called *conditional probability*. To illustrate the concept, I continue with the horse-race example. Suppose, as before, that there are three horses in the race: A, B, and C. The first two horses are from Kentucky, whereas C is from Tennessee. The probability distribution is given by $p(A) = 1/2$, $p(B) = 1/4$, and $p(C) = 1/4$. If I ask you about the chance that horse A will win, you will tell me that A wins with probability $1/2$. However, what if I tell you that the race has been run and that I heard from a reliable source that a horse from Kentucky won. (That is, C did not win.) On the basis of this new information, what would you assess the probabilities of A and B to be?

Your updated assessment should be A with probability $2/3$ and B with probability $1/3$. Here is why. Initially, you believed that A is twice as likely to win as is B. Knowing that C lost does not contradict this statement; it merely implies that the probability of C has been reduced to 0. Your updated probability distribution must assign probability 1 to the set of all states. Thus, because the probability of C has been reduced to 0, the probabilities of A and B have to be raised so that they sum to 1. Maintaining that A is twice as likely as is B, you conclude that A and B must now have probabilities of $2/3$ and $1/3$, respectively. These are the only two numbers that sum to 1 and have a ratio of 2.

In general, conditional probability can be figured by reducing to 0 the probabilities of states that have been ruled out and scaling up the probabilities of states that are left. The "scaling up" operation is done by using the same scaling factor for all states and it makes the sum of the probabilities equal to 1. In more technical language, take a state space X, a probability distribution p, and two events $K \subset X$ and $L \subset X$. Suppose you learn that the state is definitely in the set L and you want to know the updated probability that the state is in K. Because the state cannot be outside of L, the only way in which the state can be in K is if it is an element of $K \cap L$. To maintain the proportions between the states in L, you need to scale up the probabilities by the factor Prob[L]. Thus, the probability of event K conditional on event L is defined as

$$\text{Prob}[K \mid L] \equiv \frac{\text{Prob}[K \cap L]}{\text{Prob}[L]}.$$

Note that this expression is undefined if $\text{Prob}[L] = 0$. That is, if you initially believe that L cannot occur and then I ask you about the probability of K given that L has occurred, your initial assessment is of little guidance in calculating the conditional probability. In this case, the conditional probability of K can be arbitrarily defined, as long as $K \cap L \neq \emptyset$.

Rearranging the definition of conditional probability yields the following expression:

$$\text{Prob}[K \cap L] = \text{Prob}[K \mid L]\text{Prob}[L].$$

In words, the probability that $x \in K \cap L$ equals the probability that $x \in L$ times the probability that $x \in K$ conditional on $x \in L$. Reversing the roles of K and L, we have

$$\text{Prob}[L \cap K] = \text{Prob}[L \mid K]\text{Prob}[K].$$

Note that the set $K \cap L$ is exactly the same as the set $L \cap K$. Thus, the left sides of the two preceding equations are the same. Equating the right sides and rearranging yields

$$\text{Prob}[K \mid L] = \frac{\text{Prob}[L \mid K]\text{Prob}[K]}{\text{Prob}[L]}.$$

This equation is known as *Bayes' rule*.[4] It is very useful for applications of conditional probability. Note that, as with the definition of conditional probability, Bayes' rule does not constrain the assessment $\text{Prob}[K \mid L]$ if the probability of L is zero.

---

[4]Bayes' rule is named for Thomas Bayes, who reported the equation in "An Essay towards Solving a Problem in the Doctrine of Chances," *Philosophical Transactions of the Royal Society of London* 53, 1764.

THE MATHEMATICS OF RATIONALIZABILITY AND EXISTENCE OF NASH EQUILIBRIUM

B

This appendix concerns two of the components of rationalizability: the relation between best response and dominance and the procedure by which dominated strategies are iteratively removed. I first sketch some of the analysis not provided in Chapter 6 on the relation between dominance and best response. Then I develop the formal mathematical construction of the rationalizable set of strategies. Both of these topics are technical and are not essential material for those interested only in concepts and applications.

DOMINANCE, BEST RESPONSE, AND CORRELATED CONJECTURES

To understand the formal relation between dominance and best response, you have to begin with the concept of correlated conjectures. Remember that a belief, or conjecture, of player i is a probability distribution over the strategies played by the other players. In two-player games, this amounts to a probability distribution over the strategy adopted by player j (player i's opponent). In games with more than two players, though, player i's belief is more complicated. It is a probability distribution over the strategy combinations (profiles) of player i's opponents. For example, consider a three-player game in which player 1 chooses between strategies A and B, player 2 chooses between M and N, and player 3 chooses between X and Y. The belief of player 1 represents his expectations about both player 2's *and* player 3's strategies. That is, player 1's conjecture is an element of ΔS_{-1}. The belief is a probability distribution over

$$\{\text{M}, \text{N}\} \times \{\text{X}, \text{Y}\} = \{(\text{M}, \text{X}), (\text{M}, \text{Y}), (\text{N}, \text{X}), (\text{N}, \text{Y})\}.$$

Let us explore the possible beliefs.

Suppose that player 1 thinks that with probability $1/2$ player 2 will select M, that with probability $1/2$ player 3 will choose X, and that his opponents' actions are independent. The last property implies that the probability of any profile of the opponents' strategies is the product of the individual

FIGURE B.1 Possible beliefs of player 1.

(a)

2 \ 3	X	Y	
M	1/4	1/4	1/2
N	1/4	1/4	1/2
	1/2	1/2	

(b)

2 \ 3	X	Y	
M	1/6	1/6	1/3
N	1/3	1/3	2/3
	1/2	1/2	

(c)

2 \ 3	X	Y	
M	1/2	0	1/2
N	0	1/2	1/2
	1/2	1/2	

probabilities.[1] That is, player 1 thinks that (M, X) is his opponents' strategy profile with probability 1/4, (M, Y) occurs with probability 1/4, and so on. We can represent this belief by the matrix in Figure B.1(a). The *marginal* distributions appear on the outside of the matrix, on the right for player 2 and below for player 3. The marginals are the probabilities of each strategy for these players individually. Note that the probability of each strategy profile (the number in a given cell) is the product of the marginal probabilities.

Take another example of a conjecture of player 1, as depicted in the matrix in Figure B.1(b). Here player 1 thinks that with probability 1/3 player 2 will select M and with probability 2/3 player 2 will select N. He believes that player 3's strategies are equally likely. Again, he believes that his opponents' actions are independent, as evident in the matrix because the joint probabilities of the opponents' strategies are equal to the product of the marginal probabilities.

The belief represented in the matrix in Figure B.1(c) has a different flavor. Here player 1 believes that player 2's strategies are equally likely and that player 3's strategies are equally likely; the marginal probabilities are both (1/2, 1/2). However, in this case, the joint distribution over the strategies of players 2 and 3 is not defined by the product of the marginal distributions. Player 1 does *not* believe that the actions of players 2 and 3 are independent. He believes that with probability 1/2 his opponents will play (M, X) and with probability 1/2 they will play (N, Y).

To have such a belief, it is not necessary for player 1 to think that players 2 and 3 are in cahoots. That is, they do not have to be actively coordinating their strategies to support such a belief. For example, player 1's belief seems reasonable in the following kind of situation. Suppose player 1 engages in this game in a foreign country. He knows that, in some regions of the world, people

[1] If this is beginning to confuse you, think about flipping coins. The outcome of one flip is independent of the outcome of a second flip. If the probability of heads is 1/2, then the probability of getting heads in both of the two flips is $(1/2)(1/2) = 1/4$.

tend to play the strategies M and X in this game. In other regions, people tend to play N and Y. Player 1 is uncertain of the culture of the people in the foreign country and thus has the belief shown in the matrix in Figure B.1(c).

The beliefs captured by the matrices in Figure B.1(a and b) are called *uncorrelated* conjectures, whereas the belief captured by the matrix in Figure B.1(c) is a *correlated* conjecture. In a given game, a correlated conjecture for player i is any general belief from the set ΔS_{-i}. An uncorrelated conjecture has the property that the joint distributions over S_{-i} are equal to the product of the marginal distributions over each S_j, for $j \neq i$. In this case, the strategies of the other players are, probabilistically speaking, independent. In general, we shall assume that beliefs are uncorrelated. But, where a distinction must be made, I denote uncorrelated distributions by Δ^u and correlated distributions by Δ. Write as B_i the set of best responses of player i over uncorrelated conjectures; denote by B_i^c the set of best responses over all conjectures, including correlated conjectures.

As written in Chapter 6, the relation between dominance and best response for general, finite n-player games, is described by:

> **Result:** For a finite game, $B_i \subset UD_i$ and $B_i^c = UD_i$ for each $i = 1, 2, \ldots, n$.

To expose you to the methodology of formal proof, I shall prove parts of this result.

First, let us prove that $B_i^c \subset UD_i$. That is, we will prove that, if a strategy is a best response to *some* belief, then it cannot be strictly dominated.[2] Take a strategy s_i^* for player i and suppose that it is a best response to some belief $\theta_{-i} \in \Delta S_{-i}$. (The belief may be correlated.) We need to establish that there is no mixed strategy σ_i for player i such that $u_i(\sigma_i, \theta_{-i}) > u_i(s_i^*, \theta_{-i})$. Let us use a proof "by contradiction," which requires (1) assuming that such a σ_i does exist and then (2) showing that this implies that s_i^* cannot be a best response (a contradiction).

The mixed strategy σ_i is a probability distribution over player i's pure strategies.[3] The expected payoff of strategy σ_i against belief θ_{-i} is

$$u_i(\sigma_i, \theta_{-i}) = \sum_{s_i \in S_i} u_i(s_i, \theta_{-i})\sigma_i(s_i),$$

which can exceed $u_i(s_i^*, \theta_{-i})$ only if there is a pure strategy $s_i' \in S_i$ such that $u_i(s_i', \theta_{-i}) > u_i(s_i^*, \theta_{-i})$. This is because the expected payoff from σ_i is an

[2]Note that, because allowing correlation means allowing more conjectures and possibly more best responses, $B_i \subset B_i^c$.

[3]It may put probability 1 on a single pure strategy, in which case it is synonymous with that pure strategy.

average over the expected payoffs of player i's pure strategies. Because s_i' yields a higher expected payoff than does s_i^*, it cannot be that s_i^* is a best response to the belief θ_{-i}. This concludes the proof that $B_i^c \subset UD_i$.

To complete the analysis behind the result, we must establish that $UD_i \subset B_i^c$ as well. Taken together, $B_i^c \subset UD_i$ and $UD_i \subset B_i^c$ imply that $B_i^c = UD_i$; that is, with correlated conjectures, the set of undominated strategies is equivalent to the set of strategies that are best responses for some possible beliefs. Rather than prove $UD_i \subset B_i^c$ for general games, which is quite difficult and requires more sophisticated mathematics than most readers would care to see, I just prove this claim for the following special case. There are two players; player 1 has three strategies and player 2 has two strategies.

In such a game, because player 2 has only two strategies, we do not need to worry about one of player 2's strategies being dominated by a mixed strategy. Let us therefore focus on the strategies of player 1 and, for the sake of analysis, call them a, b, and c. Name the strategies of player 2 m and w. Note that, because this game has only two players, correlated conjectures are not an issue and we know that $B_i = B_i^c$, for $i = 1, 2$. To prove that $UD_1 \subset B_1$, I shall demonstrate that, if a strategy is not a member B_1, then it cannot be a member of UD_1.

Suppose strategy c is not a member of B_1. This means that, for every belief that player 1 may have about player 2's strategy, c is not a best response. In mathematical terms, for every probability $p \in [0, 1]$, $c \notin BR_1(p, 1-p)$. [Here, the belief $(p, 1 - p)$ refers to player 2's strategy m occurring with probability p and strategy w occurring with probability $1 - p$.] We must show that there is a strategy of player 1 that strictly dominates strategy c. The proof is easy if either $a \in BR_1(p, 1 - p)$ for all p or if $b \in BR_1(p, 1 - p)$ for all p, because in this case it is obvious that either a or b (or both strategies) strictly dominates c. Let us take the case in which $BR_1(p, 1 - p) = \{a\}$ for some values of p and $BR_1(p, 1 - p) = \{b\}$ for other values of p, but neither a nor b is a best response for all p. An example is pictured in Figure B.2.

As functions of p, the payoffs $u_1(a, (p, 1 - p))$, $u_1(b, (p, 1 - p))$, and $u_1(c, (p, 1 - p))$ are affine—that is, their graphs are straight lines, as pictured in Figure B.2. Furthermore, player 1's expected payoff from choosing the mixed strategy $(q, 1 - q, 0)$—the strategy that puts probability q on a, $1 - q$ on b, and zero probability on c—also is affine as a function of p. That is, the graph of $qu_1(a, (p, 1 - p)) + (1 - q)u_1(b, (p, 1 - p))$ as a function of p is a straight line. As Figure B.2 indicates, one can find a number p' between 0 and 1 such that $u_1(a, (p', 1 - p')) = u_1(b, (p', 1 - p'))$. We know that at p' the expected payoffs from choosing a and b are equal and strictly greater than the expected payoff from choosing c. Graphically, the a and b lines cross

FIGURE B.2
Dominance and best response.

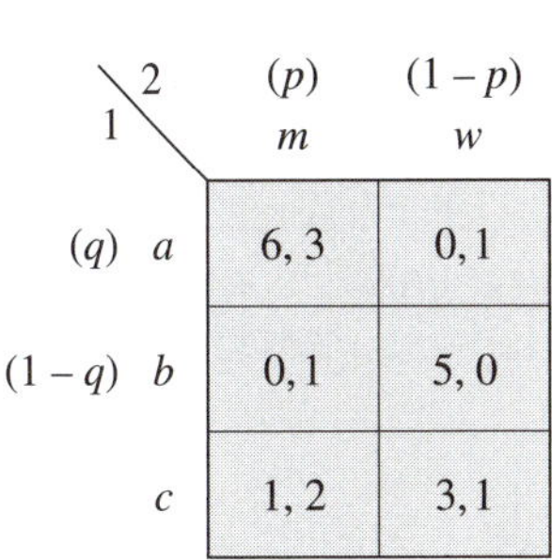

1 \ 2	(p) m	(1 − p) w
(q) a	6, 3	0, 1
(1 − q) b	0, 1	5, 0
c	1, 2	3, 1

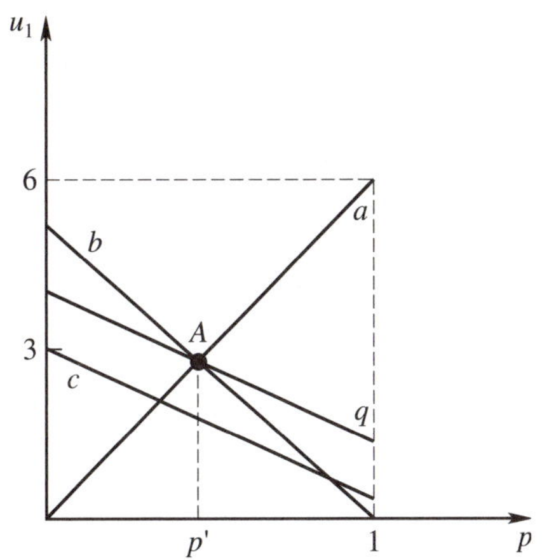

(at p') *above* line c. We also know that the expected payoff line associated with mixed strategy $(q, 1 - q, 0)$ crosses the a and b lines at p' as well. Such a line is included in Figure B.2 and is labeled q. The point at which the lines cross is designated as point A.

We need to find a number $q' \in [0, 1]$ such that the expected payoff of strategy $(q', 1 - q', 0)$ is strictly greater than the payoff of strategy c, for *all* p. It is not difficult to see that such a number exists. By varying q, one changes the slope of the q line, shifting it from the a line (where $q = 1$) to the b line (where $q = 0$). But, regardless of the value of q, the line giving the expected payoff of the mixed strategy passes through point A. We can just pick the value q' that makes this line parallel to line c. Obviously, the q' line is above the c line, which means that the expected utility from the mixed strategy $(q', 1 - q', 0)$ is strictly greater than the expected utility from strategy c, and this is true for all values of p. In particular, it is true for $p = 0$ and $p = 1$. That is, the mixed strategy $(q', 1 - q', 0)$ yields a strictly greater payoff than does strategy c against both of player 2's strategies. Thus, mixed strategy $(q', 1 - q', 0)$ strictly dominates strategy c, which establishes what we needed to prove.

RATIONALIZABILITY CONSTRUCTION

Next let us examine the iterative dominance procedure in more detail. Remember that, in each new round of deleting dominated strategies for some player, one does not consider the other players' strategies that were deleted in preceding rounds. That is, when a strategy has been "thrown out," it does not have to be considered to determine whether strategies of other players are dominated. Let us define B_i and UD_i in relation to subsets of the strategy space as follows.

Remember that n is the number of players in the game. A set of strategy profiles $X \subset S$ is called a *product set* if $X = X_1 \times X_2 \times \cdots X_n$, for some $X_1 \subset S_1, X_2 \subset S_2, \ldots, X_n \subset S_n$. For any product set $X \subset S$, we say that a strategy s_i of player i is *dominated with respect to X* if there is a mixed strategy $\sigma_i \in \Delta S_i$ such that $u_i(\sigma_i, s_{-i}) > u_i(s_i, s_{-i})$ for all $s_{-i} \in X_{-i}$. This definition is exactly the usual definition except that one does not need to compare player i's strategies against strategies of the other players that are not contained in X_{-i}. Given a product set $X \subset S$, define

$$UD_i(X) \equiv \{s_i \in S_i \mid s_i \text{ is not strictly dominated with respect to } X\}$$

and

$$B_i(X) \equiv \{s_i \in S_i \mid s_i \in BR_i(\theta_{-i}) \text{ for some } \theta_{-i} \in \Delta^u X_{-i}\}.$$

These definitions are the same as the ones in Chapter 6, restricted to the subset of strategies X. Thus, "$\Delta^u X_{-i}$" means the set of uncorrelated probability distributions over S_{-i} that assign positive probability only to strategies in X_{-i}. The definitions of $UD_i(X)$ and $B_i(X)$ have the same properties as do UD_i and B_i. For instance, in two-player games, $B_i(X) = UD_i(X)$.

I can now formally describe the procedure of iterative removal of strictly dominated strategies. Let $R^0 \equiv S$ and define

$$UD(X) \equiv UD_1(X) \times UD_2(X) \times \cdots \times UD_n(X).$$

Starting with the full game, we must delete players' strategies that are strictly dominated. Those strategies of player i that survive are given by $UD_i(R^0)$, and so $UD(R^0)$ is the set of strategy profiles that survive the first round of iterated dominance. Let $R^1 \equiv UD(R^0)$. We then continue with a second round of deleting strictly dominated strategies. In evaluating strategies in the second round, we do not need to consider strategies that were removed in the first. Thus, R^1 is the relevant set of strategy profiles, and so $UD(R^1)$ is the set of strategy profiles that survive through the second round.

This process continues. In mathematical terms, it defines a sequence of sets $R^0, R^1, R^2, \ldots$, where $R^0 \equiv S$ and, for every positive integer k,

$$R^k \equiv UD(R^{k-1}).$$

The set of strategies that survives the iterative removal of strictly dominated strategies is equal to the limit of this sequence. This may seem very complicated, and it is in general. However, for finite games the process is fairly simple, at least in application (as you can see from reading Chapter 7).

A few more comments are in order. First, it is not difficult to show that UD has the following property (which is, technically, called a *monotonicity* property): for any product sets X and Y that are subsets of S, if $X \subset Y$ then $UD(X) \subset UD(Y)$. Also, if X is finite and not empty (that is, it contains at least one strategy profile) then $UD(X)$ is also not empty. For finite games, these facts imply that, for all k, R^k is not empty and $R^k \subset R^{k-1}$. The facts also imply that the process of removing dominated strategies stops at some point, in the sense that no more strategies are deleted in future rounds. In mathematical terms, this means that there is a positive integer K (which depends on the game) such that, for all $k > K$, $R^k = R^K$. The set of strategies that survive iterated dominance is equal to R^K. A more elegant way of writing this is $\bigcap_{k=1}^{\infty} R^k$.

One can perform the same kind of routine by using the operator BR_i in place of UD_i. In formal terms, the set of *rationalizable strategies* is defined as $R = \bigcap_{k=1}^{\infty} R^k$, where BR_i is used instead of UD_i. We know that one will arrive at exactly the same set of strategies if the game has two players, because BR_i and UD_i are identical in this case. With games of more than two players, the resulting set depends on whether correlated conjectures are allowed, as noted earlier in this appendix.

EXISTENCE OF NASH EQUILIBRIUM

In Chapter 11, I stated a useful result, due to John Nash, on the existence of Nash equilibrium. Here it is again:

> **Result:** Every finite game (one that has a finite number of players and a finite strategy space) has at least one Nash equilibrium (in pure or mixed strategies).

The proof of this result is based on intuition that you can get from going over a simple example. Consider the game pictured on the left side of Figure B.3 on the next page. You will quickly observe that this game has no pure-strategy Nash equilibrium but that the mixed-strategy profile ((1/2, 1/2), (1/3, 2/3)), where player 1 selects U with probability 1/2 and player 2 selects L with probability 1/2, is a mixed-strategy Nash equilibrium. If this does not make sense to you, then I suggest that you review the definitions in Chapters 9 and 11. If you haven't read through Chapter 11 yet, I say "quit skipping ahead!"

Examine how equilibrium arises in the example of Figure B.3. Let us do this from the perspective of the players' best responses in the space of mixed

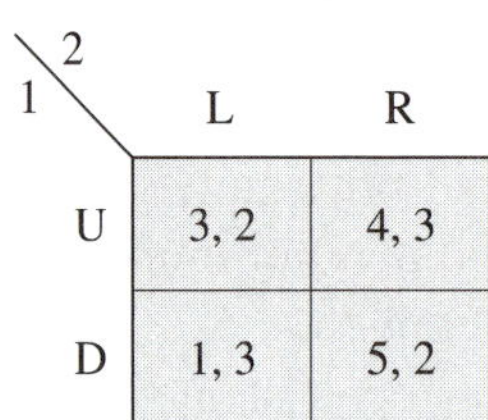

	L	R
U	3, 2	4, 3
D	1, 3	5, 2

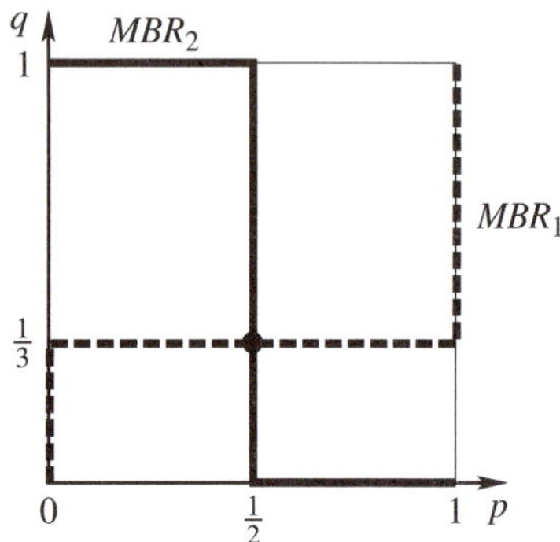

FIGURE B.3
A normal-form game and mixed best responses.

strategies. Let p denote the probability that player 1 selects U, meaning that his mixed strategy is $(p, 1 - p)$. Likewise, let q denote the probability that player 2 selects L, meaning that her mixed strategy is $(q, 1 - q)$. Note that player 1's belief about player 2's strategy is then given by q and player 2's belief about player 1's strategy is given by p.

Against player 2's mixed strategy, player 1 obtains an expected payoff of $3q + 4(1 - q)$ if he selects U, whereas he expects $q + 5(1 - q)$ if he were to select D. The payoff of U is strictly higher if

$$3q + 4(1 - q) > q + 5(1 - q),$$

which simplifies to $q > 1/3$. Thus, U is player 1's only best response if $q > 1/3$, D is the only best response if $q < 1/3$, and both U and D are best responses if $q = 1/3$. Observe how this statement translates this into mixed strategies. If $q > 1/3$, then the only mixed strategy that player 1 can rationally select is $p = 1$ (the mixed strategy that puts all probability on U). If $q < 1/3$, then the only mixed strategy that player 1 can rationally select is $p = 0$ (the mixed strategy that puts all probability on D). Finally, if $q = 1/3$, then all mixed strategies $p \in [0, 1]$ are best responses for player 1. This last point is critical. At $q = 1/3$, player 1 is indifferent between U and D, so he is indifferent between *any* mixture of the two.

We can represent player 1's mixed-strategy best responses in terms of a function MBR_1 that maps the belief q into subsets of the interval $[0, 1]$. That is, $MBR_1(q)$ is the set of mixed strategies that are best responses for player 1 to the belief q. In the example, we have $MBR_1(q) = \{1\}$ for $q > 1/3$, $MBR_1(q) = \{0\}$ for $q < 1/3$, and $MBR_1(q) = [0, 1]$ for $q = 1/3$. A graphical representation of MBR_1 appears as the dashed line in the right part of Figure B.3.[4]

[4]In technical terms, we would say that the dashed line is the graph of MBR_1 when viewed as a "correspondence," which is a special function whose domain is the set of subsets of a given set.

Repeating the analysis for player 2, you can see that the expected payoff of L is strictly higher than the payoff of R if

$$2p + 3(1 - p) > 3p + 2(1 - p),$$

which simplifies to $p < 1/2$. Thus, writing MBR_2 as player 2's set of mixed-strategy best responses, we have $MBR_2(p) = \{1\}$ if $p < 1/2$, $MBR_2(p) = \{0\}$ if $p > 1/2$, and $MBR_2(p) = [0, 1]$ if $p = 1/2$. This is represented by the solid line in the right part of Figure B.3.

The mixed-strategy Nash equilibrium occurs where the dashed and solid lines cross—that is, where the players are best responding to each other. The key intuition about the existence of a Nash equilibrium is that MBR_1 and MBR_2 have a continuity property. Specifically, MBR_1 is a mapping that applies from $q = 0$ to $q = 1$, meaning from the bottom to the top of the box in Figure B.3. Likewise, MBR_2 is a mapping that applies from $p = 0$ to $p = 1$, meaning from the left to the right of the box in the figure. These mappings are continuous in the sense that you can draw each of them without lifting your writing instrument off the page as you scan across the box. In summary, if you draw two lines through a box, one that goes from the bottom to the top and the other from the left to the right, and these lines are each continuous, then they *must* cross at some point. The crossing point is a Nash equilibrium by definition.

The general existence proof works in the same way. One starts with the best-response functions for each player (in mixed-strategy space) and collects them into a function MBR that gives the profile of best responses for a given profile of mixed strategies. One can show that MBR is well defined for finite games and has a continuity property along the lines of what we saw in the example.[5] One then employs *Kakutani's fixed-point theorem*, which establishes the existence of a mixed-strategy profile σ^* such that $\sigma^* \in MBR(\sigma^*)$.[6] The profile σ^* is a mixed-strategy Nash equilibrium because, for each player i, $\sigma_i \in MBR_i(\sigma_{-i})$. That is, each player's strategy is a best response to the strategies of the others.

EXERCISES

The following exercises are challenging and recommended only for readers with ample mathematics background.

---

[5]The continuity property is called *upper hemi-continuity*.

[6]S. Kakutani, "A Generalization of Brouwer's Fixed Point Theorem," *Duke Mathematical Journal* 8(1941):457–459.

1. Prove the following assertions made at the end of this appendix: for finite games, (a) R^k is nonempty for each k; (b) $R^k \subset R^{k-1}$, for each positive integer k; and (c) there exists an integer K such that $R^k = R^K$ for all $k > K$.

2. Consider the following game. There are n students. Simultaneously and independently, they each write an integer between 1 and 100 on a slip of paper. The average of all of the students' numbers is then computed and the student coming closest to 2/3 of this average wins $20. If two or more students tie for the closest to 2/3 of the average, then these students equally share the $20 prize. Compute the set of rationalizable strategies in this game. It may be helpful to construct the best-response function for a typical student; think of this as a function of the expected average of the other students' numbers. Calculate the strategies that are removed in the first round of deleting strictly dominated strategies by calculating the strategies that are best responses over various beliefs. That is, calculate the set $B_i(S)$. Then calculate those that are removed in the second round by calculating $B_i(B_i(S))$. Try to find a formula for R^k as a function of k.

3. Examine the game pictured here.

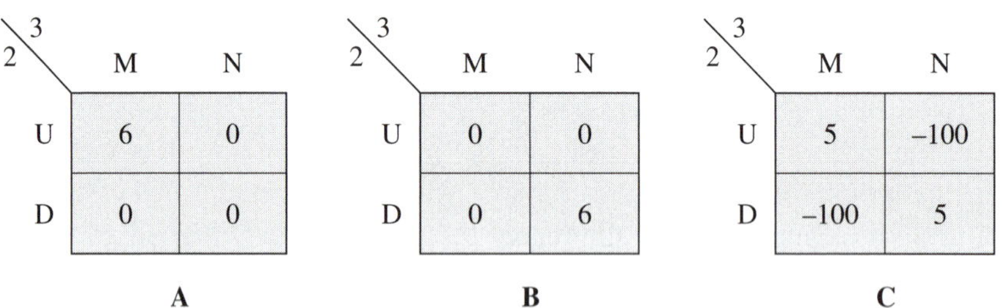

A (2 \ 3)	M	N
U	6	0
D	0	0

B (2 \ 3)	M	N
U	0	0
D	0	6

C (2 \ 3)	M	N
U	5	–100
D	–100	5

In this three-player game, player 1 selects between A, B, and C and therefore chooses which matrix will be effective. Players 2 and 3 select between U and D and between M and N, respectively. The actions of the three players are taken simultaneously. Only the payoff of player 1 appears in the cells of the matrices. This game illustrates the relation between correlated conjectures and the best-response/dominance equivalence. Show first that player 1's strategy C is not strictly dominated. Then show that C is never a best response if player 1's conjecture about the strategies of the other players is not allowed to exhibit correlation. Finally, show that C can be a best response for player 1 if his belief exhibits correlation. [Look at a belief in which player 1 thinks the other players will coordinate on (U, M) or (D, N).]

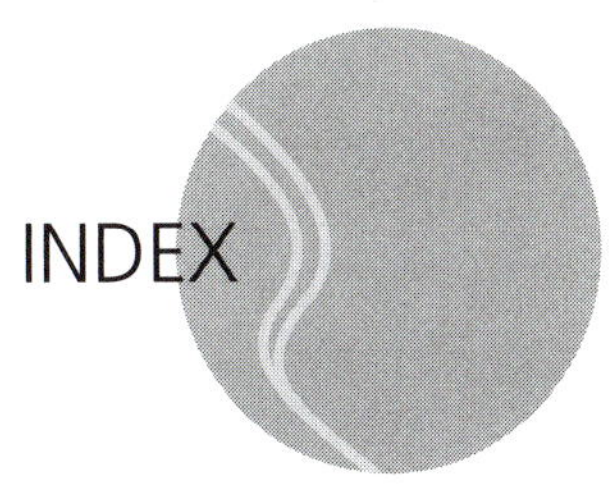

INDEX

Page numbers in *italics* refer to figures.